Graded examples in mathematics

Revision of Topics for GCSE

M. R. Heylings M.A., M.Sc.

Schofield & Sims Limited Huddersfield

0 7217 2339 x

First printed 1987
Reprinted 1987

Acknowledgements
The timetables on pages 173 and 212 and the British Rail Logo on page 48 reproduced with the permission of the British Railways Board.
The tables on pages 98, 99 and 213 reproduced with the permission of British Telecom.
The timetables on pages 96 and 179 reproduced with the permission of Merseyside Passenger Transport Executive.
The extract on page 216 reproduced with the permission of Radio Times.
The tables on pages 97 and 210 reproduced with the permission of The Post Office.
The map and timetable on page 196 reproduced with the permission of Tyne & Wear Passenger Transport Executive.
The map on page 94 reproduced with the permission of West Yorkshire Passenger Transport Executive.

The series **Graded examples in mathematics** comprises:

Fractions and Decimals	0 7217 2323 3
Answer Book	0 7217 2324 1
Algebra	0 7217 2325 x
Answer Book	0 7217 2326 8
Area and Volume	0 7217 2327 6
Answer Book	0 7217 2328 4
General Arithmetic	0 7217 2329 2
Answer Book	0 7217 2330 6
Geometry and Trigonometry	0 7217 2331 4
Answer Book	0 7217 2332 2
Negative Numbers and Graphs	0 7217 2333 0
Answer Book	0 7217 2334 9
Matrices and Transformations	0 7217 2335 7
Answer Book	0 7217 2336 5
Sets, Probability and Statistics	0 7217 2337 3
Answer Book	0 7217 2338 1
Revision of Topics for GCSE	0 7217 2339 x
Answer Book	0 7217 2340 3

Designed by Graphic Art Concepts, Leeds
Printed in England by The Bath Press, Avon

Author's Note

This series has been written and produced in the form of eight topic books, each offering a wealth of graded examples for pupils in the 11-16 age range; plus a further book of revision examples for those nearing examination in year 5.

There are no teaching points in the series. The intention is to meet the often heard request from teachers for a wide choice of graded examples to support their own class teaching. The contents are clearly labelled for easy use in conjunction with an existing course book; but the books can also be used as the chief source of examples, in which case the restrictions imposed by the traditional type of mathematics course book are removed and the teacher is free to organise year-by-year courses to suit the school. Used in this way, the topic-book approach offers an unusual and useful continuity of work for the class-room, for homework or for revision purposes.

The material has been tested over many years in classes ranging from mixed ability 11-year-olds to fifth formers taking public examinations. Some sections are useful for pupils of above average ability while other sections suit the needs of the less able, though it is for the middle range of ability that the series is primarily intended.

'Graded examples in Mathematics' and the GCSE

The advent of the GCSE has increased the demands made on the teaching and learning of mathematics. A heavy stress is laid on the content of a mathematics programme being related to applications and contexts, particularly those from everyday situations, and being such that pupils increase their self-confidence by experiencing success in their work. The exercises offered in this series are so graded as to encourage success; in addition, throughout the series, a continuing emphasis is placed on providing exercises where pupils can apply the topics being studied in a variety of contexts and situations. The main intention is to provide, for the teacher, a high degree of flexibility in the variety of exercises on offer.

Revision of Topics, intended for pupils nearing their GCSE examinations, provides a close coverage of GCSE syllabuses and will be of most use with those pupils of average or even above-average ability. A final selection of short and of longer problems from all topics is included in separate exercises; five aural tests are also provided.

Contents

Symbols

$=$	is equal to
$\neq$	is not equal to
$\simeq$	is approximately equal to
$<$	is less than
$\leqslant$	is less than or equal to
$\nless$	is not less than
$>$	is greater than
$\geqslant$	is greater than or equal to
$\ngtr$	is not greater than
$\Rightarrow$	implies
$\Leftarrow$	is implied by
$\rightarrow$	maps onto
$\in$	is a member of
$\notin$	is not a member of
$\subset$	is a subset of
$\not\subset$	is not a subset of
$\cap$	intersection (or overlap)
$\cup$	union
A'	the complement (or outside) of set A
$\mathscr{E}$	the Universal set
$\varnothing$ or $\{\ \}$	the empty set
(x, y)	the co-ordinates of a point
$\begin{pmatrix} x \\ y \end{pmatrix}$	the components of a vector

The Greek alphabet

A	α	alpha
B	β	beta
Γ	γ	gamma
Δ	δ	delta
E	ε	epsilon
Z	ζ	zeta
H	η	eta
Θ	θ	theta
I	ι	iota
K	κ	kappa
Λ	λ	lambda
M	μ	mu
N	ν	nu
Ξ	ξ	xi
O	o	omicron
Π	π	pi
P	ρ	rho
Σ	σ, ς	sigma
T	τ	tau
Υ	υ	upsilon
Φ	ϕ, φ	phi
X	χ	chi
Ψ	ψ	psi
Ω	ω	omega

Decimals

Part 1 Introduction

1 Write:
(i) as a decimal and (ii) as a fraction that part of each square which is *shaded*.

a 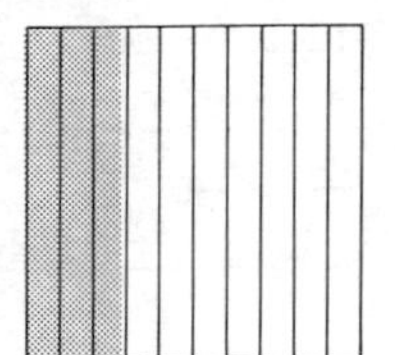 b 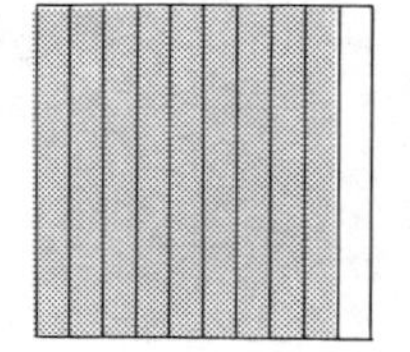 c 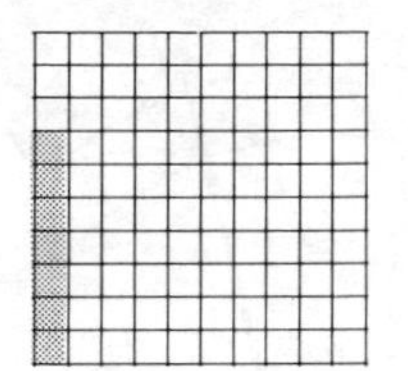 d 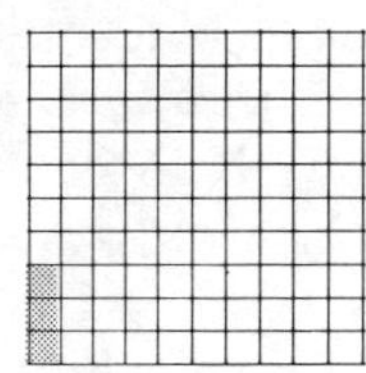

2 Write the fraction *shaded*, as in the example alongside.

Example

$\frac{3}{10} + \frac{4}{100} = \frac{34}{100} = 0.34$

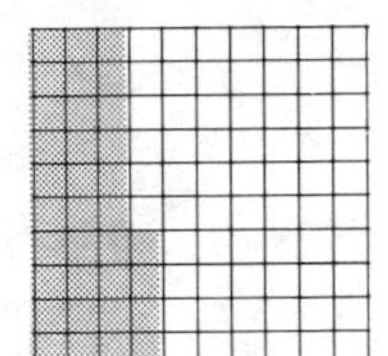

a 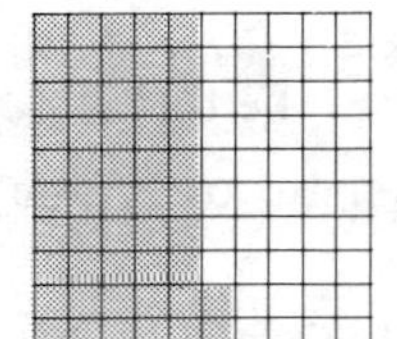 b 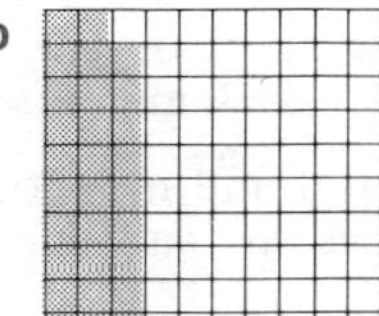 c 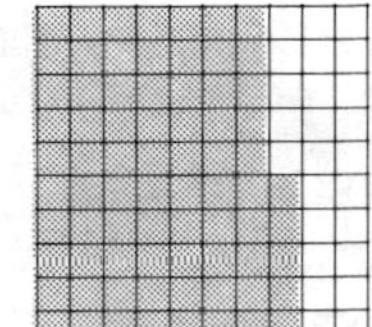 d 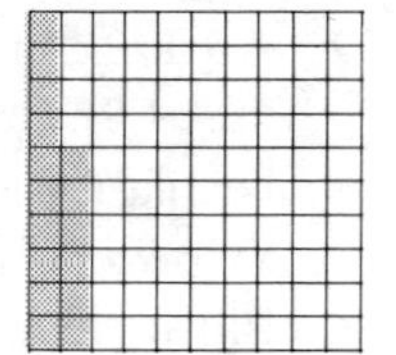

3 Write each answer as a *decimal*.

a $\frac{7}{10} + \frac{6}{100}$ b $\frac{3}{10} + \frac{8}{100} + \frac{2}{1000}$ c $\frac{8}{10} + \frac{1}{100} + \frac{7}{1000}$

d $\frac{6}{10} + \frac{4}{1000}$ e $\frac{5}{100} + \frac{3}{1000}$ f $\frac{9}{10} + \frac{2}{1000}$

4 Change these decimals to fractions.

a 0.7 b 0.07 c 0.31 d 0.89 e 0.003 f 0.031

5 X and Y are thermometers which measure temperatures in degrees Celsius.

a What is the reading shown on each thermometer?

b What temperatures are indicated by the two arrows on the thermometers?

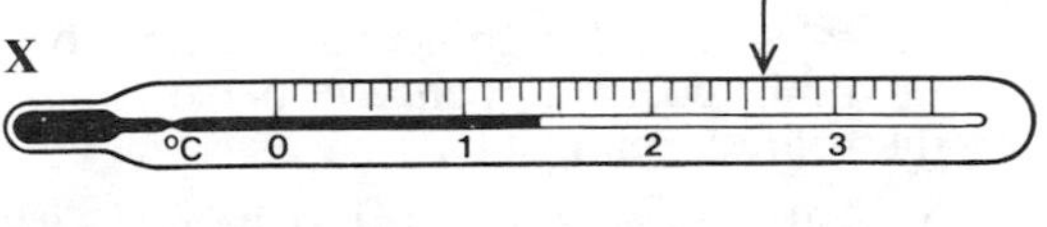

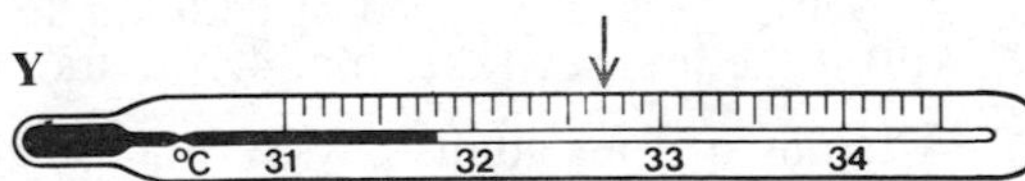

6 Write the decimal number to which each arrow is pointing on these scales.

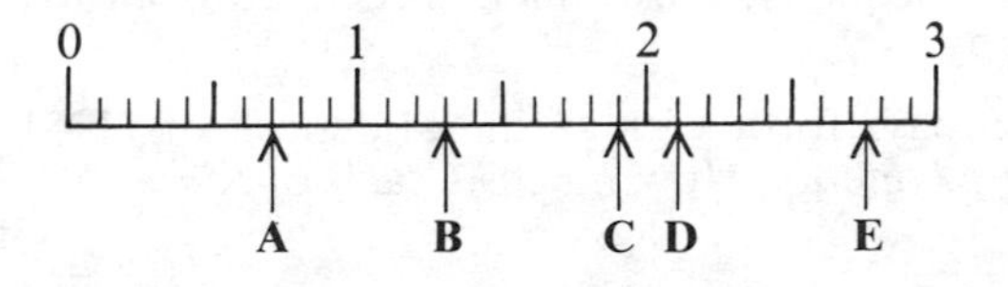

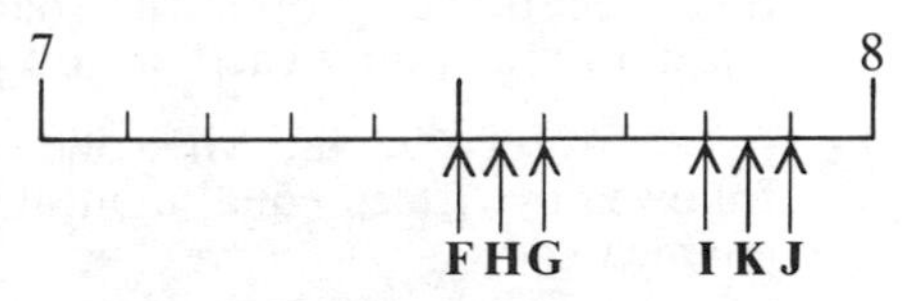

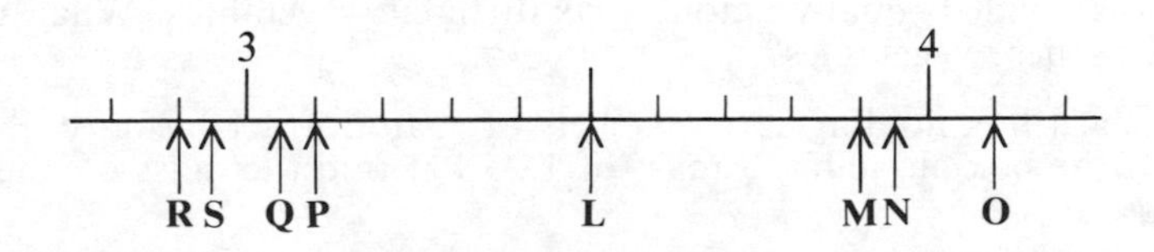

Decimals

7 What readings are indicated by the pointers on these weighing machines?

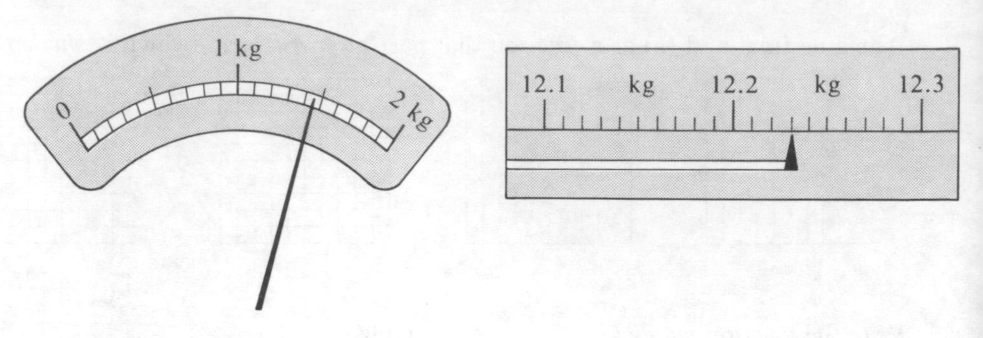

Part 2 Problems

1 Mrs Black went shopping and bought three items for £3·56, £8·43 and £6. How much did she spend altogether?

2 A garage mechanic opens a full 15-litre tin of oil and uses 4.8 litres, 2.7 litres and 3 litres in the engines of three cars. How much is left in the tin?

3 Nicola Russell buys 12 metres of material and uses four lengths of 2.75 m, 0.65 m, 3.3 m and 4 m. How much does she have left?

4 Bill Richards bought four records at £4·65 each and three books at £1·97 each. How much did he spend?

5 If one rose bush costs £2·18 and Susie Croft buys seven of them, how much change does she have out of a £20 note?

6 George Hanson earns £4·23 per hour. What is his basic weekly wage for a 36-hour week?

7 How many kilocalories are there in a loaf of twenty four slices if each slice has 72.5 kilocalories?

8 Aziz Awan makes three car journeys of 36.8 miles, 126.4 miles and 59 miles. If each mile costs 6.5 pence in petrol, find the total cost of the petrol he uses to the nearest penny.

9 A cyclist averages a speed of 24.6 km/h. How far does he go (to the nearest km) if the journey takes him 4.5 hours?

10 A family of four adults pays a total of £106·16 for a train journey. How much does it cost per person?

11 A roll of cloth should cost £47·76 but it is faulty and is sold for one-third of this price to four people who share it equally. How much is the cloth sold for and how much does each person pay?

12 A car costs £4697 and Mr Rolleson pays for it with an initial deposit of £1751 followed by fifteen equal monthly payments. How much is each of these payments?

13 A gardener has a 210 m^2 plot, one quarter of which is planted with potatoes. The rest is divided equally amongst six different vegetables. What area is given to each of these vegetables?

14 A cardboard box holding seven packets of soap powder has a total mass of 24 kg. If the box alone has a mass of 1.25 kg, find the mass of one packet of soap powder.

Decimals

15 Copy this table and write in your answers.

	1000	100	10	1	$\frac{1}{10}$	$\frac{1}{100}$	$\frac{1}{1000}$
a $17.8 \times 10 =$				.			
b $9.42 \times 100 =$				.			
c $6.7 \div 10 =$				.			
d $3.4 \div 100 =$				.			
e $0.73 \times 1000 =$				.			
f $1.25 \times 1000 =$				.			
g $672 \div 100 =$				.			
h $195 \div 1000 =$				.			

16 Write the answers only.

a Ten nails have a total mass of 42 grams. What is the mass of one nail?

b 100 cubes of sugar have a total mass of 625 grams. What is the mass of one lump?

c If one pin has a mass of 0.27 grams, what is the mass of 1000 pins?

d If one postage stamp has an area of 2.8 cm^2, what is the area of a sheet of 100 stamps?

e If 48.7 cm^3 of liquid drips into a beaker in 100 minutes, what is the average rate of flow in cm^3/min?

f One tin of beans has a mass of 87.5 grams. If ten tins are packed in a box of mass 150 grams, what is the total mass of tins and box?

17 Find the missing numbers indicated by each box $\square$.

a $42.7 \times \square = 4270$

b $3.86 \times \square = 38.6$

c $9.41 \div \square = 0.0941$

d $309 \div \square = 30.9$

e $0.67 \div \square = 0.0067$

f $0.48 \times \square = 4800$

g $2.7 \div \square = 0.027$

h $0.052 \times \square = 520$

Fractions

Part 1 Equivalence, addition and subtraction

1 Write the fraction of each shape which is (i) *blank* (ii) *shaded* (iii) *dotted*.

a
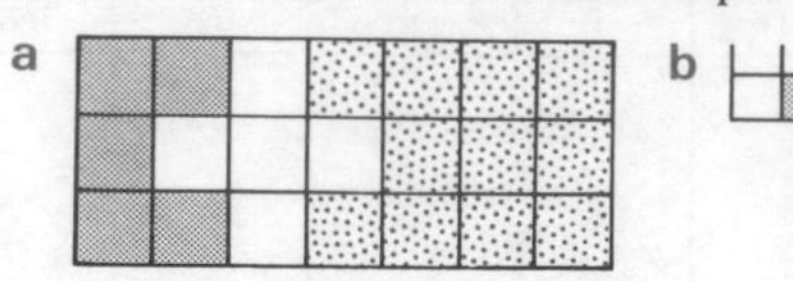

b **c** **d**
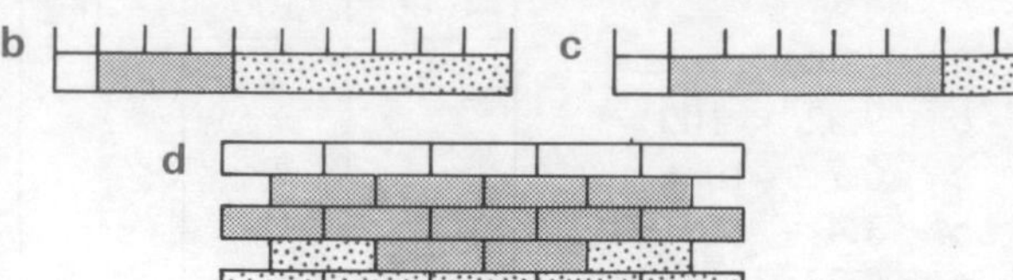

2 A workman spends two hours travelling, three hours collecting equipment and nine hours installing equipment. What fraction of the total time is spent on the installation?

3 A young girl spends £14 on clothes, £12 on shoes, £5 on cosmetics and £4 on magazines. What fraction of the total is spent on shoes?

4 A clerk types a report and, on checking it, finds that 89 lines contain spelling mistakes, 65 lines contain errors in spacing, 12 lines have incorrect numbers and 426 lines are correct. What fraction of the total number of lines contain spelling mistakes?

5 Write down the reading to which each arrow points on these diagrams.

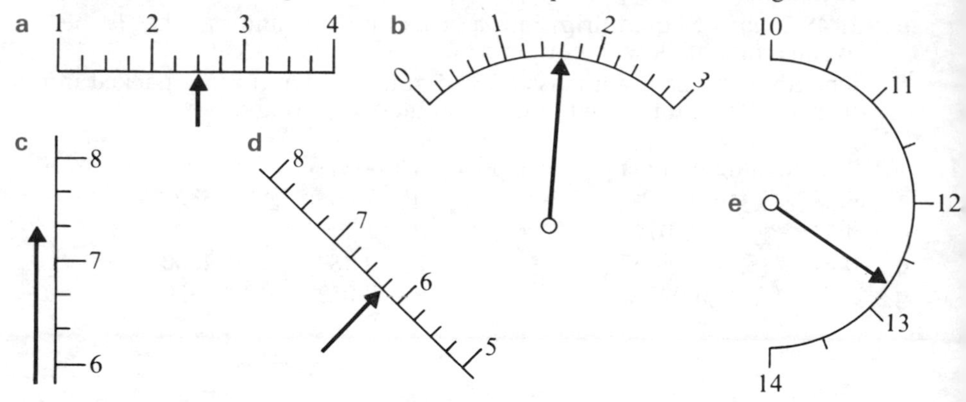

6 British industry has not yet totally converted to using metric units of measurement, and the USA has still not gone metric. Therefore the use of inches (″) rather than centimetres (cm) is still common, and this involves using fractions.

Copy and complete the following.

a $\frac{1}{2} = \frac{}{4} = \frac{}{8} = \frac{}{16} = \frac{}{32}$ **b** $\frac{1}{4} = \frac{}{8} = \frac{}{16} = \frac{}{32}$ **c** $\frac{3}{4} = \frac{}{8} = \frac{}{16} = \frac{}{32}$

d $\frac{1}{8} = \frac{}{16} = \frac{}{32}$ **e** $\frac{5}{8} = \frac{}{16} = \frac{}{32}$

7 Add or subtract these fractions.

a $3\frac{1}{2} + 1\frac{1}{2}$ **b** $4\frac{1}{2} + 2\frac{1}{4}$ **c** $5\frac{1}{4} + 3\frac{3}{4}$ **d** $2\frac{3}{4} + 7\frac{1}{4}$

e $2\frac{3}{4} + 7\frac{1}{2}$ **f** $1\frac{1}{2} + 1\frac{3}{4}$ **g** $8\frac{3}{4} - 1\frac{1}{4}$ **h** $6\frac{3}{4} - 2\frac{1}{2}$

i $5 - 1\frac{1}{2}$ **j** $4 - 1\frac{1}{4}$ **k** $3\frac{1}{4} - 1\frac{1}{2}$ **l** $6\frac{1}{2} - 2\frac{3}{4}$

m $4\frac{3}{8} + 2\frac{1}{8}$ **n** $1\frac{1}{4} + 2\frac{3}{8}$ **o** $3\frac{1}{8} + 4\frac{1}{2}$ **p** $1\frac{3}{4} + 5\frac{1}{8}$

q $2\frac{5}{8} + 4\frac{1}{2}$ **r** $\frac{7}{8} + 1\frac{1}{4}$ **s** $3\frac{7}{8} - 1\frac{1}{2}$ **t** $5 - 1\frac{3}{8}$

u $5\frac{1}{8} - 1\frac{3}{8}$

Fractions

8 Find the distance on this sheet of metal between

- **a** *A* and *C*
- **b** *A* and *D*
- **c** *C* and *E*
- **d** *A* and *E*.

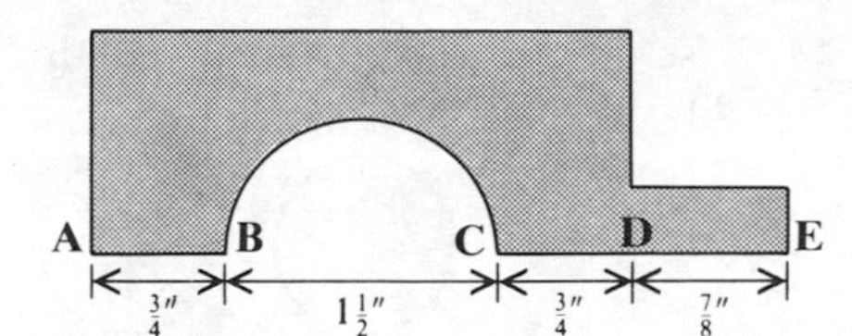

9

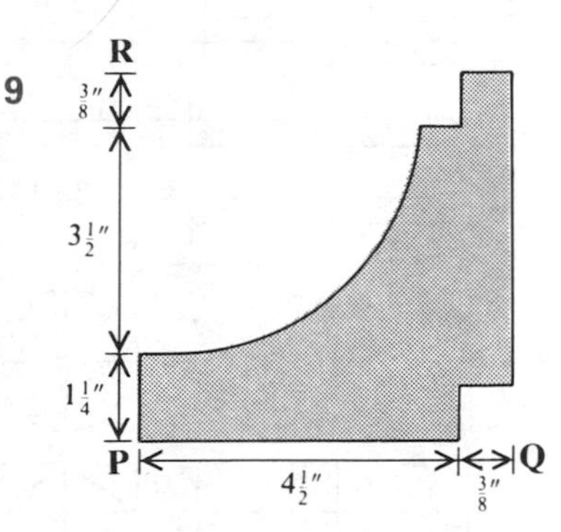

Find the distance on this sheet of hardboard between

- **a** *P* and *Q*
- **b** *P* and *R*.

10 The diagram shows an assembly of three bolts. Calculate the distance between

- **a** *L* and *N*
- **b** *L* and *O*
- **c** *M* and *P*
- **d** *L* and *P*.

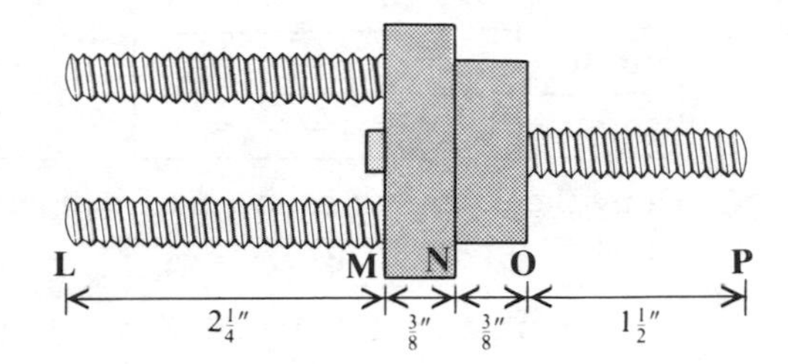

11

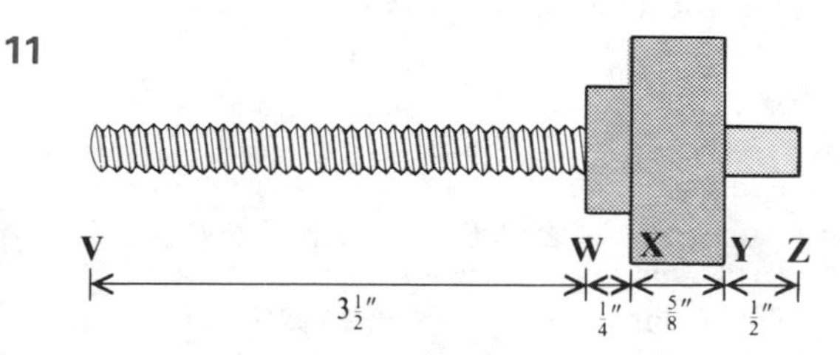

Calculate the distance on this nut and bolt between

- **a** *V* and *X*
- **b** *W* and *Y*
- **c** *W* and *Z*
- **d** *V* and *Y*
- **e** *V* and *Z*.

12 Find the distance on this bolt between

- **a** *A* and *C*
- **b** *B* and *D*
- **c** *C* and *E*
- **d** *D* and *F*
- **e** *C* and *F*
- **f** *A* and *D*
- **g** *A* and *F*.

A B C D E F

1/2″ 1 1/4″ 3/8″ 1 1/8″ 1/2″

13 These three shapes are to be cut from sheet metal. Calculate the lettered lengths *p* to *u*.

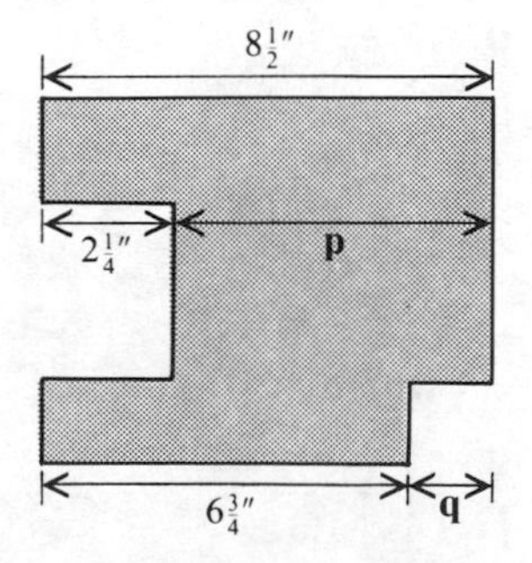

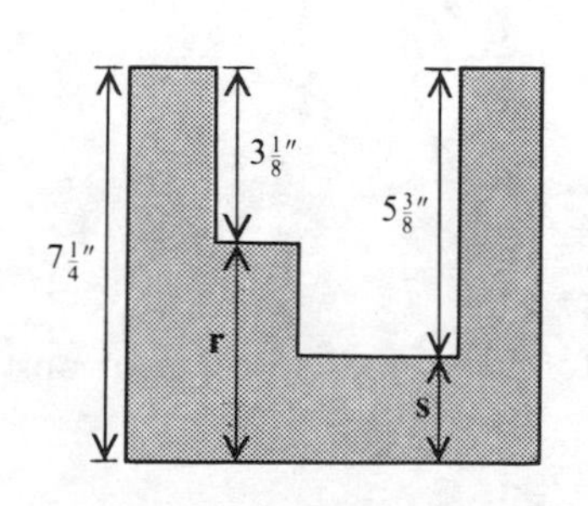

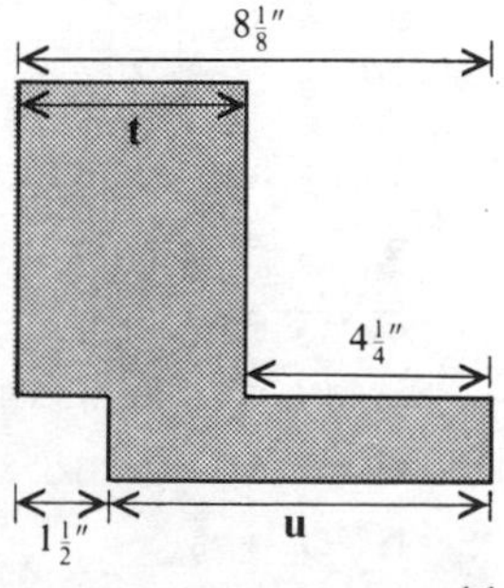

Fractions

14 Use the diagrams to help you find the missing number in each pair of equivalent fractions.

a $\frac{1}{2} = \frac{}{8}$

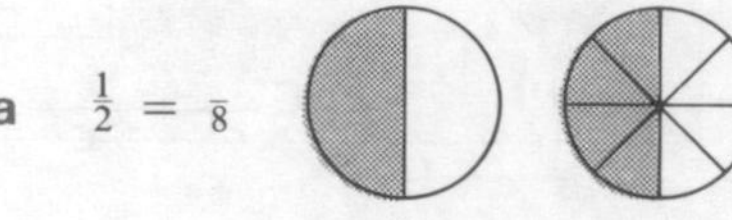

b $\frac{1}{4} = \frac{}{12}$

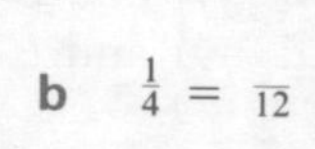

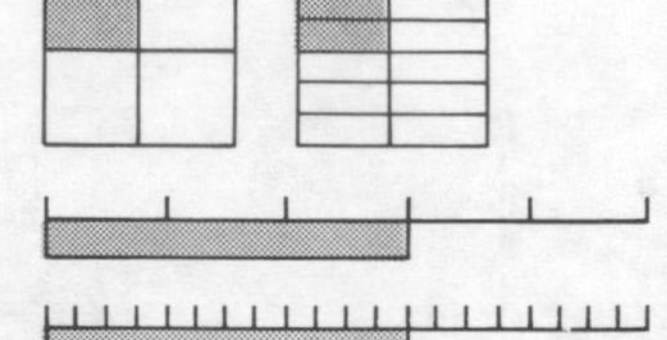

c $\frac{2}{3} = \frac{}{6}$

d $\frac{3}{5} = \frac{}{20}$

15 Copy and complete this table to give the fraction of each shape which is *shaded*.

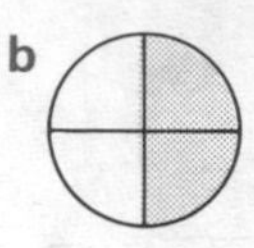

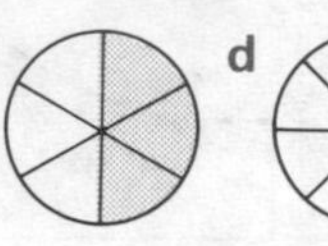

	a		b		c		d
circle		=		=		=	
square		=		=		=	

16 Copy these pairs of equivalent fractions and find the missing numbers.

a $\frac{3}{4} = \frac{}{20}$ **b** $\frac{5}{8} = \frac{}{24}$ **c** $\frac{3}{5} = \frac{}{15}$ **d** $\frac{7}{10} = \frac{}{50}$

e $\frac{2}{3} = \frac{}{15}$ **f** $\frac{5}{12} = \frac{}{60}$ **g** $\frac{1}{2} = \frac{}{18}$ **h** $\frac{5}{6} = \frac{}{48}$

17 Find equivalent fractions by cancelling these as far as possible.

a $\frac{20}{35}$ **b** $\frac{48}{64}$ **c** $\frac{280}{400}$ **d** $\frac{48}{80}$

e $\frac{75}{200}$ **f** $\frac{27}{108}$ **g** $\frac{126}{180}$ **h** $\frac{120}{144}$

18 Express your answers in their lowest terms.

a A small business has a monthly turnover of £4800 of which £1600 is spent in wages. What fraction of the turnover goes on wages?

b Mr Howroyd earned a total of £8640 last year of which £1920 was taken in tax and £720 by a pension fund. What fraction of his total earnings went (i) in tax (ii) into the pension fund?

c A commuter train takes 1 h 20 min for the whole journey of which 24 mins is spent standing at stations on the route. What fraction of the total time is spent (i) at rest (ii) in motion?

19 Copy and complete.

a $\frac{2}{3} + \frac{1}{9} = \frac{}{9} + \frac{1}{9} = \frac{}{9}$ **b** $\frac{2}{5} + \frac{3}{10} = \frac{}{10} + \frac{3}{10} = \frac{}{10}$

c $\frac{19}{24} - \frac{1}{4} = \frac{19}{24} - \frac{}{24} = \frac{}{24}$ **d** $\frac{5}{6} - \frac{5}{12} = \frac{}{12} - \frac{5}{12} = \frac{}{12}$

Use a similar method for these and cancel your answer where possible.

e $\frac{2}{3} + \frac{1}{24}$ **f** $\frac{1}{12} + \frac{5}{6}$ **g** $\frac{2}{5} + \frac{4}{15}$

h $\frac{4}{5} - \frac{3}{20}$ **i** $\frac{3}{8} + \frac{5}{24}$ **j** $\frac{2}{3} - \frac{1}{6}$

20 Copy and complete.

a $\frac{1}{4} + \frac{2}{3} = \frac{}{12} + \frac{}{12} = \frac{}{12}$ **b** $\frac{1}{2} + \frac{4}{9} = \frac{}{18} + \frac{}{18} = \frac{}{18}$

c $\frac{2}{3} - \frac{1}{2} = \frac{}{6} - \frac{}{6} = \frac{}{6}$ **d** $\frac{4}{5} - \frac{1}{3} = \frac{}{15} - \frac{}{15} = \frac{}{15}$

Use a similar method for these and cancel your answers where possible.

e $\frac{3}{4} + \frac{1}{6}$ **f** $\frac{2}{5} + \frac{1}{2}$ **g** $\frac{7}{8} - \frac{1}{5}$

h $\frac{1}{6} + \frac{2}{9}$ **i** $\frac{5}{8} - \frac{2}{5}$ **j** $\frac{1}{4} + \frac{3}{10}$

Fractions

Part 2 Multiplication and division

1 Three quarters of this rectangle is shaded and half of the shaded part is dotted.
Use the diagram to calculate $\frac{1}{2} \times \frac{3}{4}$.

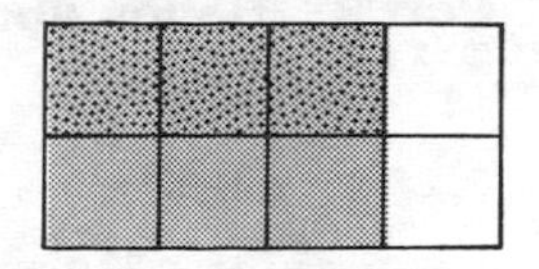

2 Multiply these fractions.

a $\frac{4}{5} \times \frac{2}{3}$ **b** $\frac{3}{4} \times \frac{5}{7}$ **c** $1\frac{1}{2} \times \frac{1}{4}$

d $\frac{2}{5} \times 2\frac{1}{3}$ **e** $2\frac{1}{2} \times \frac{1}{4}$ **f** $2\frac{1}{2} \times 1\frac{1}{2}$

Multiply these, remembering to cancel where possible.

g $\frac{5}{12} \times \frac{6}{7}$ **h** $\frac{4}{5} \times \frac{5}{12}$ **i** $\frac{1}{9} \times 1\frac{1}{2}$

j $1\frac{1}{6} \times \frac{3}{7}$ **k** $1\frac{2}{5} \times 2\frac{1}{7}$ **l** $1\frac{1}{2} \times 1\frac{7}{9} \times \frac{3}{8}$

3 Calculate the areas of these rectangles.

a

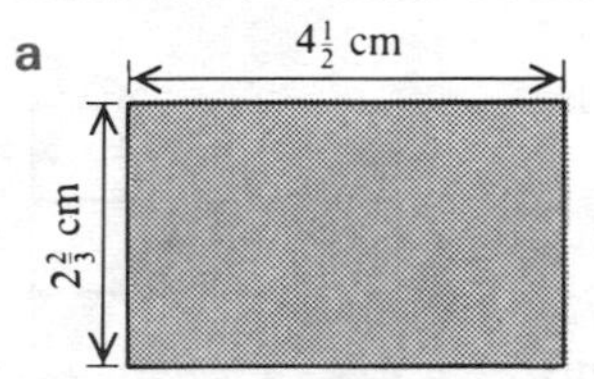

b

7½ cm

1⅗ cm

c

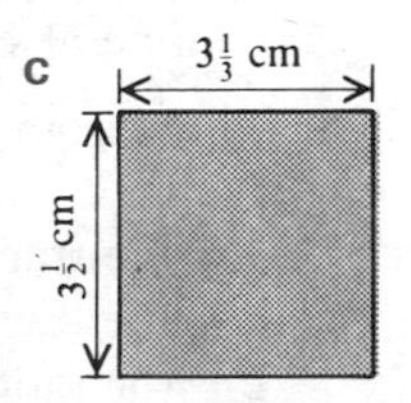

4 Calculate the volumes of these cuboids.

a

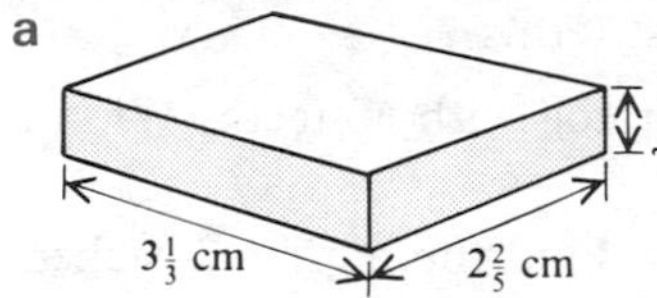

b

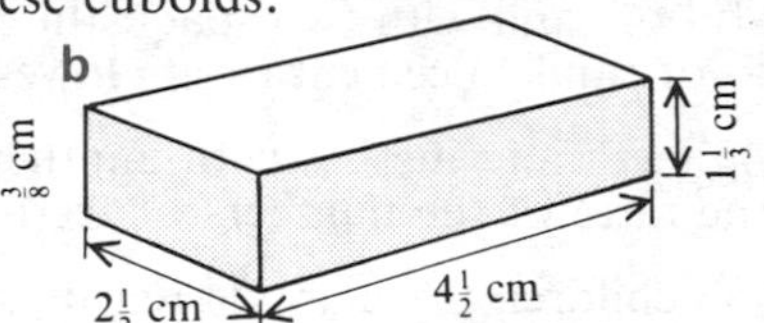

c

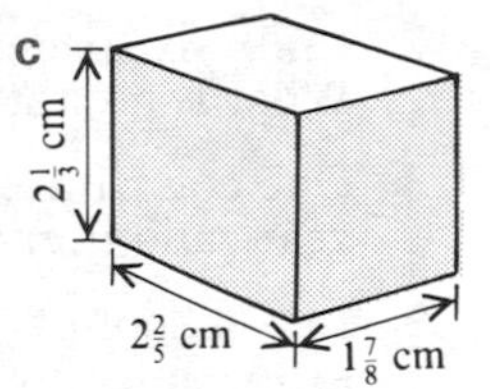

5 **a** I have a $7\frac{1}{2}$ kg bag of cement. If I use two thirds of it, what mass is left?

b My lawn has an area of $17\frac{1}{2}$ m^2. If I dig up two fifths of it for a rose bed, what is the area of (i) the rose bed (ii) the remainder of the lawn?

c I spent $1\frac{1}{2}$ hours on my homework last night, of which one third was on Geography and half of that on drawing a map. What fraction of an hour did I spend drawing the map?

6 **a** A 2 kg box of tea is used to fill several $\frac{1}{4}$ kg packets. How many packets can be filled?

b Calculate (i) $2 \div \frac{1}{4}$ (ii) $2 \div \frac{1}{3}$
(iii) $3 \div \frac{1}{2}$ (iv) $4\frac{1}{2} \div \frac{1}{2}$
(v) $1\frac{2}{5} \div \frac{1}{5}$ (vi) $3\frac{1}{2} \div \frac{1}{4}$.

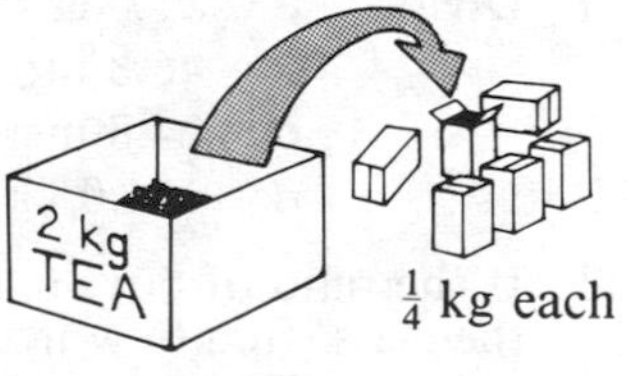

7 Work the following, remembering to cancel where possible.

a $1\frac{1}{3} \div \frac{2}{3}$ **b** $4\frac{1}{2} \div \frac{3}{4}$ **c** $2\frac{2}{5} \div 1\frac{1}{3}$ **d** $3\frac{1}{3} \div 7\frac{1}{2}$

e $\dfrac{1\frac{1}{3}}{2\frac{2}{9}}$ **f** $\dfrac{10\frac{1}{2}}{2\frac{1}{4}}$ **g** $\dfrac{4\frac{2}{3}}{1\frac{1}{6}}$ **h** $\dfrac{4\frac{1}{2}}{3\frac{3}{4}}$

8 **a** Mr Watson takes $7\frac{1}{2}$ hours to travel to Paris of which $2\frac{1}{4}$ hours is spent in an aeroplane. What fraction of the whole journey is spent in the plane?

b A $12\frac{1}{2}$ m^3 crate is partly filled with $7\frac{1}{2}$ m^3 of powder. What fraction of the volume of the crate has powder in it?

c I have $6\frac{1}{4}$ kg of oats to feed my pets. If I use $\frac{5}{8}$ kg each day, how many days will the oats last?

Fractions

Part 3 Ratios and fractions of quantities

1 Calculate

a $\frac{2}{3}$ of £9·36 **b** $\frac{3}{5}$ of £7·25 **c** $\frac{3}{4}$ of 29.2 metres

d $\frac{5}{6}$ of 3.72 metres **e** $\frac{7}{10}$ of 12.8 litres **f** $\frac{4}{5}$ of 4.65 kg.

2 If I use three-fifths of a 1.25 kg bag of flour, how much flour is left?

3 A school has 680 children of whom $\frac{5}{8}$ walk to school, $\frac{1}{4}$ come by bus and the rest come by car. How many come **a** on foot **b** by bus **c** by car?

4 Two-sevenths of Mr Auty's annual salary of £7175 is taken in tax. How much does he pay in tax each year?

5 I have 12.8 m^3 of ready-mixed concrete delivered to build a wall and lay a path. I only use $\frac{7}{8}$ of it and $\frac{3}{5}$ of what I use goes into making the path. Find the volume of concrete in the path.

6 This rectangle is made from twelve equal squares. Find

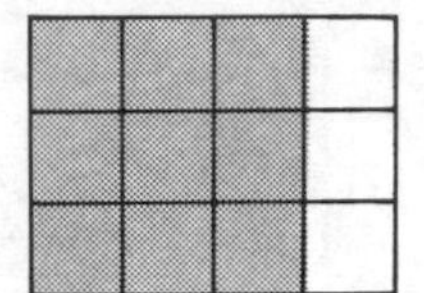

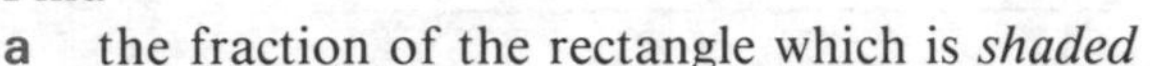

a the fraction of the rectangle which is *shaded*

b the fraction of the rectangle which is *unshaded*

c the ratio of the *shaded* part to the *unshaded* part.

7 I mix six spadesful of sand with two spadesful of cement to make concrete. What is the ratio of sand to cement (in its lowest terms)?

8 One cog with 240 teeth intermeshes with another cog with 60 teeth. What is the gear ratio (i.e. the ratio of the numbers of teeth)?

9 In a school of 432 children there are 24 teachers. Find the pupil to teacher ratio, in its lowest terms.

10 This pie chart shows the weather conditions for the days of January in a certain town. What is the ratio of

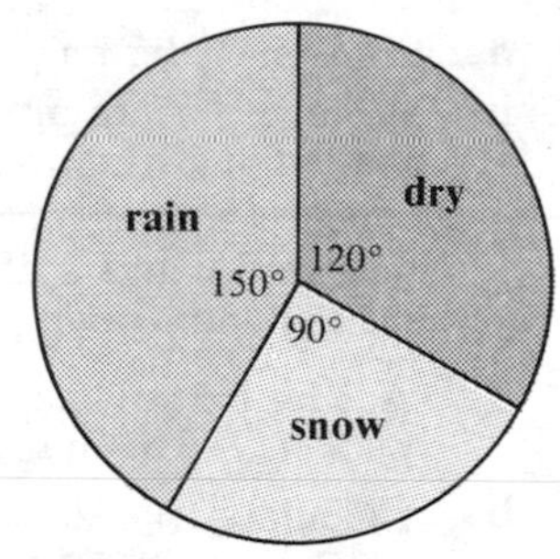

a dry days to days of snow

b dry days to days of rain?

11 Divide **a** £17·12 in the ratio 1:3

b 46.8 litres in the ratio 1:5

c 24.8 metres in the ratio 2:3

d 134.0 kg in the ratio 1:2:7

12 If the ratio of boys to girls in a class of pupils is 2:3 and there are 25 pupils in the class, find how many are **a** boys **b** girls.

13 A company posts 832 letters in a week and the ratio of first-class to second-class is 3:5. How many letters were sent by

a first-class post **b** second-class post?

14 For every hectare of land planted with oats, a farmer plants 3 hectares with barley. If he sows seed on 136 hectares altogether, how many hectares are sown with **a** oats **b** barley?

15 An alloy of metal has 5 grams of iron and 4 grams of copper for every 2 grams of nickel. In 264 grams of alloy, what mass is **a** nickel **b** iron?

Percentages

Part 1 Percentages, decimals and fractions

1 What percentage of these squares is shaded?

a 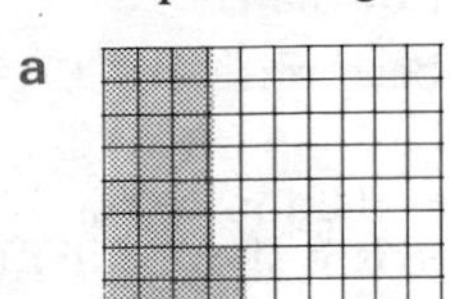 **b** 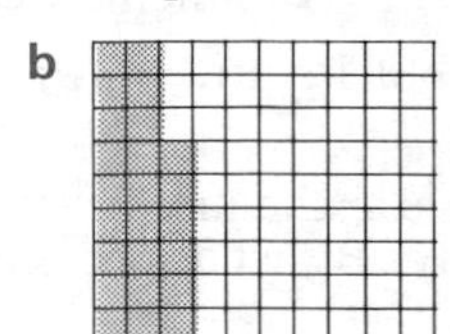 **c** 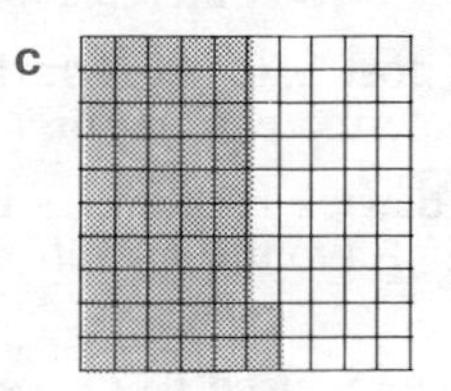 **d** 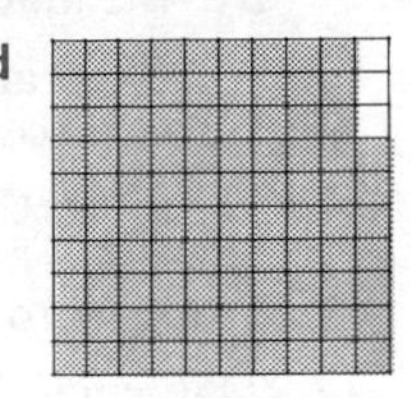

2 Change these percentages to fractions and cancel as far as possible.

a 40% **b** 80% **c** 25% **d** 35% **e** 10%
f 5% **g** 50% **h** 75% **i** 90% **j** $12\frac{1}{2}\%$

3 Change these fractions to percentages.

First type

a $\frac{13}{50}$ **b** $\frac{27}{50}$ **c** $\frac{6}{25}$ **d** $\frac{21}{25}$ **e** $\frac{3}{20}$
f $\frac{7}{10}$ **g** $\frac{3}{5}$ **h** $\frac{1}{4}$ **i** $\frac{3}{4}$ **j** $\frac{7}{200}$

Second type

k $\frac{3}{8}$ **l** $\frac{5}{8}$ **m** $\frac{1}{3}$ **n** $\frac{2}{3}$ **o** $\frac{2}{9}$
p $\frac{4}{9}$ **q** $\frac{1}{7}$ **r** $\frac{6}{7}$ **s** $\frac{5}{6}$ **t** $\frac{7}{12}$

4 Write these percentages as decimals.

a 67% **b** 32% **c** 6% **d** 2% **e** 145%

Write these decimals as percentages.

f 0.84 **g** 0.03 **h** 1.67 **i** 0.96 **j** 0.005

Part 2 Percentages of quantities

1 a *First type* A man earns £6720 in a year. If 26% of this is paid in tax, find the amount of tax he pays in the year.

b *Second type* Another man earns £7200 in a year. If he pays £1440 in tax, find the percentage of his earnings paid in tax.

The remainder of this exercise is a mixture of these two types of problem. Decide which type of problem each one is before you start your calculations.

2 I pay a 12% deposit on a television costing £250. How much do I pay as a deposit?

3 15% of the cost of a £140 holiday covers the cost of the travel. How much is this?

4 A boy saved £80 and then spent £16 of it on a holiday. What percentage of his savings did he spend?

5 A new bicycle is advertised at £75 but during a sale it is reduced by £9. What percentage reduction of the full cost is this?

6 A tank holds 175 litres of petrol, but 16% is lost through a leak. Find how many litres are lost.

7 If John Green planted 320 trees in spring but 35% of them died during the very dry summer, how many died?

8 A workman earns £240 per month but pays 24% of it in tax. Find how much tax he pays each month.

Percentages

9 A new gas cooker cost £470 last month, but its price has just been increased by 5%. Find
a the amount of this increase **b** the new cost of the cooker.

10 A refrigerator cost £56 last month, but its price has just been raised by £7. What percentage increase is this?

11 Sutton to Chadswick is 48 miles if I use the motorway but the journey is 18 miles shorter if I avoid the motorway. By what percentage is it shorter if I use ordinary roads?

12 2450 cm^3 of steel is used to make ball-bearings. If 14% is wasted in the process, find
a the volume wasted **b** the volume used.

13 A farmer has 640 hectares of land of which 96 hectares are unproductive. What percentage of the land is
a unproductive **b** productive?

14 The Forestry Commission buys 3450 hectares of land for planting trees. If trees are not planted on 8% of the land, how many hectares are unplanted?

15 A ball of string is 25.6 metres long, but 6.4 metres are cut off. What percentage is
a cut off **b** left?

16 Of 6.25 metres of dress material, 0.25 metres is wasted. What percentage is
a wasted **b** used?

17 A pair of shoes cost £12·60 but the price is reduced by £1·89 in a sale. Find the percentage reduction.

18 A building society gives 7.6% p.a. interest on money deposited with it. Find the interest on savings of £3250 in one year.

19 A magazine sold 164 250 copies in March. Its April sales increased by 2.4%. Find how many extra copies were sold in April.

20 In thc County of Elmet, out of 378 000 people of working age, 34 020 have no job. What percentage of the working population is unemployed?

21 A car travels the 540 km from Gloucester to Glasgow, but its odometer is faulty and its reading is 558 km. Find
a the error in kilometres **b** the percentage error.

22 A council spends £355 000 on all the services for its area during one month. If 42.6% of this amount is spent on education, find the cost of the education service for the month.

Part 3 Percentages in society

Discounts

1 When dealing with percentages of money, it is often easier to think of 'pennies in the pound', especially as many stores advertise their discounts in this way. Find the amount of these discounts.

a 2p in the £ on shoes costing £6
b 3p in the £ on a coat costing £20
c 5p in the £ on a carpet costing £60
d 12p in the £ on furniture costing £70
e 15p in the £ on garage repairs costing £82
f 13p in the £ on a typewriter costing £124
g 17p in the £ on a new cooker costing £436
h 12p in the £ on a dining-room suite costing £748

Percentages

2 If you pay your rates promptly, some councils will give you a refund.
Find the amount of these refunds.

- **a** 3p in the £ on rates of £120
- **b** 5p in the £ on rates of £234
- **c** 6p in the £ on rates of £172
- **d** 4p in the £ on rates of £80
- **e** 3p in the £ on rates of £140
- **f** 7p in the £ on rates of £286

Huntswich District Council

RATE DEMAND NOTE

		Property address
Reference No.		
Valuation		
District rate		
Domestic rate		
TOTAL RATE		

3 You order goods through the post from a mail-order firm which gives you a discount of 2p in the £ on the first £10 you spend. This discount rises to 4p in the £ on the next £20 spent, and rises again to 5p in the £ for every £1 spent over the first £30.
Find the discount on goods costing
a £8 **b** £12 **c** £20 **d** £30 **e** £31 **f** £64.

Cost price and selling price

4 Find the profit and the selling price when an article is bought at a cost price of

- **a** £240 with a percentage profit of 6%
- **b** £380 with a percentage profit of 15%
- **c** £764 with a percentage profit of 24%
- **d** £493 with a percentage profit of 32%.

5 Find (i) the actual profit (or loss)
(ii) the percentage profit (or percentage loss) when the cost price and selling price of an article are respectively

- **a** £150 and £180
- **b** £60 and £75
- **c** £280 and £350
- **d** £120 and £102.

6 A child's toy costs £4·50 to make and is sold for £5·40. Find
a the profit **b** the percentage profit.

7 A garden seat is made for £15·60 and is sold to make a profit of £1·95. Find
a the percentage profit **b** the selling price.

8 A shop buys a camera for £124 and intends to make a profit of 16% when it is sold.
a How much profit does the shop want to make?
b What will be the selling price?

9 A department store wants to sell out-dated stock which cost £1470. If it is prepared to make a loss of 24%, what should be the selling price?

10 A car tyre cost £32 on Monday. Tuesday's price is 10% higher than Monday's. On Wednesday, a sale is announced '10% off yesterday's price'. Find the price of the tyre on both Tuesday and Wednesday.

Inflation and Compound Interest

11 A new car cost £8400 in January 1983. Each year for the next three years, the price increased by 10% p.a. Use this table to find its price in January 1986.

	£
Price in Jan. 1983	
Increase	+ ______
Price in Jan. 1984	
Increase	+ ______
Price in Jan. 1985	
Increase	+ ______
Price in Jan. 1986	______

Percentages

12 A bank pays interest at a rate of 8% p.a. If you deposit £1250 with the bank, use this table to find the total amount after three years, to the nearest penny.

		£
Initial amount		
Interest	+	________
Total after 1 year		
Interest	+	________
Total after 2 years		
Interest	+	________
Total after 3 years		________

13 The return flight to Jamaica costs £640 at today's prices. If this price inflates by 5% p.a., find the cost of the flight after two years.

14 Mr and Mrs Crummack have their savings of £3600 in a Building Society account earning interest at 12% p.a. What are their savings worth after two years?

15 A new piano costs £4000 today. If the rate of inflation stays at 10% p.a., what will be the cost of a similar piano after four years?

16 You invest £650 in a bank account where it earns 14% interest p.a. for five years. Use a calculator to find how much is in the account at the end of the five years, showing your working to the nearest penny when necessary.

Percentage error

17 The label on a jar of jam gives the mass as 500 grams. The actual mass is 515 grams. What is

a the actual error **b** the percentage error in the label?

18 Mr Goddard thought he had £10·80 in his bank account, but when a statement arrives he finds he has £14·40. Find

a the error **b** the percentage error in Mr Goddard's estimate.

19 Annabel Riley's maths exam paper has a mark of 63 written at the top, but when she checks this she finds she should have 72. Find

a the error in her teacher's marking

b the percentage error in the marking.

20 A schoolboy reads two thermometers in a science experiment. His readings are 49.8°C on thermometer A and 25.2°C on thermometer B. Both readings are inaccurate; the actual temperatures are 48.0°C for A and 28.8°C for B. Find

a the errors **b** the percentage errors in his two readings.

Estimation

Part 1 Appropriate units

1 Would you use seconds, minutes or hours to measure the time

a to cross the Atlantic by air
b for a roll of thunder
c to boil an egg
d to cook a turkey
e of flight of an arrow
f for a train journey between York and Crewe
g for a flash of lightning
h for a sermon in church
i to count to ten
j to wallpaper a room
k to empty a dustbin
l to eat an apple?

2 Would you use grams, kilograms or tonnes to measure the mass of

a a railway engine
b a spoonful of sugar
c a piece of cake
d a bag of potatoes
e a man
f a ship
g an apple
h a table
i an egg
j an aeroplane
k a light bulb
l a load of coal?

3 Would you use millimetres, centimetres, metres or kilometres to measure the length of

a the M5 motorway
b a finger
c a garden path
d a stick of celery
e a grain of sand
f a train journey
g an airport runway
h the seed of a flower
i a motor-racing circuit
j a sitting-room
k a chair leg
l a drain-pipe?

4 Would you use mm^2, cm^2, m^2 or km^2 to measure the area of

a a leaf
b a bus ticket
c a carpet
d a house wall
e Lincolnshire
f a breadboard
g a large forest
h a pin-head
i a sheet of notepaper
j the eye of a needle
k a garden
l open moorland?

5 Would you use cm^3, litres or m^3 to measure the volume of

a a spoonful of sugar
b a ship's hold
c a bottle of wine
d a water tub
e an office block
f an egg yolk
g a glass of water
h the dome of St Paul's
i an aircraft hanger
j some concrete foundations
k a hot-water cylinder
l a reservoir?

6 In these multiple-choice questions, select the most reasonable answer.

a	Mr Bray has a height of	(i) 0.65 m	(ii) 1.85 m	(iii) 3.2 m.
b	Mr Bray's car has a length of	(i) 1.3 m	(ii) 4.2 m	(iii) 9.75 m.
c	The distance across Birmingham is	(i) 1 km	(ii) 10 km	(iii) 100 km.
d	The height of a house is	(i) 3 m	(ii) 10 m	(iii) 50 m.
e	A family car has a mass of	(i) 1 gram	(ii) 1 kg	(iii) 1 tonne.
f	A bag of flour has a mass of	(i) 1 gram	(ii) 1 kg	(iii) 1 tonne.

Estimation

g An envelope has a mass of (i) 1 gram (ii) 1 kg (iii) 1 tonne.
h An apple has a mass of (i) 2 grams (ii) 100 grams (iii) 200 grams.
i A loaf of bread has a mass of (i) $\frac{1}{10}$ kg (ii) $\frac{1}{2}$ kg (iii) 10 kg.
j The length of my middle finger is (i) 1 cm (ii) 5 cm (iii) 15 cm.
k The width of a country lane is (i) 1 m (ii) 5 m (iii) 20 m.
l A tin of motor oil holds (i) 1 litre (ii) 20 litres (iii) 100 litres.
m A vase of flowers holds (i) $\frac{1}{100}$ litre (ii) $\frac{1}{2}$ litre (iii) 5 litres of water.
n A sugar lump has a volume of (i) 1 cm^3 (ii) 10 cm^3 (iii) 50 cm^3.
o A teaspoonful of sugar has a volume of (i) $\frac{1}{2}$ cm^3 (ii) 2 cm^3 (iii) 20 cm^3.
p A kitchen cupboard has a volume of (i) 1 cm^3 (ii) 1 m^3 (iii) 1 km^3.
q A small bedroom has a volume of (i) 2 m^3 (ii) 20 m^3 (iii) 200 m^3.
r The area of a postage stamp is (i) 2 mm^2 (ii) 2 cm^2 (iii) 2 m^2.
s A coffee table has a top of area (i) 1 cm^2 (ii) 1 m^2 (iii) 1 km^2.
t The surface area of a garage door is (i) $\frac{1}{2}$ m^2 (ii) 6 m^2 (iii) 19 m^2.
u The school hall has a floor area (i) 15 m^2 (ii) 150 m^2 (iii) 1500 m^2.
v The area of an 18-hole golf course is (i) $\frac{1}{2}$ km^2 (ii) 2 km^2 (iii) 10 km^2.
w The area of a football pitch is about
(i) 15 hectares (ii) 5 hectares (iii) 1 hectare.
x The distance from Inverness to Plymouth is
(i) 100 km (ii) 1000 km (iii) 10 000 km.
y A golf ball has a surface area of about
(i) 5 cm^2 (ii) 50 cm^2 (iii) 500 cm^2.
z When I have a bath I use about
(i) 10 litres (ii) 100 litres (iii) 1000 litres of water.

7 Are these statements reasonable? Answer *yes* or *no*.
If the answer is *no*, rewrite the sentence and include a reasonable figure.

a The mass of a bag of sugar is 1 kg.
b The height of a policeman is 2 metres.
c The length of a goods train is 6 km.
d The volume of a wooden crate is 1 m^3.
e The mass of a cooking apple is $\frac{1}{2}$ kg.
f The mass of a jar of jam is 500 grams.
g The height of a telegraph pole is 135 cm.
h The area of a postage stamp is 80 cm^2.
i The area of a living-room carpet is 20 m^2.
j The volume of wine in a bottle is 1 litre.
k The capacity of a bucket is 20 litres.
l A car's petrol tank will hold 50 litres.
m The distance from Leeds to London is about 200 miles.
n It takes about 24 hours to fly across the Atlantic Ocean to New York.
o A newspaper has a mass of about 100 grams.

Estimation

p A lump of sugar has a mass of about $\frac{1}{4}$ kg.
q A daffodil will grow about 30 mm high.
r A page of a book has an area of $\frac{1}{2}$ m^2.
s A needle has a length of about 50 mm.
t A shoe box has a volume of 1 m^3.
u A bedroom has a volume of 4000 m^3.
v A forest has an area of $\frac{1}{4}$ km^2.
w A car has a mass of 20 tonnes.
x A rugby field has an area of about $\frac{3}{4}$ hectare.
y My uncle has a mass of 320 kg.
z The mass of his wife is 70 kg.

8 Copy these sentences and complete them by making a reasonable estimate.
a A room in a house is about ... metres high.
b A telegraph pole is about ... metres high.
c The pavement alongside a road is about ... metres wide.
d A bottle of milk will hold about ... litres.
e A washing-up bowl has a capacity of about ... litres.
f The fuel tank in an average car will hold about ... litres of petrol.
g A table-cloth has an area of about ... m^2.
h A school blackboard has an area of about ... m^2.
i A page of this book has an area of about ... cm^2.
j A jar of marmalade has a mass of about ... grams.
k An egg has a mass of about ... grams.
l My own mass is about ... kg.
m An ordinary family car has a mass of about ... tonnes.
n A window pane has an area of about ... m^2 of glass.
o The shortest distance across the English Channel is about ... km.
p The width of a dual-carriageway is about ... metres.
q The area of a finger-nail is about ... cm^2.
r The volume of a drop of water from a tap is about ... cm^3.
s The area of the pupil in the centre of someone's eye is about ... mm^2.
t The volume of a bowl of soup is about ... litre.
u The mass of a luxury 40-seater coach is about ... tonnes.
v I can run about ... metres in 10 seconds.
w I can cycle about ... km in 5 minutes.
x On a motorway a car travels about ... km in one hour at a cruising speed.
y The distance across the rim of a teacup is about ... cm.
z An average-sized teacup can hold about ... cm^3 of tea.

9 Answer these questions *yes* or *no*.
a If you put supplies with a mass of 80 kg into a haversack, would you be able to carry it on your back?
b If a car driver puts 4 litres of petrol into his empty petrol tank, will this be enough to fill it?
c If you have a square piece of hardboard 4 metres by 4 metres, will you be able to get it through the front door of your house?
d Is it possible to run 200 metres in one minute?
e Would Mrs Simons be able to make a full-length coat from $\frac{1}{2}$ m^2 of material?

Estimation

f Would it be possible for Mr Cook to dig a hole of 1 m^3 in his garden in five minutes?

g If your friend said his front bicycle tyre had a volume of 100 cm^3, would you believe him?

h Would you be able to lift a 5 kg bag of cement with one hand?

i Could you blow up a party balloon so that it had a volume of 2 m^3?

j Would 4 hours be long enough for an ordinary family car to travel 400 km on a motorway?

Part 2 Length and area

1 Copy this table.

	a	b	c	d	e	f	g	h
Estimated length, cm								
Measured length, cm								

This line ——— is 1 cm long.
Estimate, without any measuring, the lengths of these lines in centimetres.
Enter the estimates in your table.

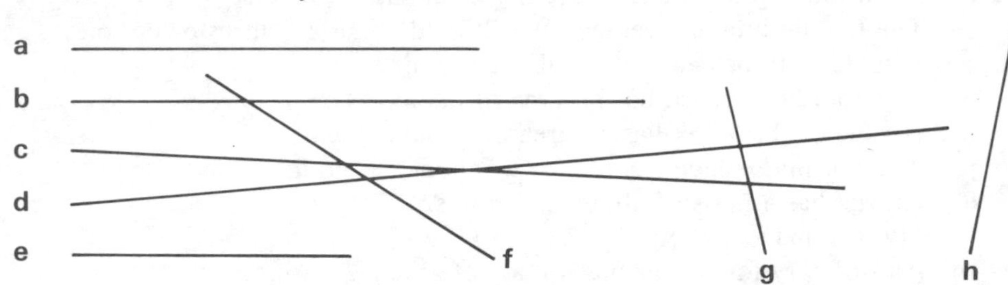

Now use a ruler to measure the lines, and enter these results in the table.

2 Copy this table.

	a	b	c	d	e	f
Estimated length, cm						
Length measured by paper, cm						
Length measured by cotton, cm						

Estimate, without any measuring, the lengths of these lines in centimetres.
Enter the estimates in your table.

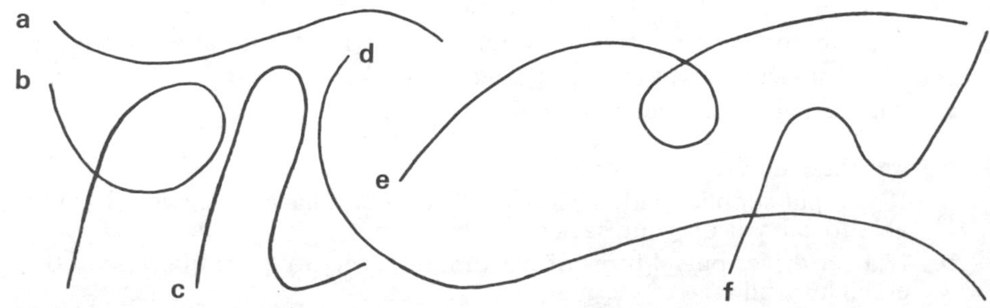

Now measure their lengths by
(i) using the edge of a piece of paper and marking off each line in sections and (ii) using a piece of cotton thread.
Enter the results in your table.

Estimation

3 This map has a scale of 1 cm = 1 km, and its edge is marked in 1 cm intervals.

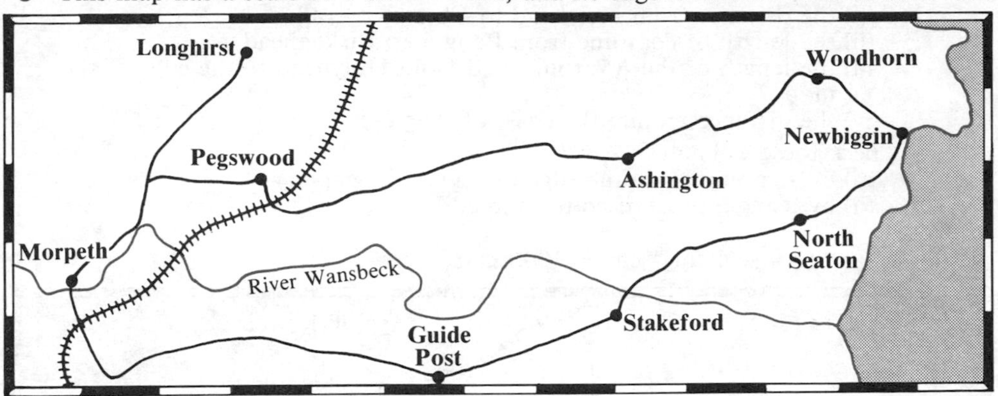

a Estimate the *shortest* distance (in kilometres) between
(i) Guide Post and Stakeford
(ii) Longhirst and Ashington
(iii) Pegswood and Guide Post
(iv) Woodhorn and Longhirst
(v) Newbiggin and Stakeford
(vi) Stakeford and Ashington
(vii) Morpeth and Longhirst
(viii) North Seaton and Woodhorn.

b Use a ruler to check these estimates. Give the answers correct to 0.1 km.

c Estimate, without measuring,
(i) the length of railway line across the map
(ii) the length of the river Wansbeck from Morpeth to where it meets the sea
(iii) the length of coastline shown on the map
(iv) the distance by road from Morpeth to Newbiggin via Pegswood and Ashington
(v) the distance by road from Morpeth to Newbiggin via Guide Post and Stakeford.

d Check these estimates
(i) by stepping along the edge of a piece of paper
(ii) by using a piece of cotton thread.

4 This map has a scale of 1 cm = 10 km.

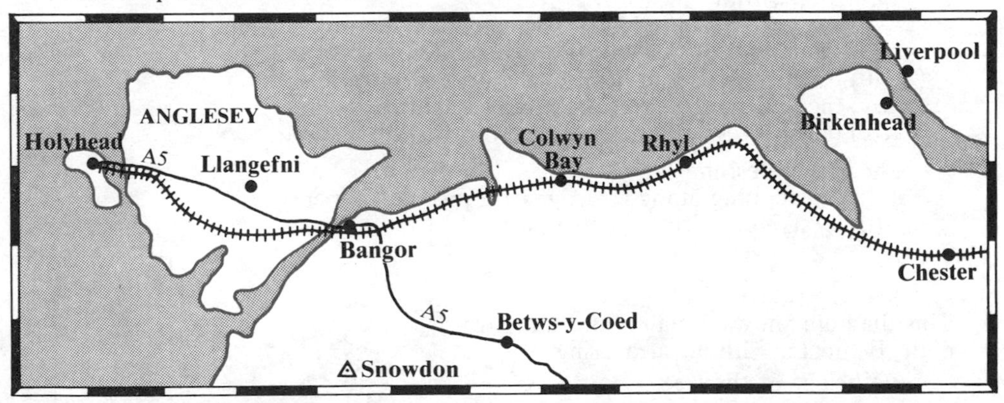

a Estimate without measuring, the *shortest* distance (in kilometres) between
(i) Holyhead and Bangor
(ii) Llangefni and Betws-y-Coed
(iii) Colwyn Bay and Chester
(iv) Betws-y-Coed and Liverpool
(v) Snowdon and Liverpool
(vi) Holyhead and Birkenhead
(vii) Chester and Bangor
(viii) Rhyl and Holyhead

b Use a ruler to check these estimates. Give the answers correct to the nearest kilometre.

Estimation

c Estimate, without measuring,
(i) the distance from Holyhead to Chester by rail
(ii) the length of coastline from Bangor to Birkenhead
(iii) the length of the A5 trunk road from Holyhead to the edge of the map
(iv) the distance around the coast of Anglesey.

d Check these estimates
(i) by stepping along the edge of a piece of paper
(ii) by using a piece of cotton thread.

5 This map has a scale of 1 cm = 100 km.

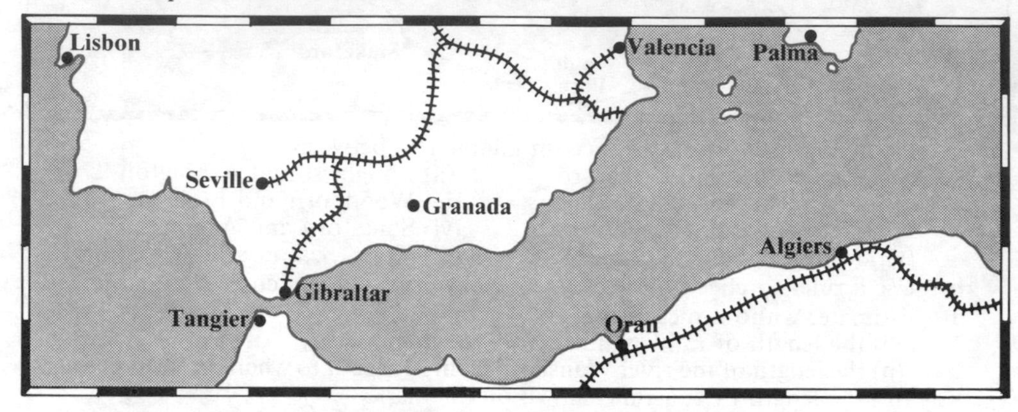

a Estimate, without measuring, the *shortest* distance (in kilometres) between
(i) Valencia and Palma
(ii) Lisbon and Palma
(iii) Granada and Seville
(iv) Tangier and Algiers
(v) Gibraltar and Oran
(vi) Valencia and Oran
(vii) Gibraltar and Palma
(viii) Lisbon and Algiers.

b Use a ruler to check these estimates. Give the answers correct to the nearest 10 km.

c Estimate, without measuring,
(i) the length of the North African coastline on this map
(ii) the length of the European coastline on this map
(iii) the distance from Seville to Valencia by rail
(iv) the length of the Algerian railway shown here
(v) the shortest distance by sea from Lisbon to Oran.

d Check these estimates
(i) by stepping along the edge of a piece of paper
(ii) by using a piece of cotton thread.

6 This diagram shows a man about 2 metres high. Estimate, without measuring,

a the height of the tree
b the height of the street light
c the height of the office block
d the length of the fence.

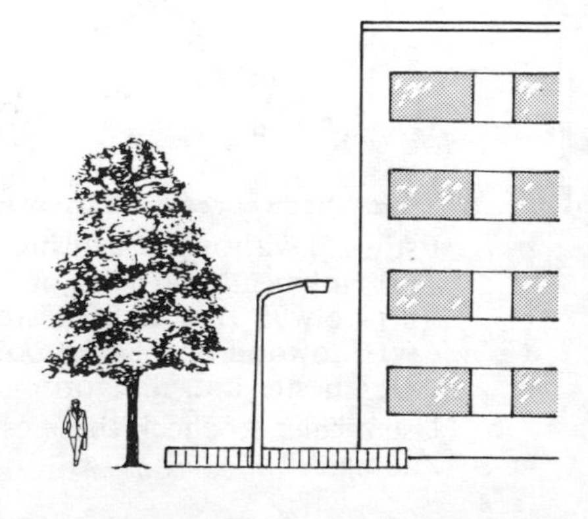

Estimation

7 This diagram shows a lorry 14 metres long.
Estimate, without any measurement,

a the length of the car
b the height of the lorry
c the height of the telegraph pole
d the height of the pylon.

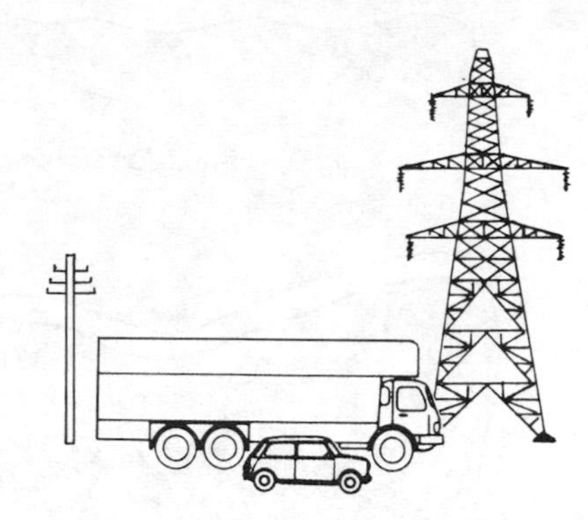

8 Copy this table.

	a	b	c	d	e	f
Estimated area, cm^2						
Measured area, cm^2						

This square [ ] has an area of 1 cm^2.

Estimate, without measuring, the areas of these rectangles and squares.

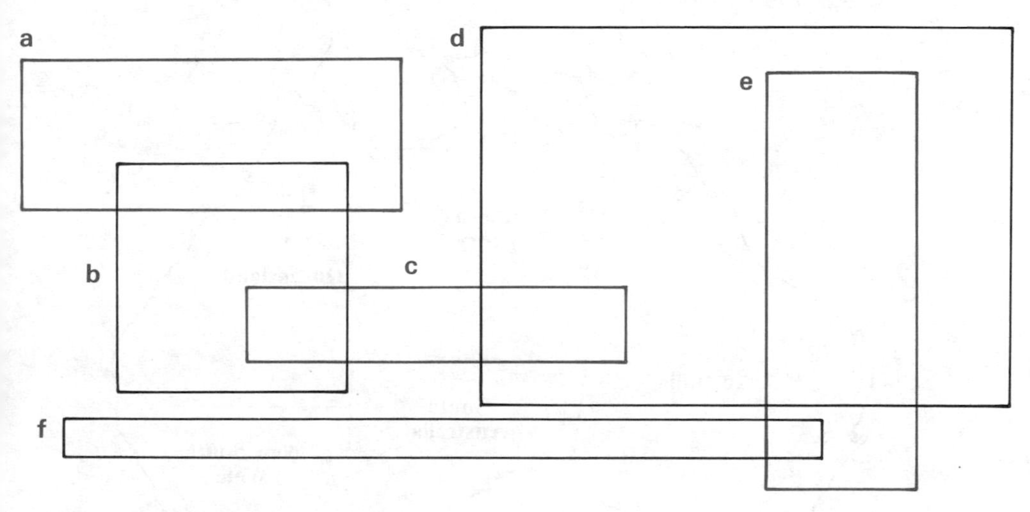

Now use a square grid on tracing paper to find the actual areas and enter the results in your table.

9 Copy this table.

	Field X	*Field Y*	*Field Z*
First estimate of area, cm^2			
Second estimate using tracing paper, cm^2			
Actual area, cm^2			

Estimation

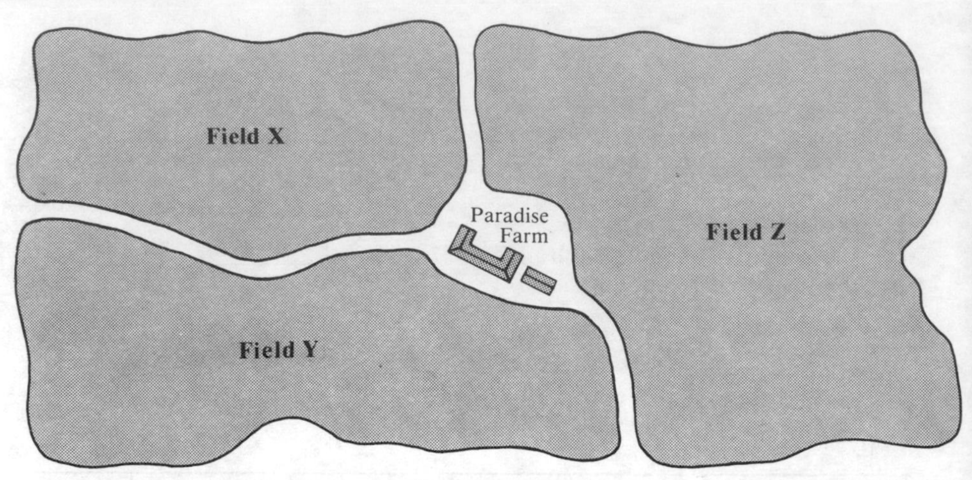

Three fields X, Y and Z, are separated by lanes which lead to Paradise Farm.

a Estimate, by looking and without any measurement, the areas of the three fields on this map in cm^2. Enter your estimates in the table.

b Find a better estimate for each field by using a square grid on tracing paper. Count a square centimetre if *more* than half of it is inside a field; but do *not* count it if *less* than half is inside. Enter your results in the table.

c Ask your teacher for the actual areas of the fields on the map and enter them in the table. See how close your estimates were.

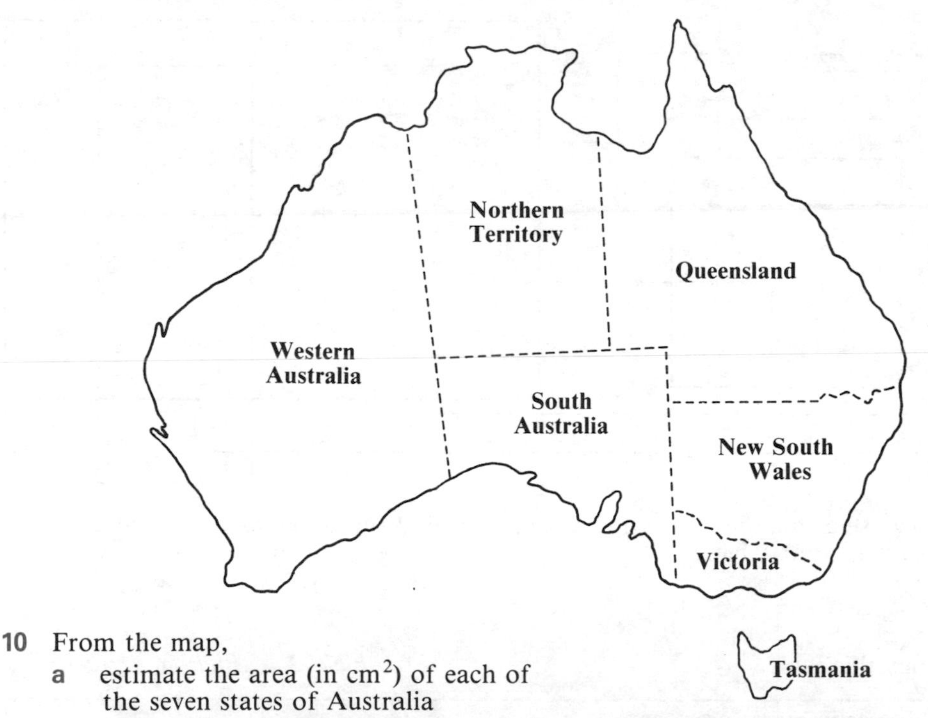

10 From the map,

a estimate the area (in cm^2) of each of the seven states of Australia

b estimate the length (in cm) of the coastline of mainland Australia

c If the length scale factor of the map is 1 cm = 400 km and the area scale factor is $1\ cm^2 = 160\ 000\ km^2$, convert your answers in **a** and **b** to give approximate values in km^2 and km.

Estimation

Part 3 Making reasonable estimates

In these multiple-choice questions, choose the answer which is the most reasonable.

1 If 1 metre is about 3 feet, then 4.2 metres are about
a 7 feet **b** 13 feet **c** 25 feet.

2 If 1 litre is rather less than 2 pints, then 12 litres are about
a 10 pints **b** 15 pints **c** 21 pints.

3 If 1 kg is a little more than 2 lb (pounds), then 7.8 kg are about
a $3\frac{1}{2}$ lb **b** 17 lb **c** 30 lb.

4 If 5 miles is about 8 km, then 21 miles are about
a 25 km **b** $2\frac{1}{2}$ km **c** 34 km.

5 If 1 gallon is about $4\frac{1}{2}$ litres, then $2\frac{1}{2}$ gallons are about
a 2 litres **b** 20 litres **c** 10 litres.

6 If 1 inch is about $2\frac{1}{2}$ cm, then 1 foot is about
a $24\frac{1}{2}$ cm **b** 40 cm **c** 30 cm.

7 If 1 hectare is about $2\frac{1}{2}$ acres, then 4.2 hectares are about
a 2 acres **b** 10 acres **c** 20 acres.

8 If 1 metre is about 1.1 yards, then 4.9 metres are about
a 2 yards **b** 5 yards **c** 10 yards.

9 If 1 ounce is about 28 grams, then 6 oz are about
a 4 grams **b** 100 grams **c** 180 grams.

10 If 1 stone is about $6\frac{1}{3}$ kg, then 8 stones 3 lb are about
a $1\frac{1}{2}$ kg **b** 20 kg **c** 50 kg.

11 If 1 square inch is about $6\frac{1}{2}$ cm^2, then 14 square inches are about
a 60 cm^2 **b** 100 cm^2 **c** 200 cm^2.

12 A tin of chopped meat costs £1·23. Seven tins will cost about
a £16 **b** £80 **c** £8.

13 A one-metre strip of metal has a mass of 52 grams. A $4\frac{1}{2}$-metre strip has a mass of about
a 20 grams **b** 200 grams **c** 2000 grams.

14 A sack of potatoes has a mass of 5.5 kg. The mass of 15 similar sacks is about
a 3 kg **b** 50 kg **c** 80 kg.

15 At an average speed of 345 miles per hour, in 3 hours an aeroplane will travel about
a 100 miles **b** 1000 miles **c** 10 000 miles.

16 A car travelling for $6\frac{1}{2}$ hours on a motorway at an average speed of 84 km/h covers a distance of about
a 45 km **b** 4800 km **c** 500 km.

17 If a rectangle is 22.8 cm long and 9.7 cm wide, its area is about
a 200 cm^2 **b** 2000 cm^2 **c** 32 cm^2.

18 Is 384 nearer to **a** 400 **b** 300 or **c** 38?
19 Is 28.9 nearer to **a** 20 **b** 2.9 or **c** 30?
20 Is 7.24 nearer to **a** 9 **b** 1 or **c** 0.72?
21 Is 0.87 nearer to **a** 1 **b** 0.01 or **c** 0.09?

Estimation

Decide which of the choices offered in each of these questions is the nearest to the actual answer. But do *not* work out the calculation accurately.

22	157 + 32 + 19	**a** 100		**b** 200		**c** 500	
23	302 + 114 + 239	**a** 60		**b** 400		**c** 600	
24	753 + 419 + 642	**a** 1500		**b** 1800		**c** 20 000	
25	46 + 561 + 37	**a** 640		**b** 1200		**c** 64	
26	6 + 12 + 352 + 84	**a** 400		**b** 450		**c** 490	
27	224 − 158	**a** 70		**b** 100		**c** 170	
28	804 − 246	**a** 450		**b** 650		**c** 550	
29	1082 − 367	**a** 820		**b** 76.5		**c** 720	
30	1416 − 551	**a** 90		**b** 900		**c** 1100	
31	621 − 93	**a** 430		**b** 480		**c** 530	
32	7.15 × 3.04	**a** 18		**b** 22		**c** 29	
33	6.92 × 5.17	**a** 30		**b** 35		**c** 42	
34	12.3 × 4.8	**a** 17		**b** 48		**c** 60	
35	21.5 × 9.4	**a** 30		**b** 180		**c** 210	
36	38.7 × 6.2	**a** 24.4		**b** 180		**c** 240	
37	42.1 ÷ 5.07	**a** 0.8		**b** 8		**c** 200	
38	24.8 ÷ 6.11	**a** 2		**b** 4		**c** 6	
39	62.7 ÷ 7.9	**a** 8		**b** 19		**c** 20	
40	146 ÷ 3.1	**a** 5		**b** 30		**c** 50	
41	872 ÷ 5.9	**a** 12		**b** 120		**c** 1200	

42 These calculations have been given approximate answers. Are the answers reasonable? Say *yes* or *no*; if *no*, give a reasonable approximate answer. Check your approximations using a calculator.

a $12.7 \times 4 \simeq 50$ **b** $25.2 \times 8.1 \simeq 200$

c $36.1 \times 1.97 \simeq 72$ **d** $103.2 \div 4.1 \simeq 2.5$

e $614.7 \div 60.1 \simeq 10$ **f** $50.3 \div 1.9 \simeq 2.5$

g $0.64 - 0.19 \simeq 0.04$ **h** $1.83 - 0.98 \simeq 0.8$

i $\dfrac{47.2 + 13.7}{2.04} \simeq 30$ **j** $\dfrac{81.6 + 38.9}{6.12} \simeq 20$

k $\dfrac{19.4 + 7.55}{12.8} \simeq 20$ **l** $\dfrac{142 - 67.8}{4.3} \simeq 17$

m $2.13 \times 5.04^2 \simeq 50$ **n** $4.7 \times 3.1^2 \simeq 144$

o $(5.2 \times 1.93)^2 \simeq 100$ **p** $(0.51 \times 18.3)^2 \simeq 80$

Part 4 Problems

Estimate the answers without any written work.
Check your estimates using a calculator.

1 If I want to buy six tins of soup at 21 pence each, will I have enough with £1?

2 Mrs Barraclough takes her three children on a bus. She pays 64 pence for herself and half price for each child. Will the £2 she offered the conductor be enough?

3 I usually average a speed of about 50 miles per hour on a motorway. How much time should I allow for the 235 miles from Exeter to Manchester?

4 Jane Askwith needs 3 metres of material to make herself a dress. If her choice costs £3·45 per metre, will she have to pay more than £10?

Estimation

5 A teacher collects in 28 exercise books for marking and spends an average of $5\frac{1}{2}$ minutes on each book. Will she take more than 2 hours to mark them all?

6 A farmer starts to plough his 7-hectare field and takes 34 minutes for the first $\frac{1}{2}$-hectare. About how many hours will he take for the whole field?

7 A room is 3.82 metres long and 5.04 metres wide. What is the approximate area of its floor?

8 A brick wall 13.2 metres long and 2.1 metres high, needs 48 bricks for each square metre.

a What is the approximate area of the wall?

b About how many bricks will be needed to build it?

9 Carpet for a bedroom costs £8·12 per m^2. If the bedroom is 5.2 m by 2.9 m, estimate

a the area of the floor **b** the cost of the carpet.

10 Mr Risby intends turfing his vegetable plot to make a lawn. If the plot is 12.3 m by 6.8 m and turf costs 52 pence per m^2, estimate

a the area of the plot **b** the cost of the turf.

11 A packet of eight bars of chocolate costs 85 pence. What is the approximate cost of one bar?

12 Mrs Minnet paid 86 pence for a bag of nine apples. About how much is this per apple?

13 Four people use one car to go to work and they share the cost of the fuel. If they pay a total of £12·63 in one week, about how much is this per person?

14 A 5-litre tin of oil costs £4·62. What is the approximate cost per litre?

15 A packet of six tubes of toothpaste costs £1·35. About how much is this per tube?

16 A teacher organises an outing for 32 pupils costing a total of about £170. Find the approximate cost per pupil.

17 A large office is to be fitted with carpet costing £9·43 per m^2. If the office has a rectangular floor 19.7 m by 6.2 m, estimate

a the area of the floor **b** the cost of the carpet.

18 A supermarket floor is 37.3 m long and 11.3 m wide. If it is laid with tiles costing £4·12 per m^2, estimate

a the area of the floor **b** the cost of the tiles.

19 A car averages 8.7 miles per litre of petrol.

a About how many litres does it use over 65 miles?

b About how much will it cost, if each litre costs 34.8 pence?

20 A commercial van averages 12.4 km per litre of diesel. Estimate

a the number of litres it consumes in 100 km

b the cost of this diesel, if one litre costs 32.3 pence.

21 A new car is advertised as using 7.34 litres of petrol every 100 km. Estimate

a how many litres it will need for a 543 km journey

b the cost of the petrol on this journey, if each litre costs 35.7 pence.

22 Members of a church youth group have a sponsored hymn-singing marathon to raise money for charity. If they take about 3 minutes per hymn, approximately how many will they sing in $4\frac{1}{2}$ hours?

Estimation

23 If I travel the 192 miles from Leeds to London, about how long will the journey take at an average speed of

a 50 miles per hour **b** 40 miles per hour **c** 30 miles per hour?

24 The formula $A = \pi r^2$ can be used to calculate the area A of a circle with a radius r, where π has an approximate value of 3.
Estimate the area A when the radius r is

a 1.07 cm **b** 2.15 cm **c** 4.98 cm
d 7.24 cm **e** 9.87 cm **f** 12.2 cm.

25 The energy E consumed by a resistor in an electrical circuit is given by the formula $E = Ri^2t$.
Estimate the value of E when

a $R = 5.13,\ i = 2.04,\ t = 3.92$ **b** $R = 12.2,\ i = 2.95,\ t = 10.1$
c $R = 25.6,\ i = 1.12,\ t = 6.4$.

26 Before Janice used her calculator to work out the volume V of a cylinder, she made a rough estimate in her head first.
Estimate the volume using the formula $V = \pi r^2 h$, taking π as 3, when

a $r = 5.2$ cm, $h = 9.97$ cm **b** $r = 7.1$ cm, $h = 1.94$ cm
c $r = 3.2$ cm, $h = 6.15$ cm.

27 Andrew calculated the volume of these three cones on a calculator using the formula $V = \frac{1}{3}\pi r^2 h$. His answers are given below.

a

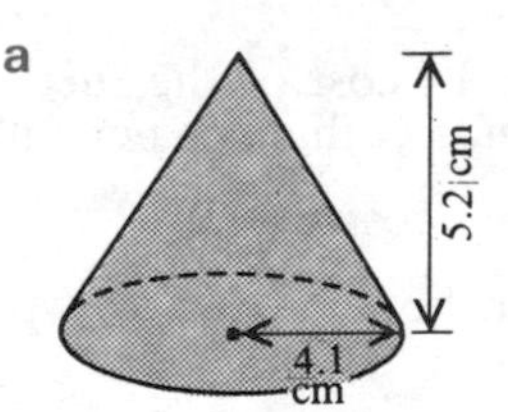

Volume = 91.5 cm³

b

7.3 cm
6.6 cm

Volume = 333 cm³

c

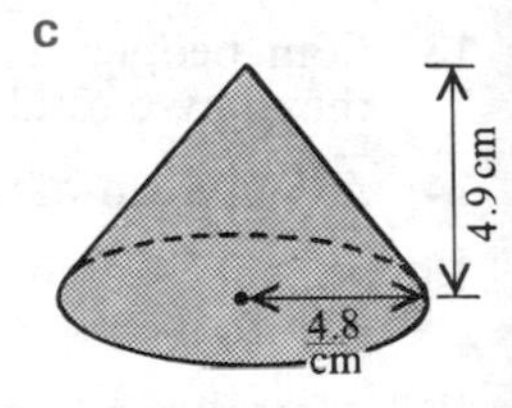

Volume = 229 cm³

Estimate the volumes and find which one of his three answers is not correct.

28 Given the area A of a circle, its radius r can be found from the formula

$r = \sqrt{\dfrac{A}{\pi}}$. Take π as approximately 3 and estimate the value of r when

a $A = 28.6$ cm² **b** $A = 47.2$ cm² **c** $A = 157$ cm² **d** $A = 243$ cm².

29 The time t taken for a car with an acceleration a to travel a distance s from rest is given by $t = \sqrt{\dfrac{2s}{a}}$. Estimate the value of t when

a $s = 26.5,\ a = 6.3$ **b** $s = 102.1,\ a = 7.97$ **c** $s = 12.6,\ a = 23.1$.

30 David used his calculator to find three values of y from the formula

$y = \dfrac{a + x^2}{b}$ taking $a = 16.2$ and $b = 2.15$ each time.

a When $x = 9.17$, he found $y = 46.6$
b When $x = 2.88$, he found $y = 11.4$
c When $x = 0.27$, he found $y = 3.01$.

Estimate the values of y for these three x-values, and so find which one of his y-values is incorrect.

Estimation

31 In these calculations, each coloured dot represents a missing operation; either +, −, × or ÷. Choose any method to find the missing operations and write out each calculation in full. Look for any clues and use a calculator to help you.

a 154 ● 289 = 443
b 847 ● 592 = 255
c 106 ● 12 = 1272
d 32 ● 85 = 2720
e 851 ● 23 = 37
f 56 ● 281 = 337
g 389 ● 472 ● 256 = 605
h (109 ● 17) ● 18 = 7
i (364 ● 4) ● 21 = 1435
j (56 ● 12) ● 3 = 204
k (112 ● 96) ● 4 = 64
l (678 ● 518) ● 92 = 13
m 36 ● 14 ● 3 = 168
n 976 ● 2 ● 8 = 61
o 4 ● (406 ● 31) = 1748
p (64 ● 85 ● 121) ● 4 = 112
q (894 ● 6) ● 14 = 163
r (366 ● 16 ● 24) ● 168 = 76

32 In these calculations, a missing digit is shown by a coloured rectangle. Choose any method to find the missing digits and write out each calculation in full. Look for any clues and use a calculator to help you.

a ■2 × 34 = 74■
b 23 × ■4 = 32■
c 42 × 5■ = 22■8
d 8■ × 47 = 39■1
e 76 × 5■ = 4■04
f 3■ × ■4 = 17■8
g 4■ × ■3 = 28■5
h ■2 × 2■ = 2■78
i ■4 × 7■ = 4■58
j 5■ × 7■ = 381■
k 8■ × 6■ = ■■89
l 7■ × 9■ = ■■54
m 1■3 × 5■ = ■8■8
n ■3■ × 24 = 7■■2
o ■5 × 23 × 6■ = 23115

Approximation

Part 1 Rounding off

1 a Use a ruler to find the perimeter of this shape in centimetres to the nearest millimetre.

b Round off your answer to the nearest whole centimetre.

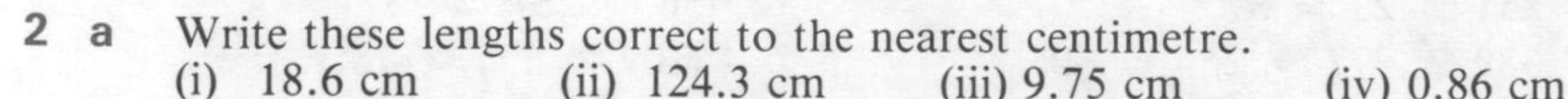

2 a Write these lengths correct to the nearest centimetre.
(i) 18.6 cm (ii) 124.3 cm (iii) 9.75 cm (iv) 0.86 cm

b Write these masses correct to the nearest kilogram.
(i) 46.2 kg (ii) 369.4 kg (iii) 3.07 kg (iv) 12.6 kg

c Write these amounts correct to the nearest ten pence.
(i) £6·19 (ii) £37·42 (iii) £0·59 (iv) £3·95

3 A square has sides 4.7 cm long. Find the length of its perimeter correct to the nearest centimetre.

4 A regular hexagon has sides of 3.64 cm. Calculate the length of its perimeter correct to the nearest millimetre.

5 A children's playground is a rectangle 67.4 metres long and 48.7 metres wide. Calculate its perimeter correct to the nearest metre.

6 Write out each of these numbers
(i) correct to two significant figures and (ii) correct to two decimal places.
a 1.647 **b** 1.063 **c** 0.0948 **d** 0.3302

7 A tin of peas has a mass of 0.475 kg. Find the total mass of 86 similar tins correct to
a the nearest kilogram **b** three significant figures
c one decimal place.

8 It is 6.47 km from my home to school. I make this journey twice a day for five days each week. What distance do I cover in a school term of fourteen weeks, correct to
a the nearest km **b** the nearest 10 km **c** the nearest 100 km?

9 A dining-room table is 2.15 m long and 0.96 m wide. Calculate its area, correct to
a the nearest m^2 **b** the nearest $\frac{1}{10}$ m^2
c one decimal place **d** three significant figures.

10 In a science experiment, I find that the mass of 6 cm^3 of white powder is 31.6 grams. Find the mass of 1 cm^3, correct to
a the nearest gram **b** the nearest $\frac{1}{10}$ gram
c two decimal places **d** two significant figures.

Part 2 Limits of accuracy

1 What are the least and greatest possible prices of an article which costs
a £8 to the nearest pound **b** £12 to the nearest pound
c £190 to the nearest ten pounds **d** £6·70 to the nearest ten pence?

2 What are the least and greatest possible lengths of a piece of string which has a length of
a 13 m to the nearest metre **b** 120 m to the nearest ten metres
c 6.4 m to the nearest $\frac{1}{10}$ metre **d** 2.94 m to the nearest $\frac{1}{100}$ metre?

Approximation

3 On three days of my holiday, I travelled 67 miles, 84 miles and 42 miles. If these distances were measured correct to the nearest mile, find

a the greatest possible value of the total distance travelled

b the least possible value of the total distance travelled.

4 A chemist pours three quantities of liquid into one jar. If the quantities are 2.7 litres, 1.5 litres and 0.8 litres, all measured correct to the nearest $\frac{1}{10}$ litre, find

a the greatest possible volume of liquid in the jar

b the least possible volume of liquid in the jar.

5 Angle p is measured to the nearest degree and found to be 148°. Find

a the greatest possible value of angle q

b the least possible value of angle q.

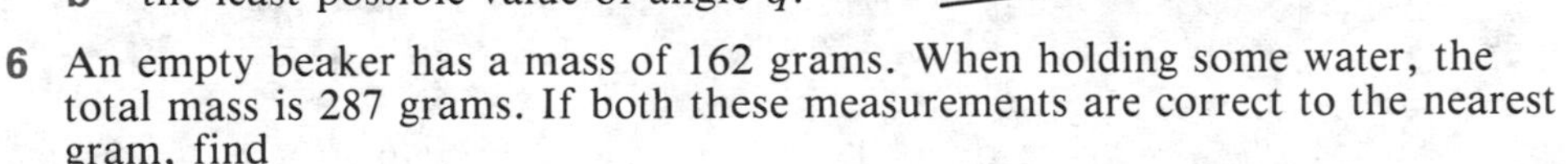

6 An empty beaker has a mass of 162 grams. When holding some water, the total mass is 287 grams. If both these measurements are correct to the nearest gram, find

a the greatest possible value of the mass of water in the beaker

b the least possible value of the mass of water in the beaker.

7 A window pane is 46 cm wide and 57 cm high. If both these measurements are correct to the nearest centimetre, find the range of possible values of its area, giving your answers correct to two significant figures.

8 An electric motor has an angular speed of 46 rps (correct to the nearest revolution). If it runs at this speed for 8.4 seconds (correct to the nearest 0.1 second), find the range of possible values for the number of revolutions through which it has turned. Give your answer correct to two significant figures.

9 6.8 m^2 of hardboard have a rectangular shape 3.2 metres long, where these measurements are given correct to two significant figures. Calculate the range of possible values for the width of the rectangle, to the same accuracy.

10 If $7 \leqslant a \leqslant 9$ and $1 \leqslant b \leqslant 2$ and $3 \leqslant c \leqslant 5$, find the range of possible values of x, y and z where $x = a^2 + bc$, $y = 8b - c$ and $z = \dfrac{4a - c}{b}$.

Indices

Part 1 Indices

1 **a** What are the first eight prime numbers?
b Write each of these numbers as a row of prime numbers multiplied together.
(i) 180 (ii) 168 (iii) 990 (iv) 33 696
c Write your answers to part **b** using *indices* (or powers).

2 Find the values of
a 2^6 **b** 6^2 **c** 2×6 **d** 3^4 **e** 4^3
f 4×3 **g** $2^3 \times 5^2$ **h** $3^3 \times 10^2$ **i** $1^7 \times 4^2$.

3 Simplify these. Do not work out the values, but give your answers using indices.
a $2^5 \times 2^6$ **b** $3^2 \times 3^8$ **c** $5^2 \times 5^4 \times 5^5$
d $\frac{2^9}{2^4}$ **e** $\frac{3^{10}}{3^4}$ **f** $\frac{4^2 \times 4^7}{4^5}$
g $\frac{8^6 \times 8^9}{8 \times 8^7}$ **h** $(4^3)^2$ **i** $(6^4)^3 \times 6^3$

4 Find the value of n when
a $7^2 \times 7^n = 7^6$ **b** $\frac{4^n}{4^5} = 4^2$
c $\frac{3^6 \times 3^4}{3^2 \times 3^n} = 3^7$ **d** $\frac{5^2 \times 5^n}{5^7} = 5^4$.

5 If $3^8 = 6561$, find the values of **a** 3^9 **b** 3^{10} **c** 3^7.

6 These numbers are written in standard form. Write them in full without using indices.
a 4.8×10^3 **b** 3.92×10^5 **c** 1.05×10^4
d 4.8×10^8 **e** 5.67×10^4 **f** 9.7×10^5

7 Write these numbers in standard form.
a 37 000 **b** 640 000 000 **c** 52 400
d The area of England is 130 000 km^2.
e The average distance between the Earth and the Moon is 384 000 km.
f The average distance from Mars to the Sun is 142 000 000 miles.

8 Copy and complete this table and use it to work these multiplications and divisions.
a 16×64 **b** 128×32
c $8 \times 8 \times 32$ **d** $4 \times 16 \times 16$
e $\frac{256}{32}$ **f** $\frac{1024}{64}$
g $\frac{64 \times 512}{2048}$ **h** $\frac{256 \times 128}{1024}$

Power of 2	Its value
0	$2^0 = 1$
1	$2^1 = 2$
2	$2^2 = 4$
3	$2^3 =$
4	$2^4 =$
5	
6	
7	
8	
9	
10	
11	
12	

Indices

9 This slide-rule is based on powers of two. Write down seven multiplications and their answers which can be found with the slide rule in the position shown here.

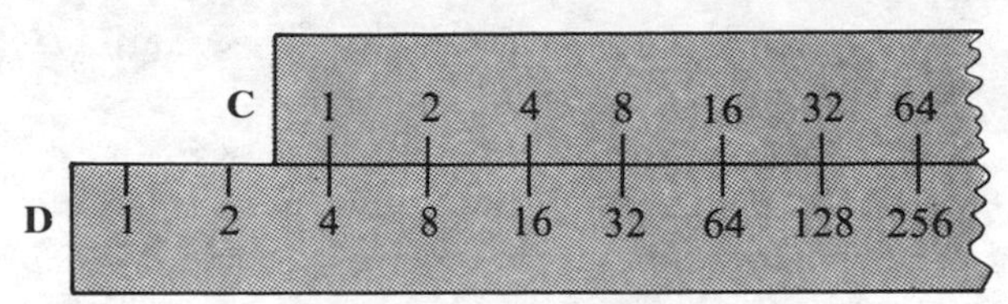

Part 2 Roots and indices

1

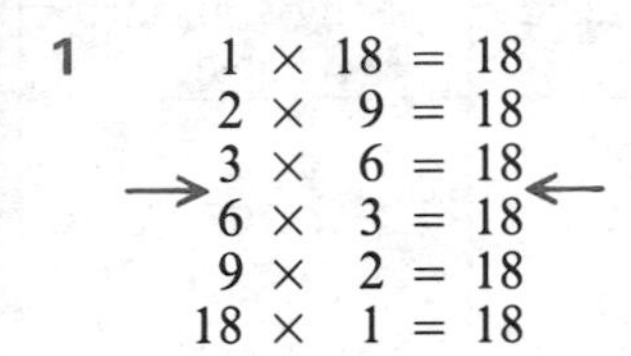

The factors of 18 as arranged here show that there is a number n between 3 and 6 such that $n \times n = 18$.
n is the *square root* of 18 i.e. $n = \sqrt{18}$
Use a calculator to find it (to two decimal places) using only the ⊠ and ⊟ operation buttons.

2 **a** Copy and complete this table of the factors of 24. Repeat the method of **1** on a calculator to find $\sqrt{24}$.

$1 \times 24 = 24$
$2 \times 12 = 24$
$3 \times .. = 24$
$4 \times .. = 24$
....
....

b Use the same method to find $\sqrt{30}$.

3 *Without* using a calculator, write down the values of
a $\sqrt{9}$ **b** $\sqrt{16}$ **c** $\sqrt{49}$ **d** $\sqrt{81}$ **e** $\sqrt{100}$ **f** $\sqrt{400}$.

4 **a** A square has an area of 25 cm^2. How long is each of its edges?
b A photograph has the shape of a square of area 36 cm^2. How long is each of its edges?
c A square postage stamp has an area of 4 cm^2. What is the length of each of the edges of the stamp?
d A gardener sows a square lawn of area 64 m^2. He wants to edge right round it with a plastic strip. What length of strip does he need?

5 Use a calculator for these problems.
a A square window pane has an area of 3.5 m^2. How long are its edges, correct to two decimal places?
b A coffee table has a square top of area 1500 cm^2. What is the length of each edge to the nearest whole centimetre?
c A garden shed has a square flat roof of area 5.6 m^2. What length is the gutter along one of the edges of the roof (to the nearest whole centimetre)?
d A square carpet of area 35 m^2 is laid in a large office. If all its edges are bound by tape, what length of tape is needed?

6

$1 \times 1 \times 1 = 1$
$\rightarrow 2 \times 2 \times 2 = 8 \leftarrow$
$3 \times 3 \times 3 = 27$
$4 \times 4 \times 4 = 64$

This pattern shows that between 2 and 3 there is a number n such that $n \times n \times n = 20$.
n is the *cube root* of 20 i.e. $n = \sqrt[3]{20}$
Use a calculator to find it (to two decimal places) using only the ⊠ operation button.

7 Repeat the method of **6** to find $\sqrt[3]{30}$ and $\sqrt[3]{100}$.

8 **a** *Without* using a calculator, write down the values of
(i) $\sqrt[3]{8}$ (ii) $\sqrt[3]{27}$ (iii) $\sqrt[3]{64}$ (iv) $\sqrt[3]{125}$ (v) $\sqrt[3]{1000}$.
b A child's set of building blocks has cubes with volumes of 27 cm^3. How long is the edge of each cube?
c A cubical water tank has a capacity of 8 m^3. What is the height of the tank and the area of its base?

Indices

9 **a** Work out (i) $4^3 \times 4^{-2}$ and (ii) $\frac{4^3}{4^2}$

b What is the value of 4^{-2}?

Write the values of the following as fractions.

c 3^{-2} **d** 5^{-2} **e** 2^{-3} **f** 10^{-3} **g** 2^{-5} **h** 3^{-4} **i** 4^{-3}

10 Copy and complete this table:

		Without standard form	Using standard form
a	the thickness of a sheet of paper	0.0095 cm	....
b	the area of a pin head	....	1.5×10^{-4} cm^2
c	the diameter of a molecule of oxygen	0.000000034 cm	....
d	the wavelength of sodium light	....	5.9×10^{-7} cm
e	the mass of a sugar lump	0.0025 kg	....

11 **a** Work out (i) $\sqrt{9} \times \sqrt{9}$ and (ii) $9^{\frac{1}{2}} \times 9^{\frac{1}{2}}$.

b Work out (i) $\sqrt[3]{8} \times \sqrt[3]{8} \times \sqrt[3]{8}$ and (ii) $8^{\frac{1}{3}} \times 8^{\frac{1}{3}} \times 8^{\frac{1}{3}}$.

c Write down the values of

(i) $9^{\frac{1}{2}}$ (ii) $8^{\frac{1}{3}}$ (iii) $36^{\frac{1}{2}}$ (iv) $81^{\frac{1}{2}}$

(v) $1000^{\frac{1}{3}}$ (vi) $16^{\frac{1}{4}}$ (vii) $32^{\frac{1}{5}}$ (viii) $81^{\frac{1}{4}}$.

Proportion

Are these pairs of variables *directly* or *inversely* proportional? As one variable increases, does the other variable *increase* or *decrease*?

a bottles of milk bought – cost of milk
b length of a roll of cloth – the weight of the cloth
c number of pieces cut from a pie – the size of each piece
d number of men loading a barge – the time taken to do the job
e time taken for a certain journey – the average speed during the trip
f number of cars on a ferry – the amount of space for each car

Part 1 Direct proportion

1 The variables in these tables are *directly* proportional.
(i) Copy and complete each table.
(ii) Draw a graph of the values in the table.
(iii) Find the equation connecting the two variables.

a

Area of rose bed, x m^2	1	2	3	4	5	10	15	20
Number of roses planted, y	4							

b

Flights to Majorca, x	1	2	3	4	5	10	15	20
Cost of flights, £y		160						

c

Volume of liquid, x cm^3	1	2	3	4	5	10	15	20
Mass, y grams			3.6					

2 If four apples cost 36 pence, how much will five similar apples cost?

3 If twelve women pack 600 boxes of biscuits in one hour, how many boxes would ten women pack in the same time?

4 If £8 can be exchanged for 13.6 Canadian dollars, how many dollars would you get for £14?

5 A lorry uses 12 litres of diesel fuel in travelling 160.8 km. How far would 10 litres take it?

Part 2 Inverse proportion

1 The variables in these tables are *inversely* proportional.
(i) Copy and complete each table.
(ii) Draw a graph of the values in the table.
(iii) Find the equation connecting the two variables.

a

Time for a journey, x hours	1	2	3	4	5	6	10
Average speed, y mph	120						

b

Number of workmen, x	1	2	3	4	5	6	10
Time to complete job, y hours	60						

c

Number of pieces in a cake, x	1	2	3	4	5	6	10
Angle of piece at centre, y°		180					

Proportion

2 Jessica Aston has enough food to feed her litter of five kittens for twelve days. How long would the food have lasted had there been six kittens?

3 Four pumps will empty the school's swimming pool in fifteen hours. If one pump is out of order, how long will the other three pumps take?

4 At an average speed of 120 km/h, a train takes 35 minutes for a certain journey. How long would it take if the average speed was 70 km/h?

A mixture

5 A freezer uses three units of electricity in 150 minutes. How long does it take to use two units?

6 It takes my three horses eight days to eat a sack of bran. If I sell one of the horses, how long would a sack last with the remaining two horses?

7 If £6 is changed for 63 French francs, how many francs would you get for £8?

8 A 24-kg sack of onions costs £7·20. What would be the cost of an 18-kg sack?

9 A van can carry forty two boxes of tins when each box holds eighteen tins. If a new size of box holds twenty eight similar tins, how many boxes of this new size can the van carry?

10 Travelling at a steady speed of 56 mph, I took 25 minutes to drive between two service areas on the motorway. If the next day the same distance took me 35 minutes, what was my average speed?

Part 3 Other types of proportion

1 For each of these, find
(i) the constant of proportionality (ii) the equation connecting x and y.

a $y \propto x^2$ and $y = 12$ when $x = 2$
b $y \propto x^3$ and $y = 54$ when $x = 3$
c $y \propto \sqrt{x}$ and $y = 10$ when $x = 4$
d $y \propto \frac{1}{x}$ and $y = 3$ when $x = 5$
e $y \propto \frac{1}{x^2}$ and $y = 1$ when $x = 3$
f $y \propto \frac{1}{\sqrt{x}}$ and $y = 8$ when $x = 9$

2 When an object falls from rest, the distance y which it falls is proportional to the square of the time, t taken in falling. If $y = 32$ when $t = 4$, find
a an equation connecting t and y **b** the value of y when $t = 5$.

3 The energy E of a rotating flywheel is proportional to the square of its angular speed ω. If $E = 36$ when $\omega = 3$, find
a an equation connecting E and ω **b** the value of E when $\omega = 2$.

4 When a car goes round a corner, its maximum speed v is proportional to the square root of the radius r of the corner. If $v = 20$ when $r = 100$, find v when $r = 25$.

5 The force F between two magnets is inversely proportional to the square of the distance x between them. If $F = 9$ when $x = 2$, find F when $x = 3$.

Changes of units

Part 1 Foreign exchange

1 Malcolm Cartwright and his friends decide on a holiday in Switzerland.
Use this graph to find how many Swiss francs he will get for

a £2 **b** £3·20
c £4·80 **d** £2·60.

When he changes francs back into pounds, how many pounds £ will he get for

e 10 Swiss fr.
f 8.50 Swiss fr.
g 6.50 Swiss fr.
h 11.50 Swiss fr?

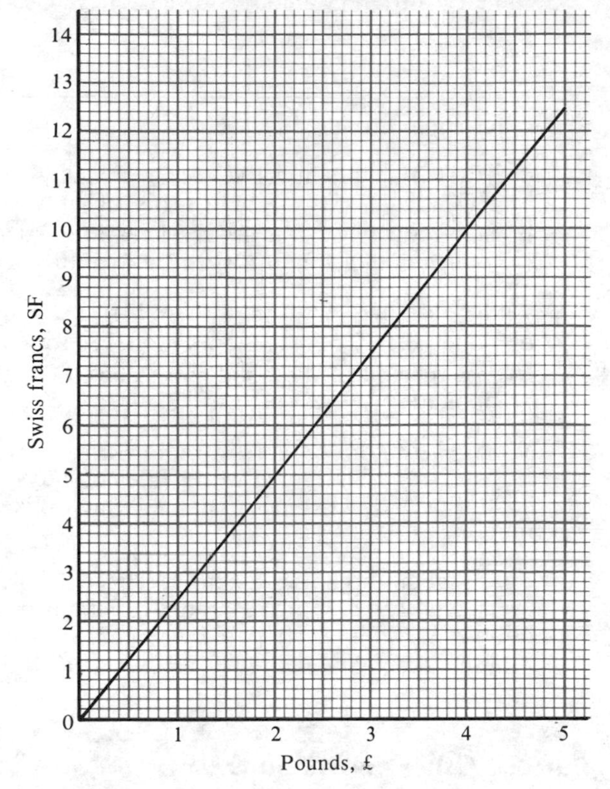

2 You go on a holiday to Portugal.
Use this graph to change pounds sterling (£) into Portuguese escudos, (esc).
How many escudos will you get for

a £2 **b** £3·50
c £4·20 **d** £8?

How many pounds £ will you get for

e 500 esc. **f** 420 esc.
g 650 esc. **h** 940 esc?

i If the exchange rate alters so that £1 = 250 escudos, draw your own conversion graph for amounts up to £5 at this new rate.

j Is this new rate of £1 = 250 escudos better or worse than the rate shown on the graph for a British tourist visiting Portugal?

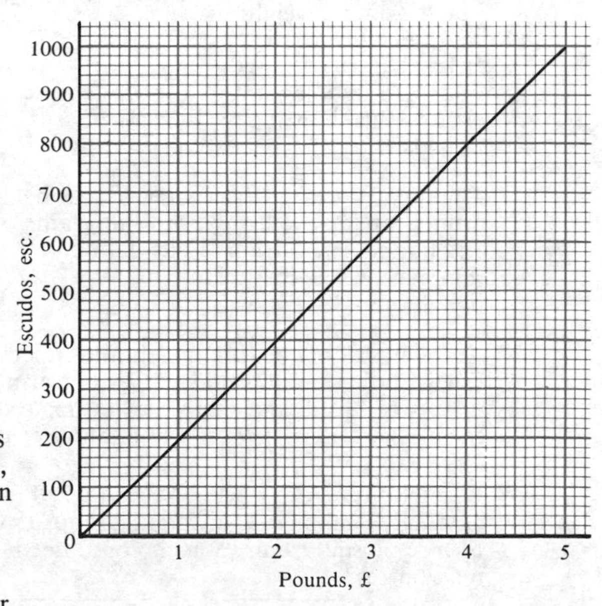

Changes of units

3 The Anglo-Icelandic exchange rate is £1 = 60 kronur. Copy and complete this table for different amounts of currency. Use the table to draw a conversion graph.

Pounds sterling, £	1	2	3	4	5	8	10
Icelandic kronur, kr	60						

a Erik Magnusson arrives at Prestwick Airport near Glasgow with 450 kronur. How much is this amount worth in pounds £?

b On returning home a week later, he arrives in Reykjavik with £8·50. How much is this amount worth in kronur?

4 Janice Pritchard went on a school trip to Yugoslavia where £1 = 588 dinar.

a She takes £35 in spending money. How much is this worth in dinar?

b At the end of the holiday, she has 4370 dinar to change back into pounds £. How much will she receive?

5 When visiting his parents in India, Ravi Gowda buys a single air ticket in England for £225. Some weeks later he buys the return ticket in Bombay for 4500 rupees. If £1 = 18.8 rupees, which ticket was the better buy and by how much?

6 Rev. Ingham organises a church trip to Jerusalem. He is able to arrange everything for £745 per person. An Israeli travel agent can organise a similar holiday for 1780 shekels. If £1 = 2.30 shekels, which is the cheaper price and by how many pounds £ per person is it cheaper?

Part 2 Other conversions

1 a The maximum weight of baggage allowed for one airline has a mass of 20 kg. Use this flow diagram to express this limit in pounds (lb).

Mass, kg → Divide by 5 → Multiply by 11 → Mass, lb

b The unladen weight of a lorry has to be printed on its side using both kilograms and pounds. A workman has painted this sign on one lorry. How many pounds should it show?

TARE WT.
4400 kg
lb

c An international boxing match announces the mass of each contestant in English using pounds and in German using kilograms. Jimmy Higgens has a mass of 164 lb and Hans Müller has a mass of 73 kg. Write the mass of each boxer to the nearest whole number of pounds and kilograms.

2 a A hotel buys from a wholesaler a large metal cannister of washing-up liquid holding $2\frac{1}{2}$ gallons. A different brand of washing-up liquid is available in 12-litre plastic drums. Which container has the greater capacity and by how many litres?

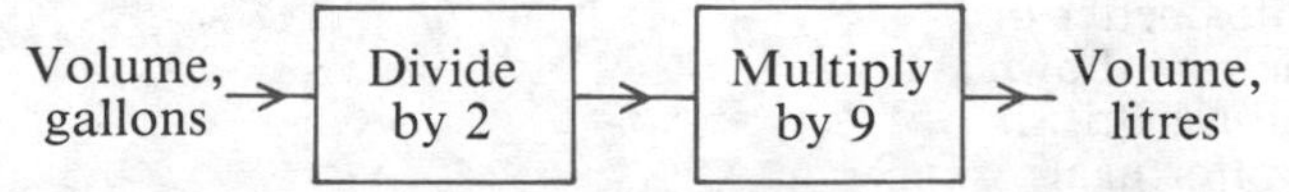

Changes of units

b An English company imports paint in 10-litre tins from a French manufacturer and sells it to the general public at £9 per gallon. How many gallons are in each tin and at what price should each tin be sold?

c A garage sells petrol at £1·85 per gallon. Another garage on the opposite side of the road sells petrol at 40.2 pence per litre. Which petrol is the better buy?

3 A Spanish company is selling plots of land on the Costa del Sol. Advertisements are printed in British newspapers.

a Two adjacent plots have areas of 2 hectares and 3.4 hectares. If the advertisements give these areas in acres, calculate their size in acres. *(1 hectare = $2\frac{1}{2}$ acres).*

b An English family is looking for a plot of about $1\frac{1}{2}$ acres. How many hectares is this?

4

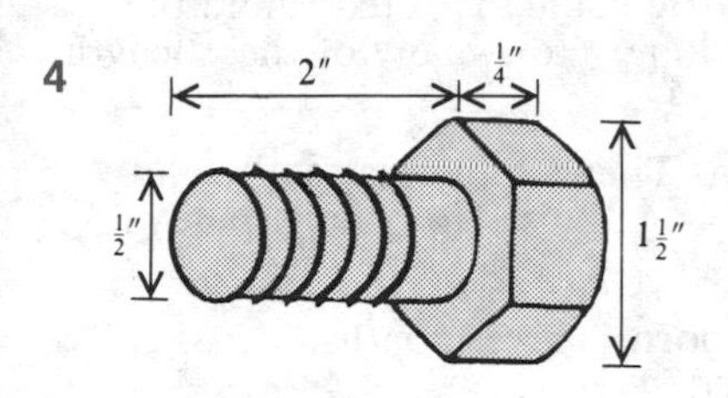

a This bolt has the measurements given in inches. Sketch the bolt and convert the measurements to the nearest millimetre. *(1 inch = 2.54 centimetres)*

b A packet contains nails of two different lengths. The label says that they are either 76 mm or 140 mm long. Convert these lengths to the nearest $\frac{1}{10}$th of an inch.

5 a The Crawshaw family are to drive across France to holiday in Biarritz. This map shows three routes in kilometres from three channel ports. How long are these journeys, correct to the nearest mile? *(1 km = 0.621 miles)*

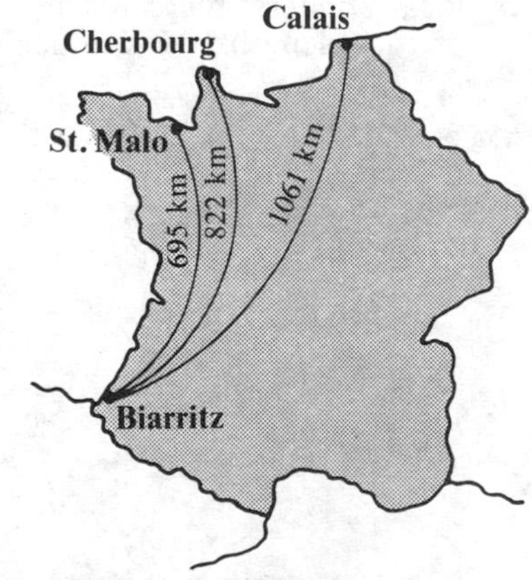

b The conversion 1 km = 0.621 miles is accurate to three significant figures. A more common conversion is 1 km = $\frac{5}{8}$ mile. What is the difference in miles between these two conversion factors for a journey of 500 kilometres?

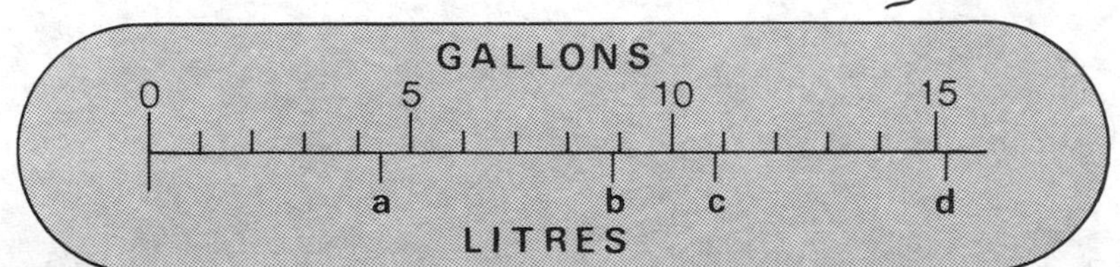

6 A garage has a number line to convert between litres and gallons. Customers buying petrol can then quickly convert between the two measures. The numbers indicating litres are missing on the number line shown here.
If 1 gallon = 4.55 litres, what numbers should be printed in the positions labelled *a, b, c* and *d,* to the nearest 10 litres?

Changes of units

Part 3 Imperial and other units

12 inches = 1 foot	16 ounces (oz) = 1 pound (lb)	8 pints = 1 gallon
3 feet = 1 yard	14 pounds = 1 stone	60 minutes = 1 hour

1 How many strips of brown sticky paper 4 inches long can be cut from a roll of sticky paper $8\frac{1}{2}$ feet long? Will there be any paper left over?

2 A recipe for a casserole to serve four people uses $1\frac{1}{2}$ lbs of meat. If Mrs Oxley makes a casserole just for herself and her husband, how many ounces of meat does she need?

3 Charles Roberts visits his cousin Alden in the USA. When asked how much he weighs, he says $11\frac{1}{2}$ stone. "Gee", replies Alden, "how many pounds is that?".

4 It took the Rickett family 5 h 20 min to make the car journey to their holiday guest-house in Torquay. If they stopped for 1 h 35 min for a meal on the way, how long were they actually driving?

5 The perimeter of a lawn is 32 yards long. Plastic edging is fitted round the lawn. Edging can be bought in lengths 2 feet long. How many of these lengths are needed?

6 Jeremy Spencer keeps two goats in his garden. They supply his family and friends with milk. In one week they produce $4\frac{1}{2}$ gallons. How many pint milk bottles will this amount fill?

7 Jane Hardcastle was 7 lb 5 oz when she was born. Three months later, her mass was 9 lb 2 oz. How much had she gained?

8 The roof of your house has a gutter 15 ft 4 in long. It needs replacing. The local builders' merchant has new guttering in lengths of 8 ft 6 in. You buy two of these lengths. How much of the second length do you *not* need to use?

Perimeter

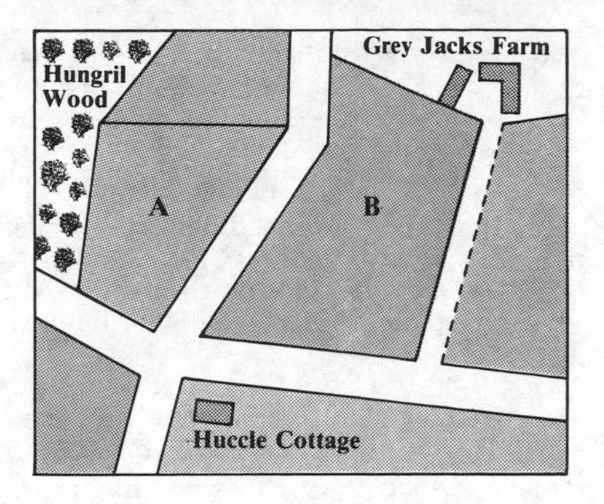

1 **a** Use your ruler to find the perimeter (in centimetres) of the two fields *A* and *B* on this map.

b If the scale of the map is 1 mm = 5 metres, find the actual perimeters of the two fields in metres.

2 This is a plan of the sitting-room of a house drawn to a scale of 1 cm = 2 metres. A fitted carpet is to be laid and kept in position by spiked strips fixed round the edge of the floor.

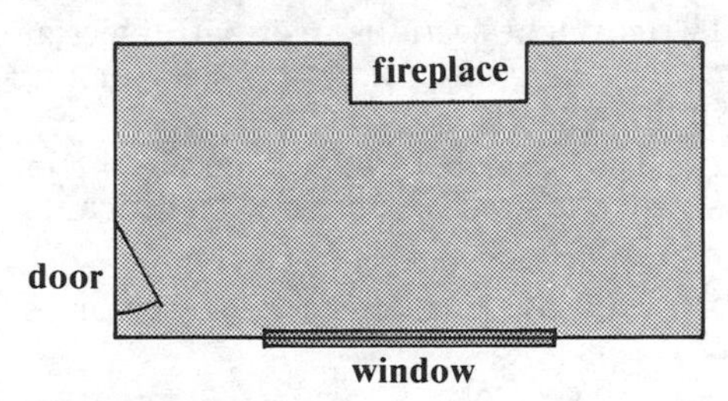

Find

a the perimeter of the room in centimetres on the plan by using a ruler

b the actual perimeter of the room in metres

c the cost of the strips if a one-metre length costs 46 pence.

3 This is a map of a reservoir which has formed behind a dam wall. The scale of the map is 1 cm = 250 metres.

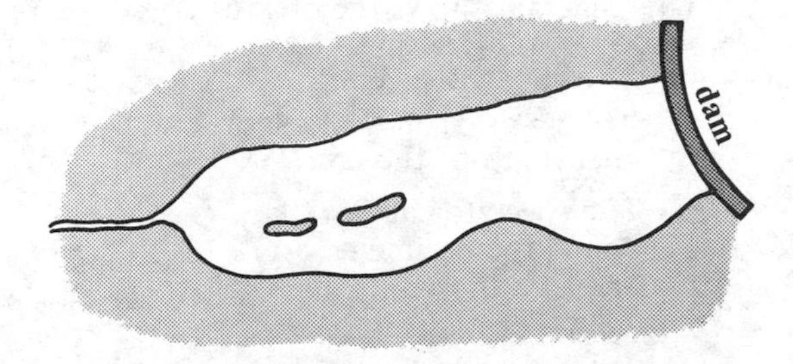

a Use a piece of cotton to find the perimeter of the reservoir on the map in centimetres.

b Calculate the actual perimeter in metres.

c A workman from the water board walks round the reservoir inspecting the perimeter at a speed of 2 km/h. How long will the tour of inspection take him?

4 Calculate the perimeter of

a a square of side 3.5 cm

b a rectangle 6.8 cm by 4.3 cm

c a regular pentagon of side 2.8 cm

d a regular hexagon of side 7.5 cm

e a 50-pence piece if the distance between two adjacent corners is 1.4 cm.

5 Calculate the perimeter of

a a square of side $2\frac{1}{3}$ cm

b a rectangle $3\frac{1}{4}$ cm long and $1\frac{3}{8}$ cm wide

c a regular pentagon of side $4\frac{3}{5}$ cm

d a regular hexagon of side $4\frac{3}{5}$ cm.

Perimeter

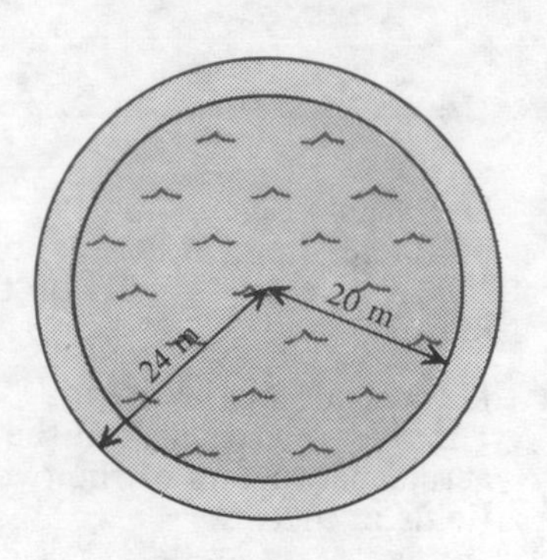

6 a A factory's private reservoir has a circular shape of radius 20 metres. Take π as 3.14 and calculate the perimeter, or *circumference*, of the reservoir to three significant figures.

b It is surrounded by a path 4 metres wide as shown. Calculate the circumference of the outside of this path to three significant figures.

7 Calculate the circumferences of the circles with these radii. Take π as 3.14 and give your answers to three significant figures.

a 4 cm b 12 cm c 5.6 cm

Now take π as $\frac{22}{7}$

d 21 cm e $3\frac{1}{2}$ cm f $10\frac{1}{2}$ cm

8 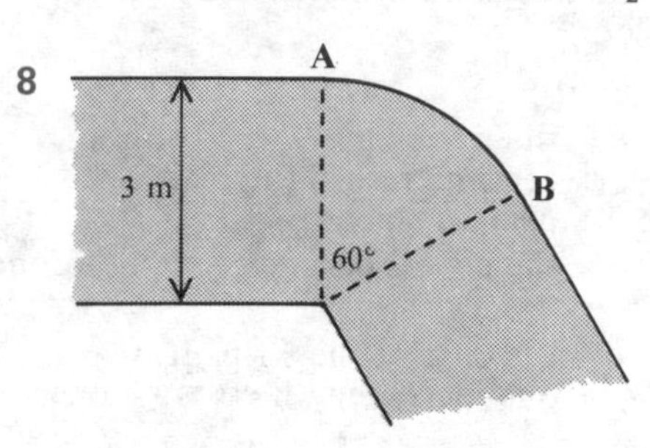

Two straight sections of a path meet at a corner. They are joined by a sector of a circle of angle 60° as shown.

a Take π as 3.14 and calculate the length of the arc AB, to three significant figures.

b Find the perimeter of this sector.

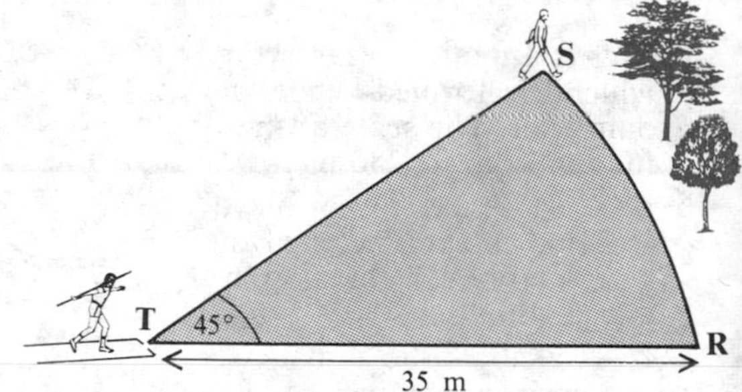

9 This diagram shows the area of a school sports field marked out for throwing the javelin.

a Take π as $\frac{22}{7}$ and calculate the length of the arc RS.

b If a spectator at S walks round the perimeter of the area, how far does he walk?

10 Calculate the lengths of the arcs of the circles with these radii and angles. Take π as 3.14 and give your answers to three significant figures.

a 4 cm and 90° b 5 cm and 45° c 2.5 cm and 72°

Now take π as $\frac{22}{7}$.

d 14 cm and 60° e $7\frac{1}{2}$ cm and 120° f $3\frac{1}{2}$ cm and 45°

Area

Part 1 Introduction

1 These four pieces *A*, *B*, *C*, *D* will fit together like a jigsaw to make the rectangle with corners numbered 1 to 4.
If piece *A* is fitted in corner 1, which pieces fit into corners 2, 3 and 4?

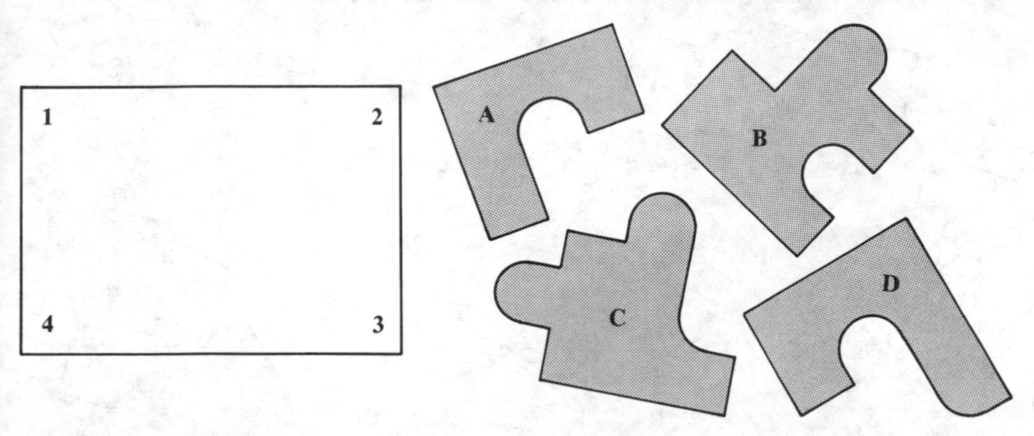

2 Which of these shapes can be used to make a tessellation?

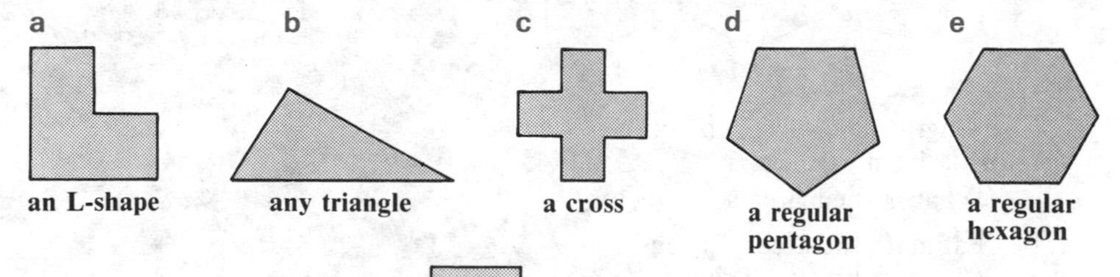

3 This is one square centimetre

How many square centimetres will fit into these two shapes?

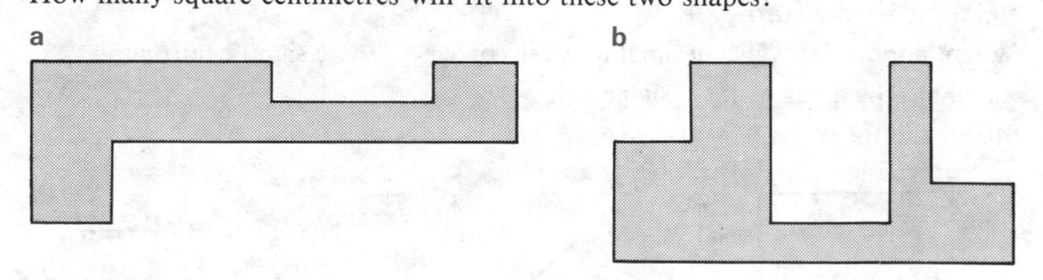

4 Use a piece of tracing paper with a centimetre grid on it to find the area (in cm^2) of these three shapes.

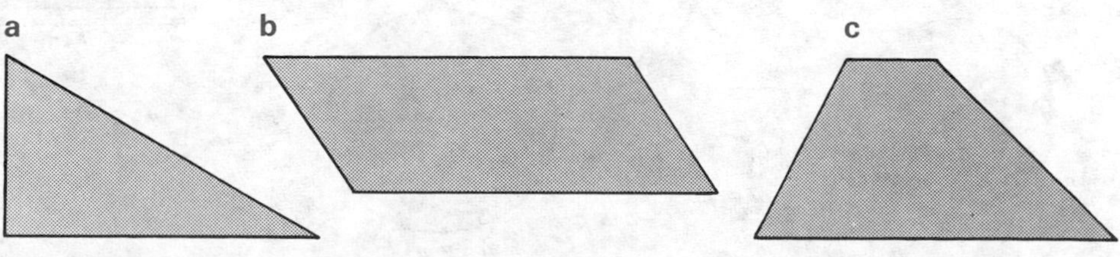

Area

5 Here are two maps, one of Anglesey and the other of the Isle of Man.
Use three methods to estimate their areas.
Method I Use tracing paper with a centimetre grid on it.
Method II Use dotted tracing paper with dots one centimetre apart.
Method III Use tracing paper with strips ruled one centimetre apart.

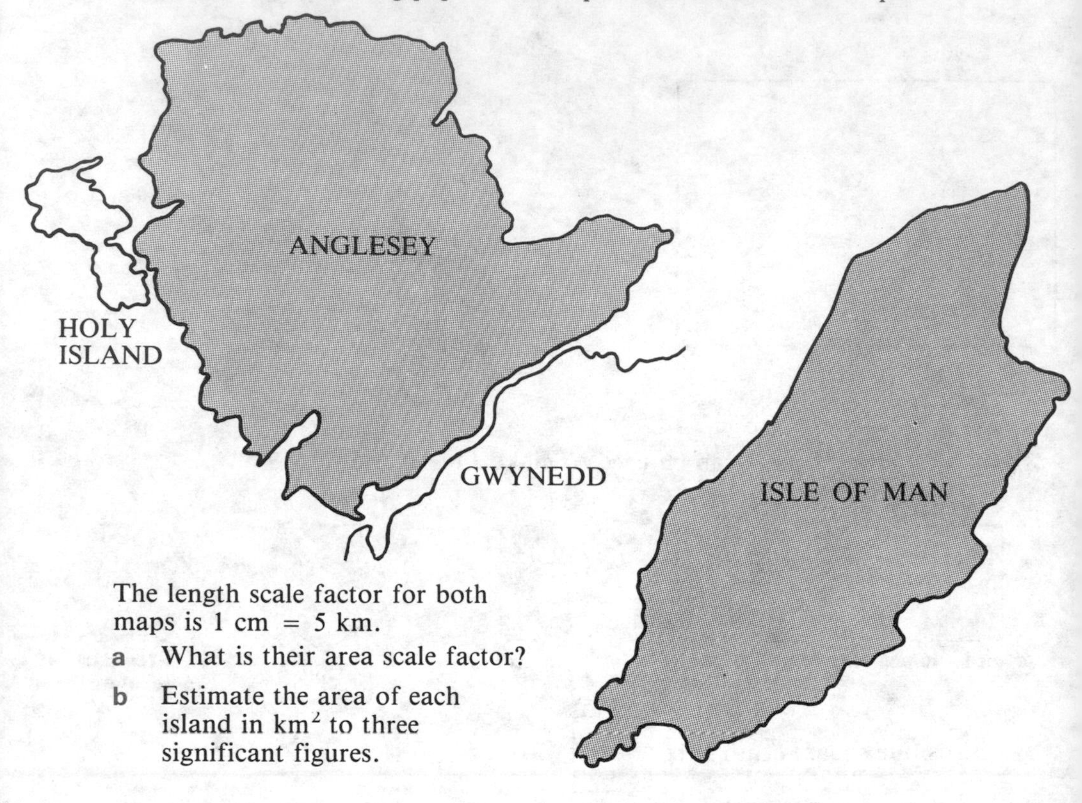

The length scale factor for both maps is 1 cm = 5 km.

a What is their area scale factor?

b Estimate the area of each island in km^2 to three significant figures.

Part 2 Using formulae

Where appropriate, give decimal answers correct to three significant figures.

1 Find the areas of these shapes.

Integer lengths

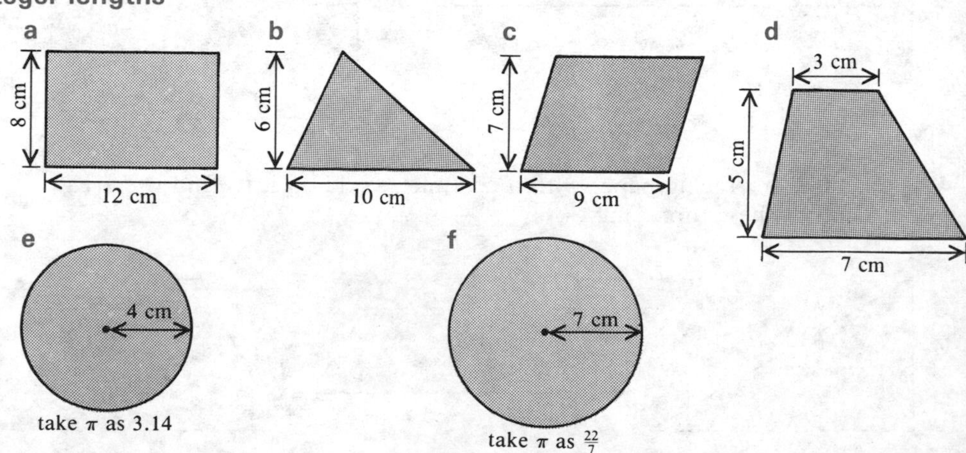

Area

Decimal lengths

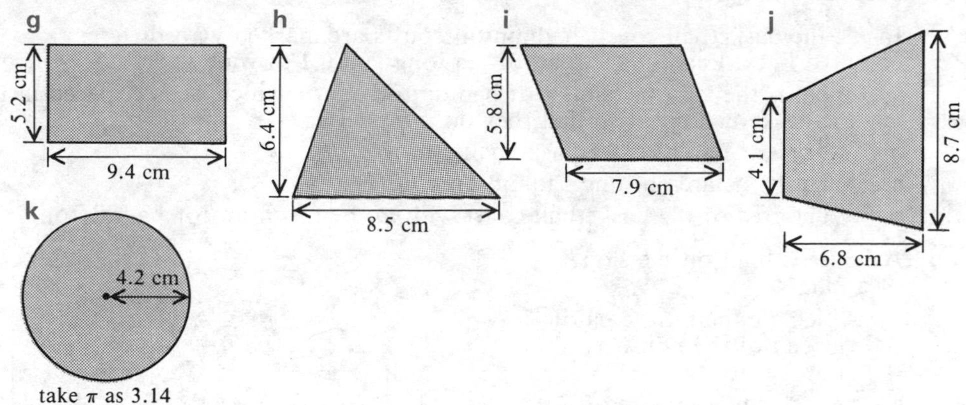

take π as 3.14

Fractional lengths

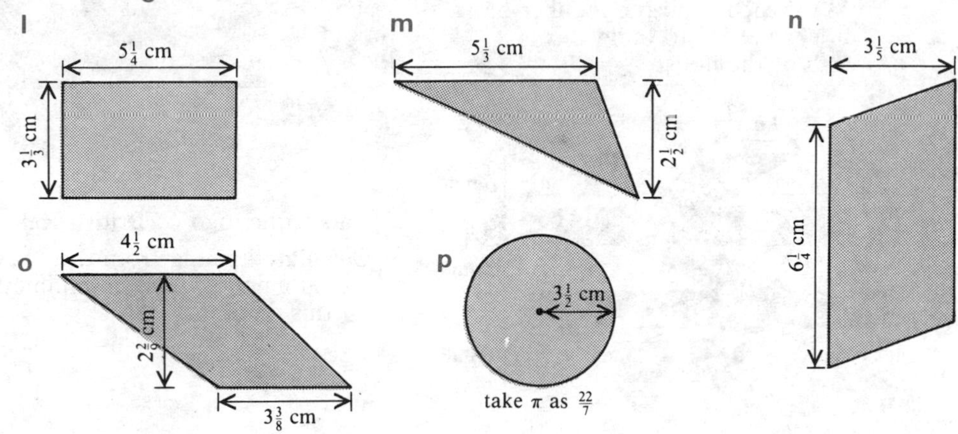

take π as $\frac{22}{7}$

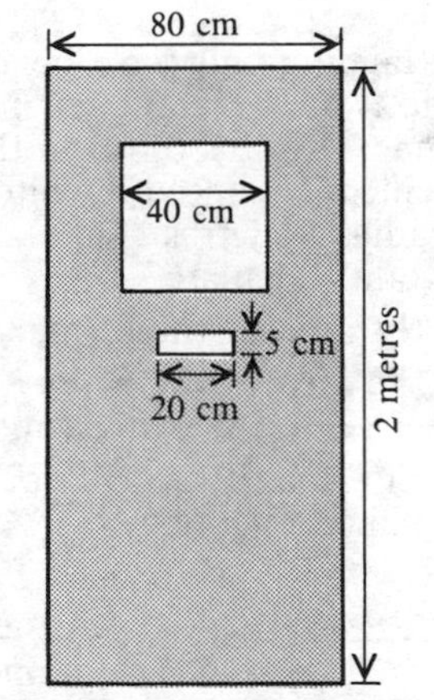

2 The front door of a house is made from one rectangular piece of wood 2 m by 80 cm. Two holes are cut in it; a square hole of side 40 cm for a window, and a rectangular hole 20 cm by 5 cm for a letter-box.

Calculate

a the area of the original rectangular piece of wood

b the area after the two holes have been cut out of the wood.

3 a A rectangular carpet has a length of 6.5 m and a width of 4.2 m. What is its area?

b It is laid in a room 7.4 m long and 5.5 m wide. Find the area of the floor of the room.

c What area of the floor is *not* covered by carpet?

Area

4 A sports hall is just large enough to take one basketball court 25.6 m by 14.5 m.
Inside the basketball court, badminton courts are marked with different coloured lines. Each of them is 13.4 m long and 6.1 m wide.

- **a** What is the largest number of badminton courts which can be placed inside the basketball court so that they do not overlap?
- **b** What is the area of the basketball court?
- **c** What is the area of one badminton court?
- **d** What area of the basketball court will *not* be taken up for badminton?

5 A house is built on the slope of a hill.

- **a** Calculate the total area of the end of the house as shown here.
- **b** If each square metre of wall needs 50 bricks and the two windows shown have an area of $1\frac{1}{2}$ m^2 each, find the number of bricks needed to build this wall of the house.

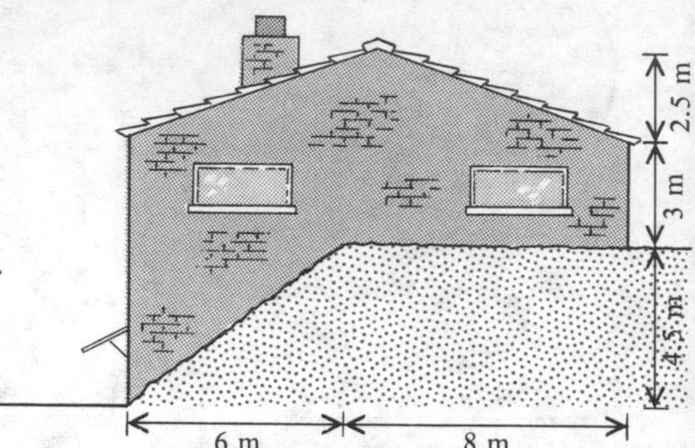

6

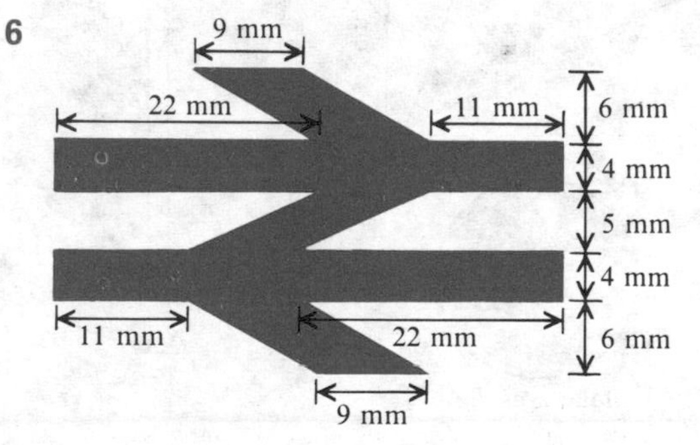

This is the logo of British Rail.
Calculate its area from the measurements given in millimetres on this diagram.

7 The landing pad for a helicopter is made from a rectangular piece of tarmac 24 m by 18 m, in the centre of which is painted a white circle of radius 8 metres. Take π as 3.14 and calculate

- **a** the area of the tarmac
- **b** the area of the white circle to the nearest m^2
- **c** the area of tarmac *not* painted white.

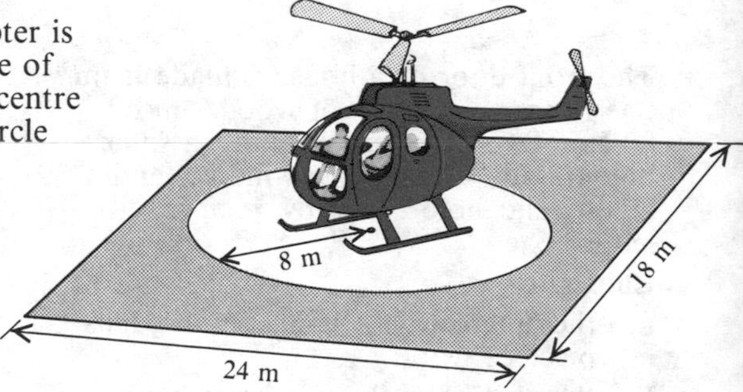

8

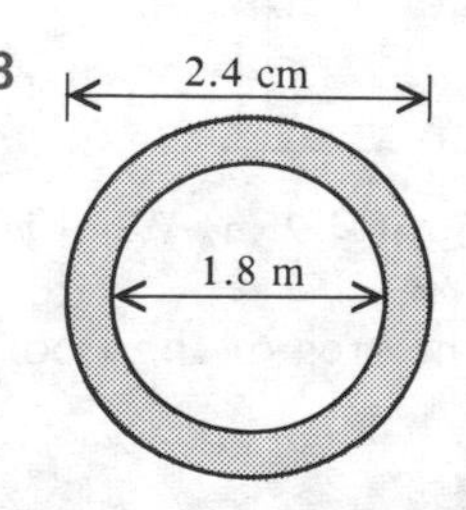

A metal washer is made with external and internal diameters of 2.4 cm and 1.8 cm respectively.

Calculate

- **a** the external and internal radii
- **b** the area of the hole in the washer
- **c** the cross-sectional area of the metal.

Area

9 A cardboard carton just holds twelve tins as shown, where each tin has a diameter of 14 cm. Take π as $\frac{22}{7}$ and calculate

a the length and width of the top of the box

b the area of the circular top of one tin

c the area of the box *not* occupied by the tins.

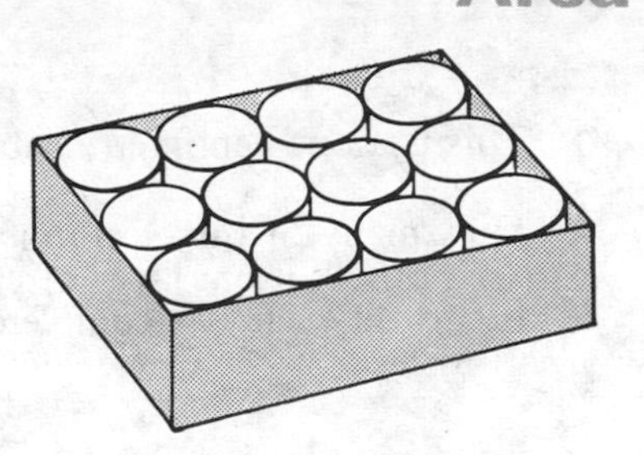

10 Calculate the areas of these sectors of circles.
Take π as 3.14 and give your answers to three significant figures.

a

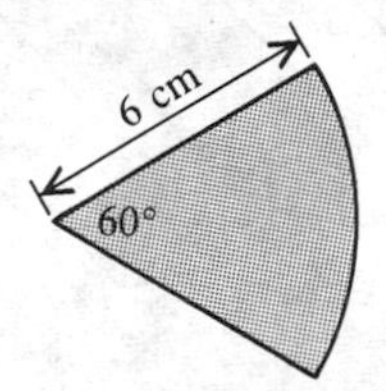

b

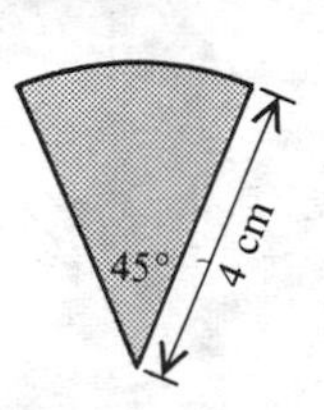

c

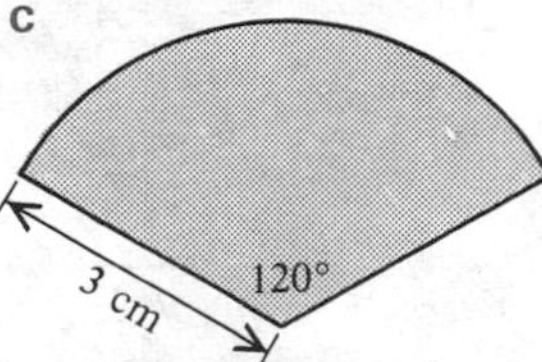

Take π as $\frac{22}{7}$ for these three sectors.

d

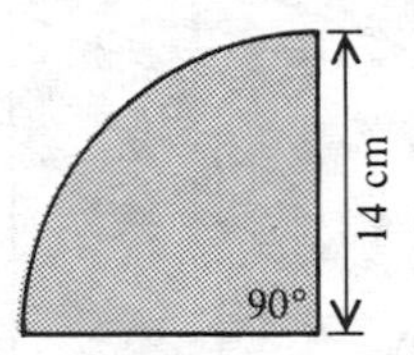

e

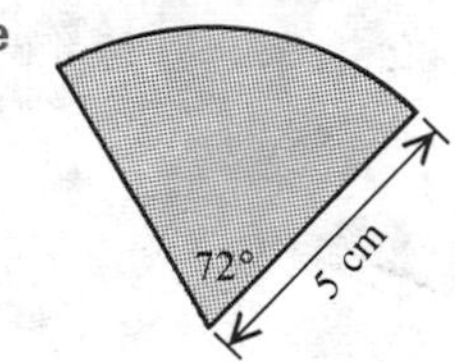

f

$3\frac{1}{2}$ cm

144°

11 This pie chart shows the proportions of the populations of the countries of the United Kingdom. This table gives the angles at the centre of the pie chart.

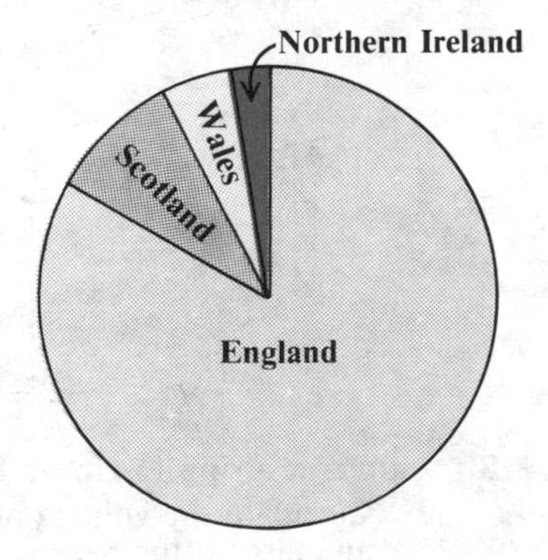

England	Scotland	Wales	Northern Ireland
300°	33°	18°	9°

If the chart has a radius of 2 cm, calculate the area of the sector for each country in cm^2 correct to one decimal place.

12 A search light can shine over a distance of 2.45 km and through an angle of 210°. What area of land can it illuminate?

Volume

Part 1 Cuboids

1 This is a one centimetre cube, 1 cm^3 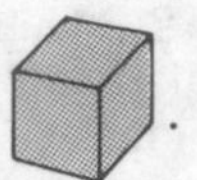.

How many of these centimetre cubes are needed
(i) in the bottom layer of each box to cover its base
(ii) to fill each box completely?

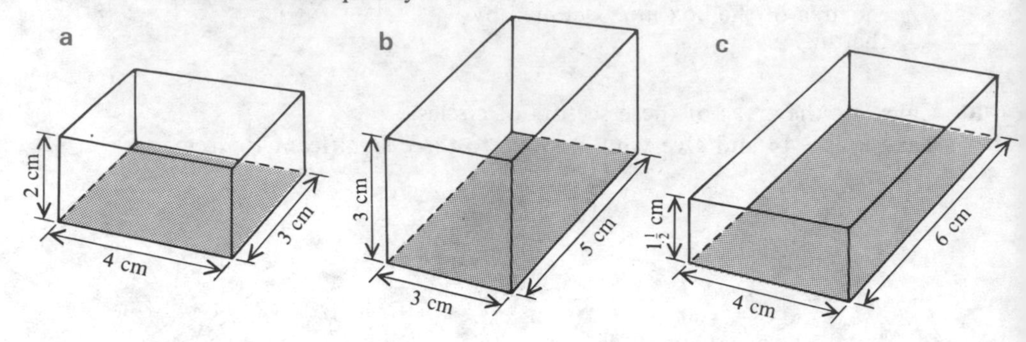

2 Find (i) the area of the base in cm^2
(ii) the volume in cm^3 of each of these cuboids.

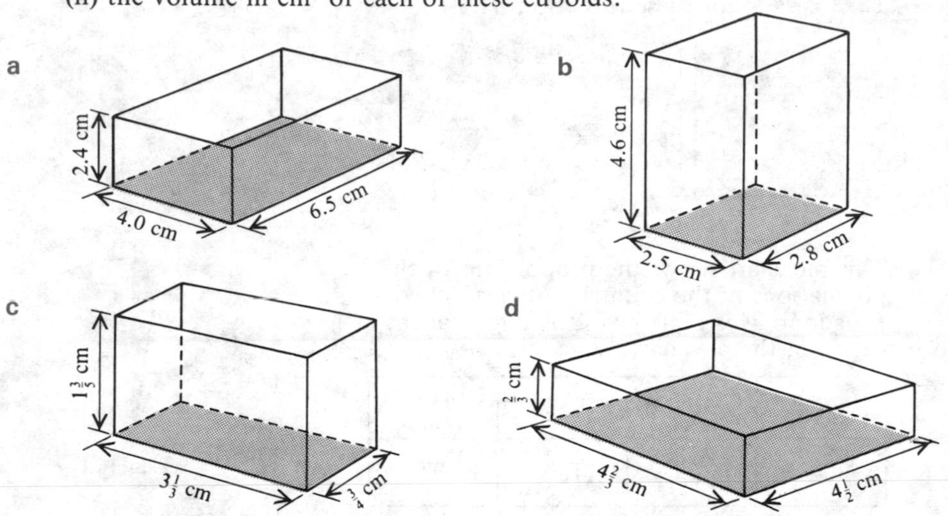

3 A tin measuring 15 cm by 16.2 cm by 20.2 cm is full of cooking oil.
- **a** Calculate the volume of the oil in the tin. Give your answer both in cm^3 and also to the nearest litre.
- **b** How many such tins would be needed to fill a drum with a capacity of 22.5 litres?

4 A number of paving slabs are made from 48 litres of concrete. If the slabs have a square cross-section of side 26 cm and are 3.7 cm thick, calculate
- **a** the volume of one slab
- **b** the number of complete slabs which can be made.

5 Water 9.5 cm deep is running down a rectangular duct 14.2 cm wide at a speed of 55.6 cm/s.
- **a** How much water is flowing from the end of the duct each second?
- **b** How long would it take to fill a tank of capacity 300 litres?

Volume

Part 2 Prisms and cylinders

1 For each prism, calculate (i) the area of the shaded base (ii) the volume.
All measurements are given in centimetres.

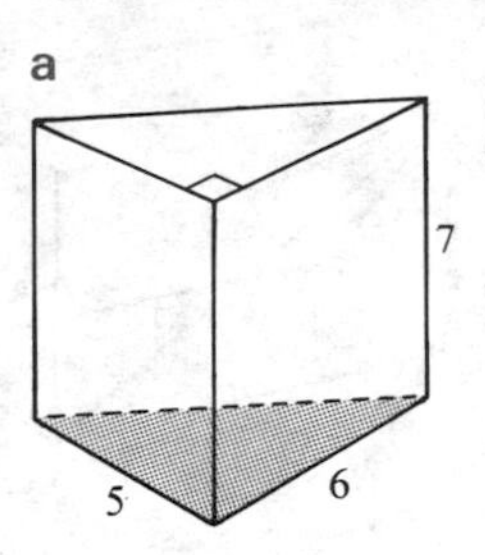

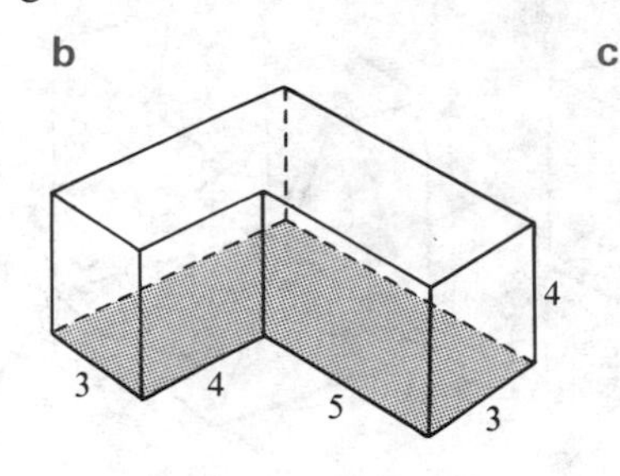

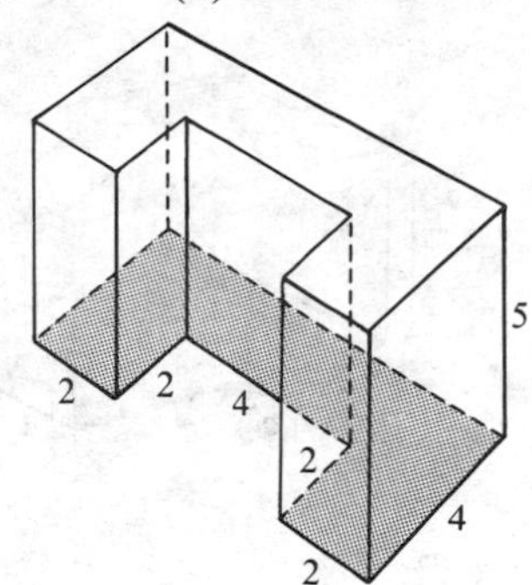

2 Find the end area and the volume of each prism.
All measurements are in centimetres.

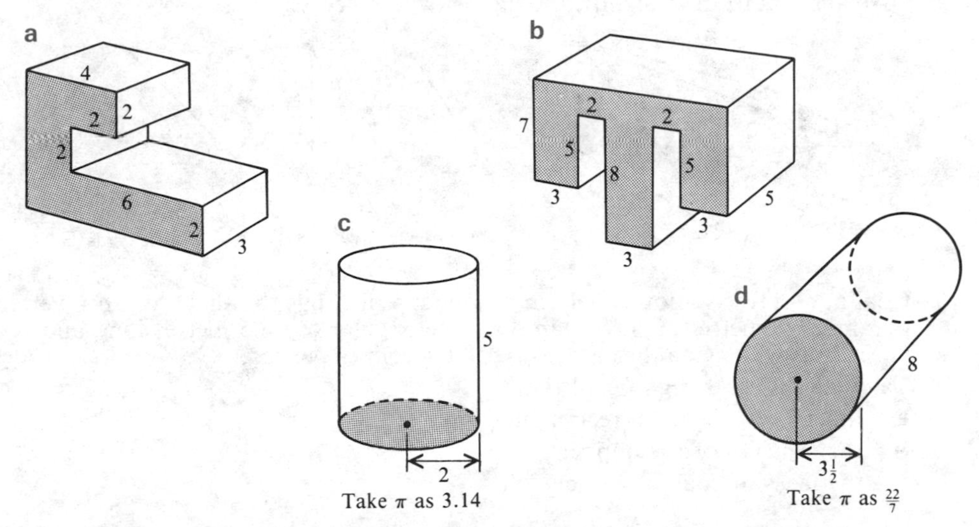

Take π as 3.14

Take π as $\frac{22}{7}$

3 A cylindrical tin of fruit juice has a radius of 4.1 cm and a height of 9.5 cm. The label on it says it has a volume of $\frac{1}{2}$ litre. Find

a the actual volume of the tin

b whether you should complain about the labelling.

4 A lump of iron is placed in a cylindrical beaker holding water. If the diameter of the beaker is 8 cm and the water level rises 2.6 cm, calculate the volume of the iron to three significant figures.

5 A cube of edge 10 cm has a cylindrical hole of diameter 6 cm, drilled 3 cm deep.
Calculate, to the nearest whole cm^3, the volume of the cube.

a before the hole is drilled

b after it has been drilled.

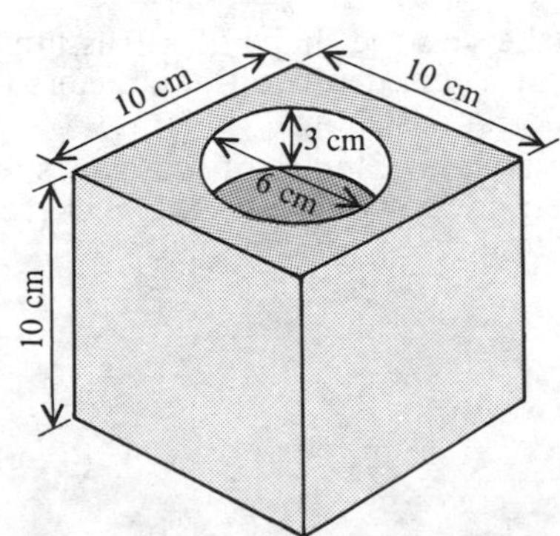

Volume

Part 3 Pyramids and cones

1 Find the volumes of these cuboids and the pyramids inside them.

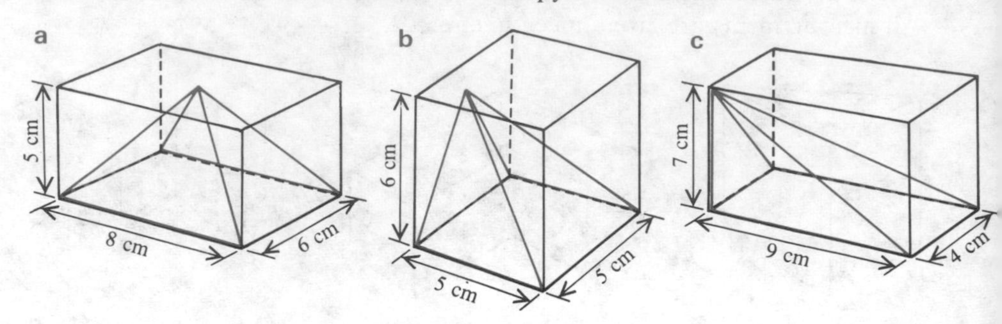

2 For each pyramid, find (i) the area of the base (ii) the volume.
Give answers to three significant figures where necessary.

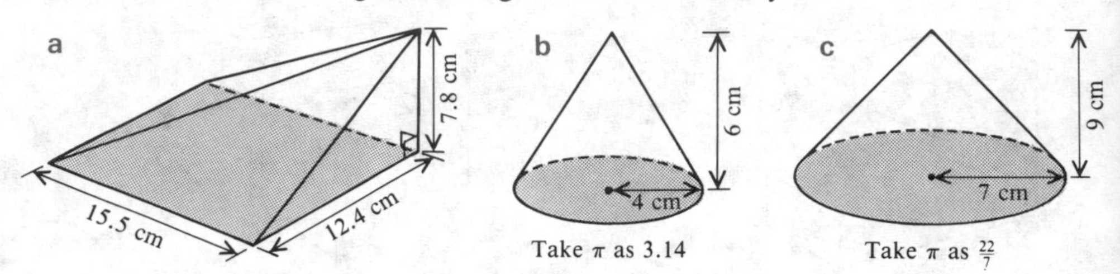

3 Coal at a power-station is held in a hopper which has the shape of an inverted pyramid 8.5 metres high with an open rectangular top 6.5 metres long and 5.6 metres wide. Coal has a density of 1.4 tonnes per m^3.
Calculate, to three significant figures,

- **a** the area of the open rectangular top
- **b** the capacity of the hopper
- **c** the mass of coal in it when full.

4 Sand and cement are mixed in the ratio of 4:1 to make concrete. The sand to be used lies in a conical pile 0.7 m high with a base of diameter 1.6 m.
Calculate, to three significant figures,

- **a** the volume of this sand
- **b** the volume of the cement needed.

5 A bucket has the shape of a frustum of a cone of 35 cm depth with diameters top and bottom as shown.
Calculate the volumes of two cones, and hence find the volume of the bucket in litres to three significant figures.

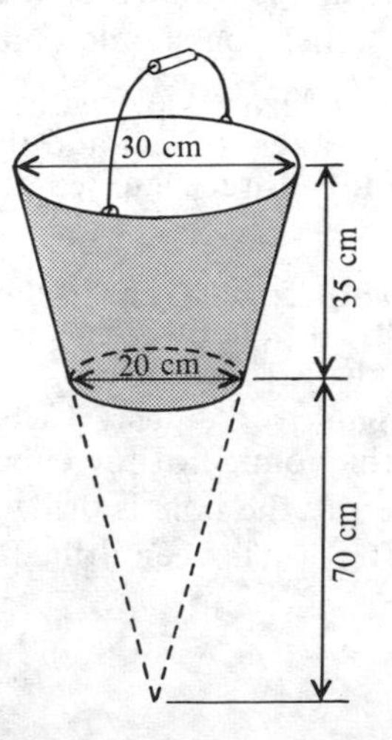

Volume

Part 4 Spheres

1 Find the volumes of the spheres with these radii,

a taking π as 3.14 and giving answers to three significant figures

(i) 2 cm (ii) 5 cm (iii) 4 cm

b taking π as $\frac{22}{7}$

(i) $3\frac{1}{2}$ cm (ii) $\frac{1}{2}$ cm (iii) $1\frac{1}{2}$ cm.

2 A school globe has a diameter of 32 cm. Calculate its volume to three significant figures.

3

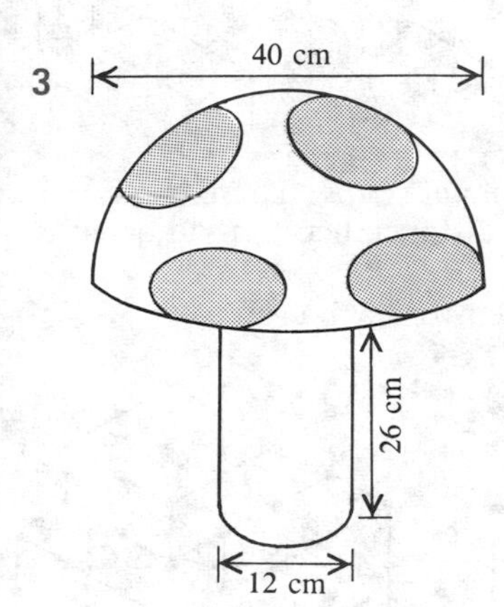

This concrete toadstool is made as a garden ornament from a solid hemisphere of diameter 40 cm and a solid cylinder of diameter 12 cm and height 26 cm. Concrete has a density of 2.7 grams per cm^3.

Calculate, to three significant figures,

a the total volume of the mushroom

b its mass in kilograms.

Solids and nets

Part 1 Isometric drawings

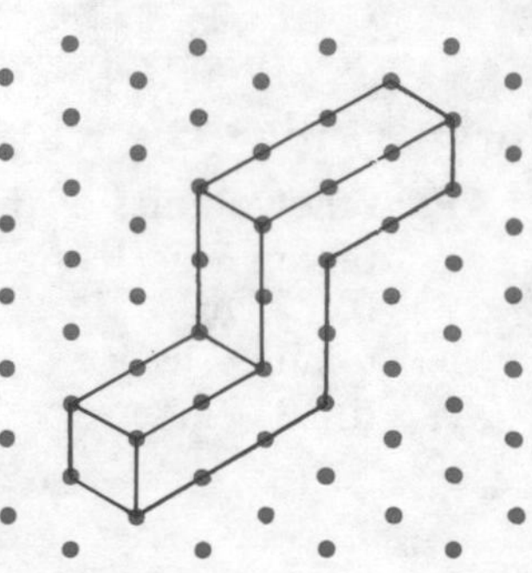

1 The first solid drawn here is on a dotted or isometric grid. Copy it and draw the other solids on a similar grid.
For each solid, count the number of faces F, vertices V and edges E.
Calculate the value of $F + V - E$ in each case. What do you notice?

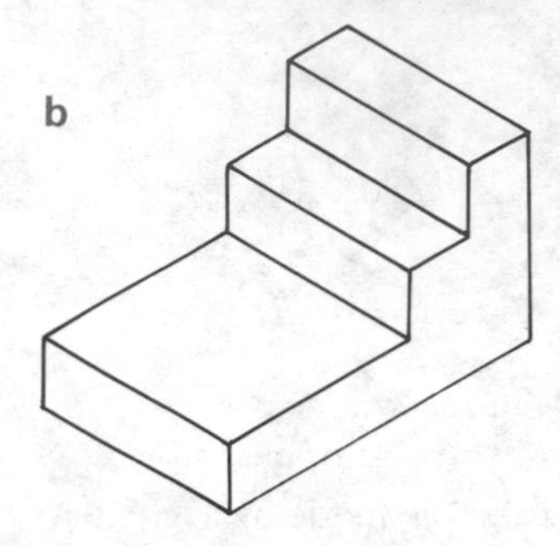

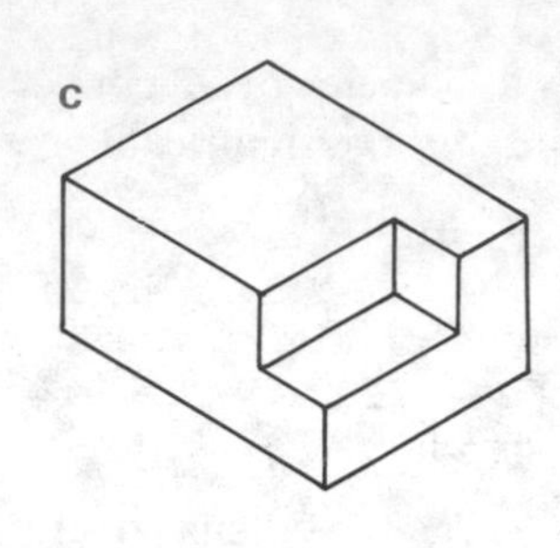

2 Three crates are to have their edges reinforced by metal strips. These three diagrams show the *skeletons* of the crates with the distance between adjacent dots representing 10 cm.
Calculate the total length of the metal edging needed for each crate.

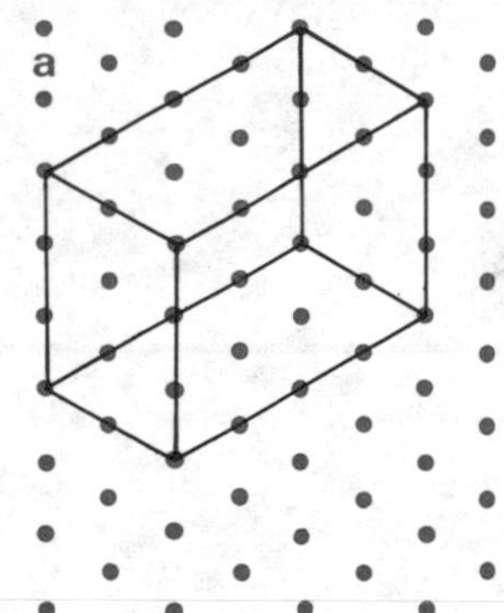

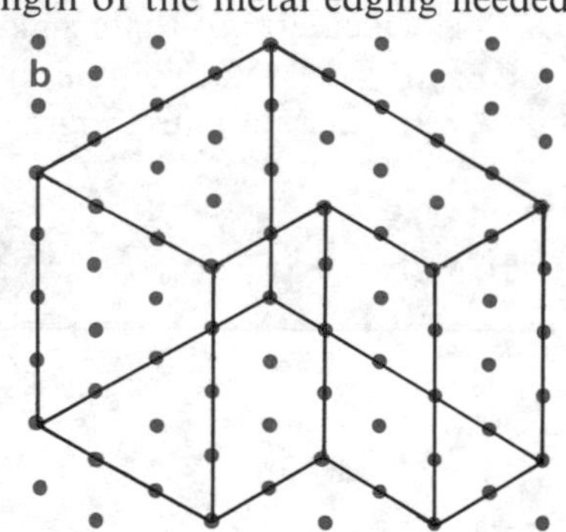

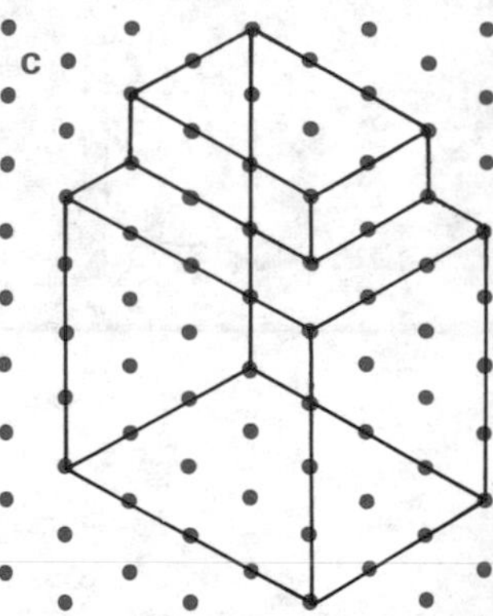

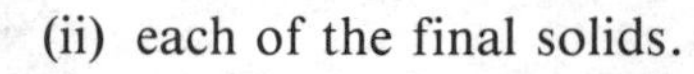

3 These solids are made from a child's interlocking cubes with edges 1 cm long. The shaded cubes are then removed.

a Draw on dotted paper the solids which are left after the shaded cubes are removed.

b Find the surface area of
(i) each of the original solids (ii) each of the final solids.

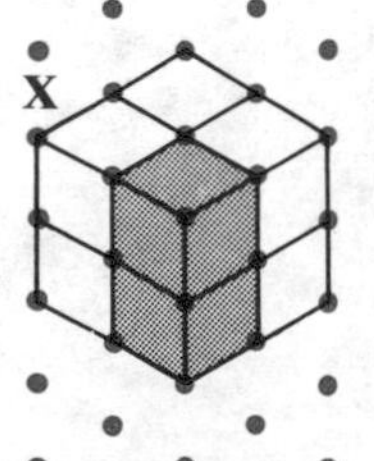

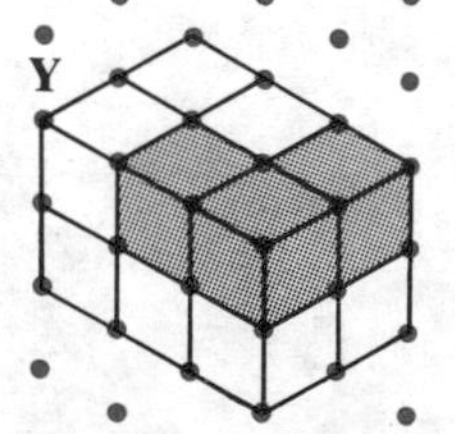

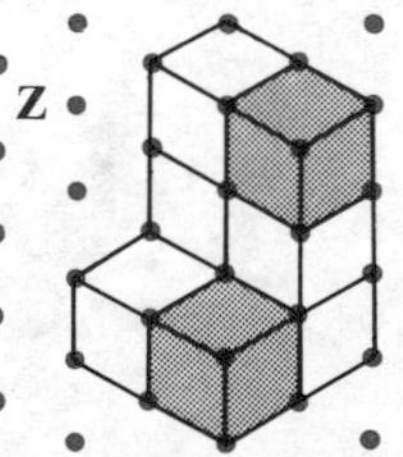

Solids and nets

4 Three solids are to be drawn on dotted paper. Only *half* of each solid is shown here, and each shaded face is a plane of symmetry of the whole solid.
Copy each diagram and complete the whole solid.

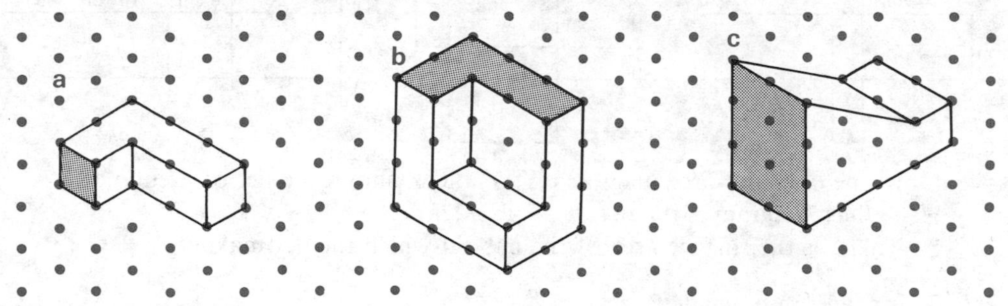

5 Three solids have rotational symmetry. Only *half* of each solid is shown here.
The dotted lines are axes of rotational symmetry of order 2.
Copy each diagram on dotted paper and complete the whole solid.

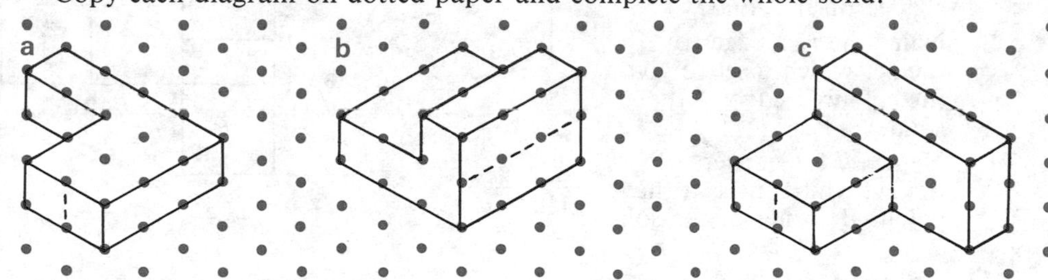

Part 2 Nets and surface area

1

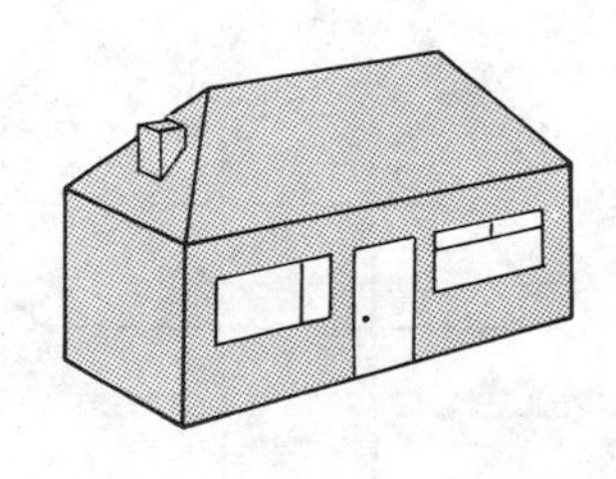

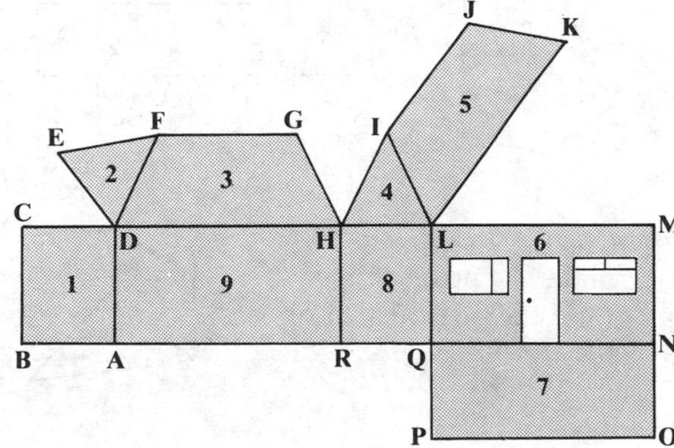

This model of a bungalow can be made from the net shown here.
Give the number of the face of the net

a which forms the base of the model
b on which the chimney should stand.

When the model is made, which edge of the net will join with each of these edges?

c *RQ* **d** *AR* **e** *FG* **f** *EF*

List *all* corners of the net which coincide with these corners when the model is made.

g *G* **h** *R* **i** *B* **j** *K*

Solids and nets

2

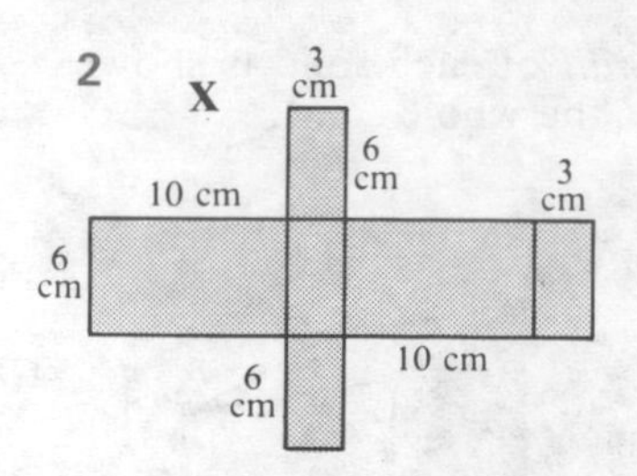

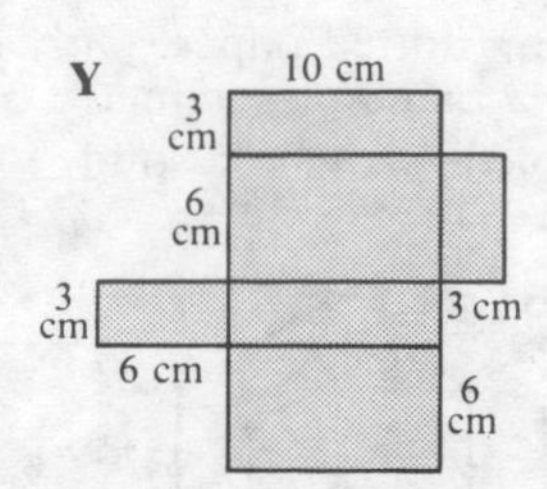

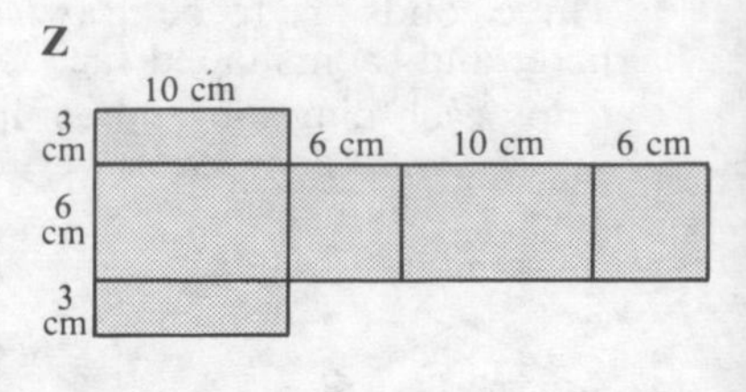

Only one of these three diagrams *X*, *Y* and *Z* shows the net of a cuboid.

a Which diagram is the net of a cuboid?

b What is the surface area of the cuboid which the net makes?

3 This solid can be made from this net.

a What is the number of the face of the net which is shaded on the solid?

b Name the *two* edges of the net which meet to give the coloured edge of the solid.

c Name *all* the corners of the net which meet at the coloured corner of the solid.

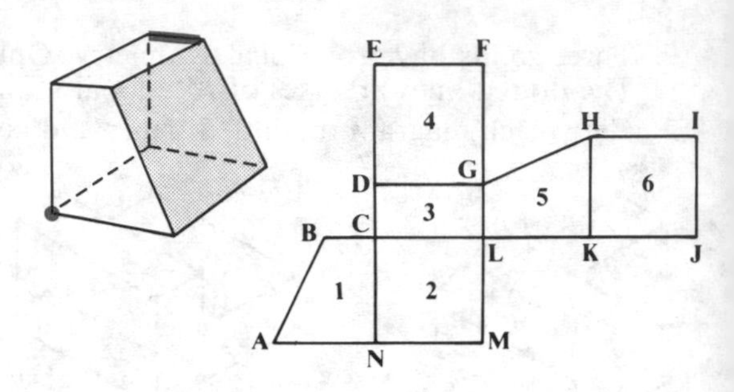

4

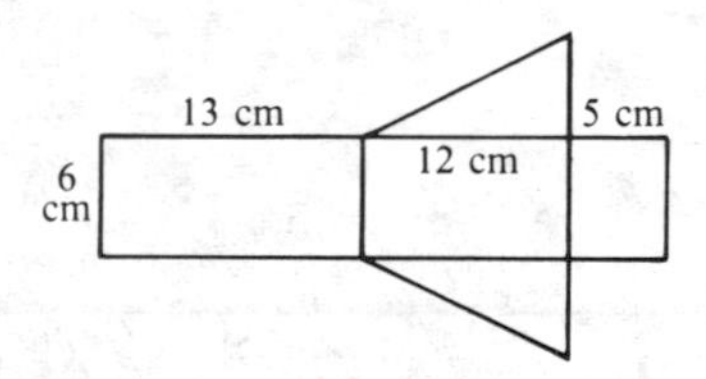

This net is used to make a triangular prism.
Calculate the surface area of the prism.

5 The net given here will make the prism shown.
Find the surface area of the prism.

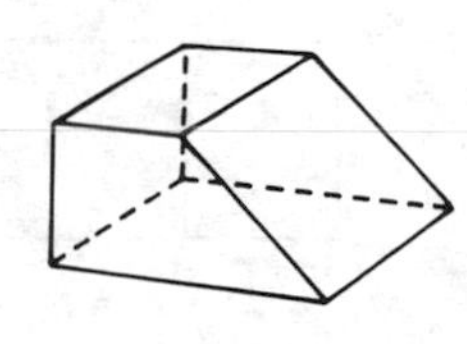

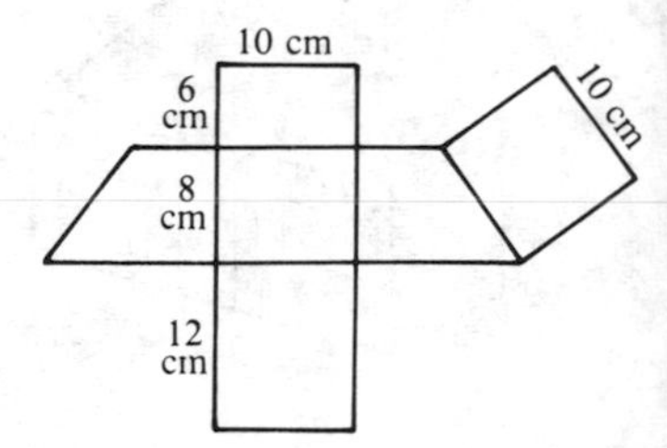

6

A breakfast-cereal box is made from one piece of cardboard. The bottom and top of the packet have a double thickness of two overlapping flaps.

a Draw a net from which the packet can be made. Label each face of the net with one of these words.

top	bottom	front	back	side

b Calculate the surface area of an unopened packet.

c Calculate the area of cardboard which is needed to make one packet.

Solids and nets

7 A shoe box has a tight-fitting lid as shown.

a Draw a net for the lid accurately, scaling down all lengths by a half. (Ignore any tabs required for gluing the net.)

b If a net for the lid is cut from a rectangular sheet of cardboard, what is the smallest possible size of rectangle and what area of the rectangle has to be cut away and wasted?

c Calculate the total area of cardboard used to make the box and the lid (not including any wastage).

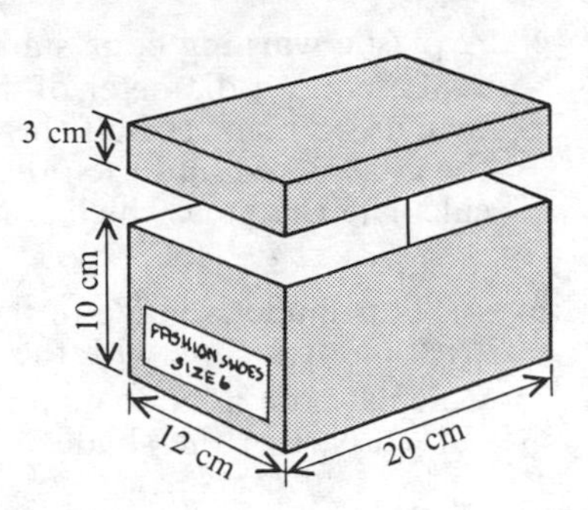

8 A cylindrical tube, open at both ends, is made from a rectangular sheet of paper. If the tube has a height of 10 cm and a radius of 5 cm, calculate its surface area to the nearest whole cm^2.

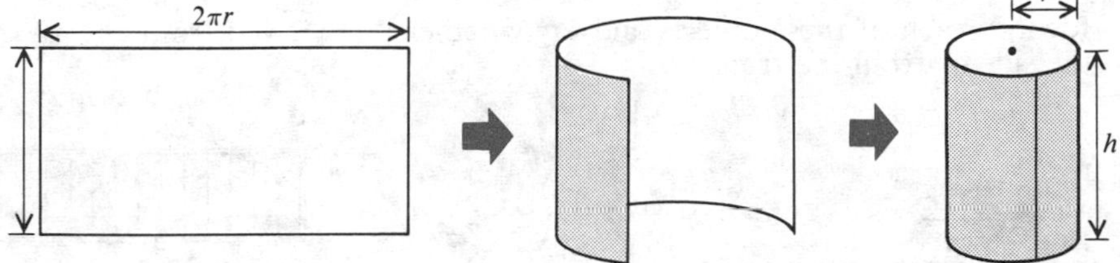

9 a A hollow tube has a height of 20 cm and a radius of 7 cm. Take π as $\frac{22}{7}$ and calculate its surface area.

b If two circles are used to close the two ends of the tube, what is its total surface area?

10

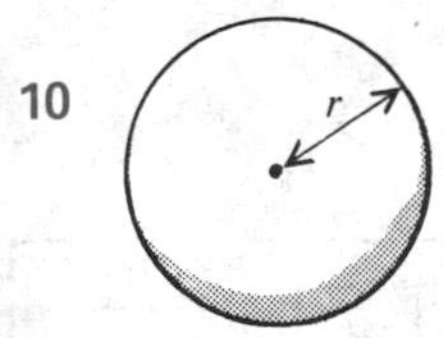

Calculate the surface area of a sphere with

a a radius of $3\frac{1}{2}$ cm (taking π as $\frac{22}{7}$)

b a radius of 3 cm (taking π as 3.14).

11 Calculate (to the nearest whole cm^2) the surface area of

a a school globe with a diameter of 40 cm

b a tennis ball with a diameter of 6.4 cm.

12 A garden toadstool is made from a concrete hemisphere on top of a concrete cylinder having the dimensions shown. Calculate its total surface area to the nearest cm^2.

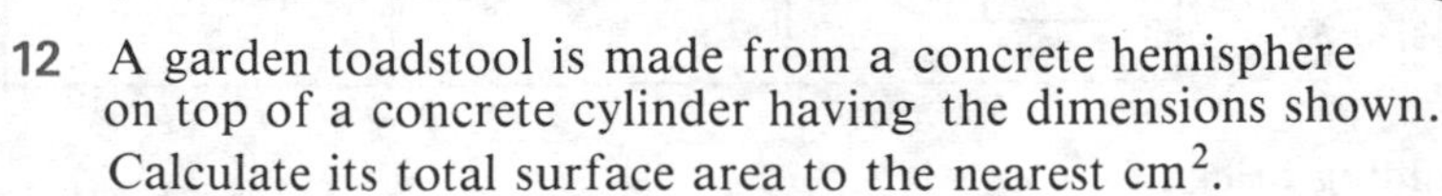

13

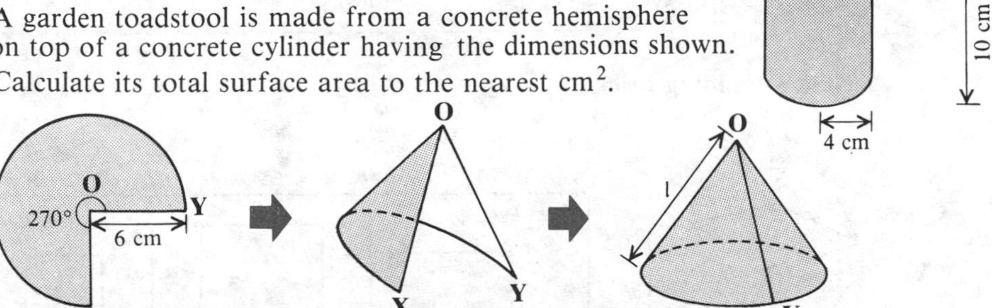

This sector of a circle is made into a hollow cone by bringing together the two edges OX and OY. Take π as 3.14 and find

a the area of the curved surface of the cone

b the slant height l of the cone.

Solids and nets

14 A plastic warning cone stands at the roadside. Its base has a diameter of 40 cm and its slant height is 50 cm. If half the curved surface of the cone is red and the other half is white, calculate the area which is red.

15 A lamp shade is made in the shape of a frustum of a cone with the dimensions shown. Calculate the area of material needed for the curved part of the shade.

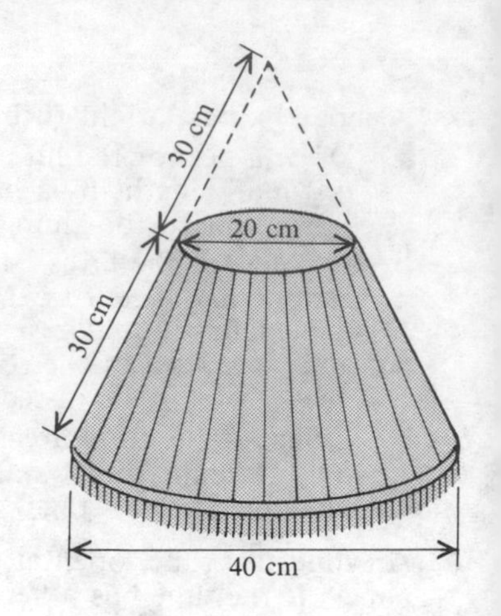

Part 3 Plans and elevations

1 Identify each of these objects and say whether it is viewed from above, from the side or from the front.

a

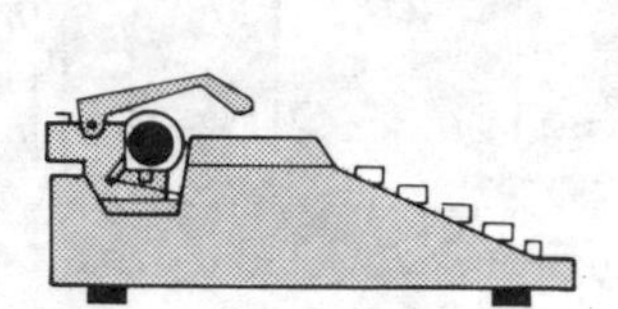

b

c

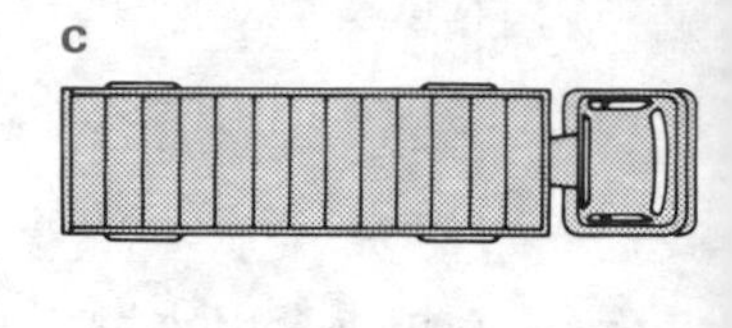

2 Each of these objects is viewed from three directions *A*, *B* and *C*. Match the directions *A*, *B* and *C* with the views *1*, *2* and *3*.

a a drop-leaf table

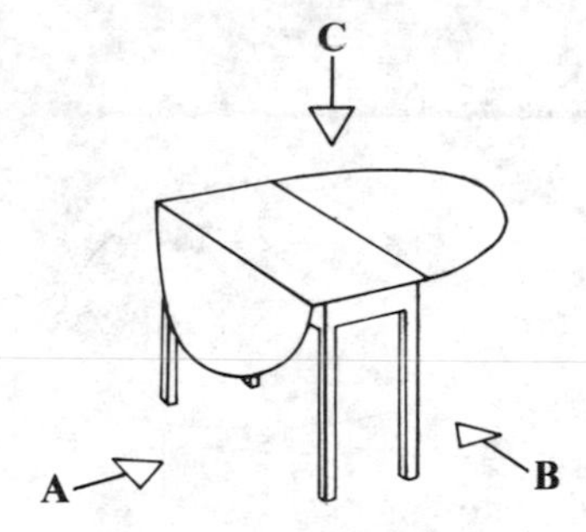

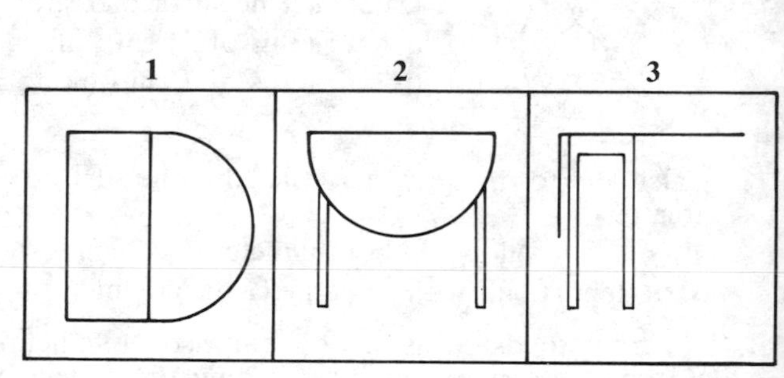

b a child's building brick

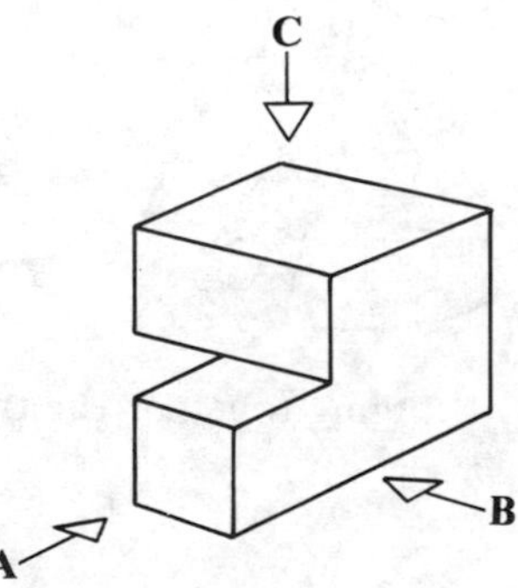

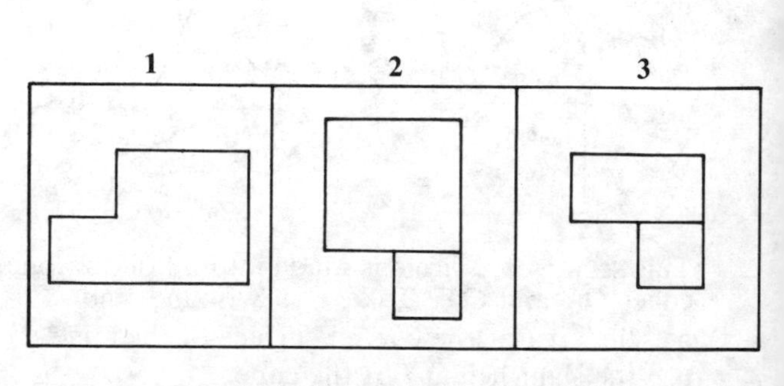

Solids and nets

3 The *plan* of an object is that view seen from *above*.
The *elevation* of an object is that view seen from one *side*.
Draw the plan and elevation of each of these objects in the directions given by the arrows.
Any hidden edges should be shown on your diagrams as broken lines.

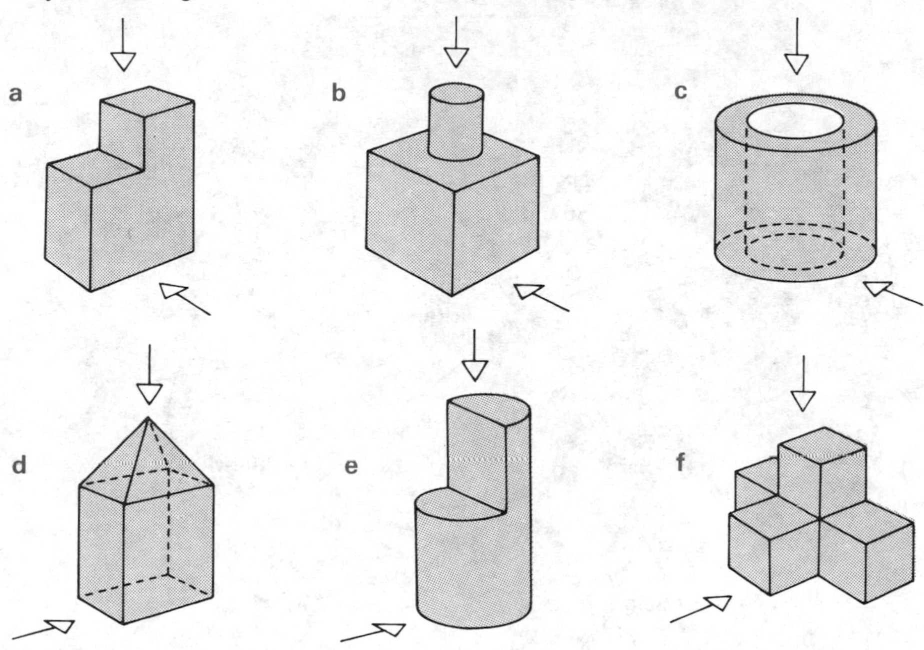

4 Sketch the three solids which have the plans and elevations given here.

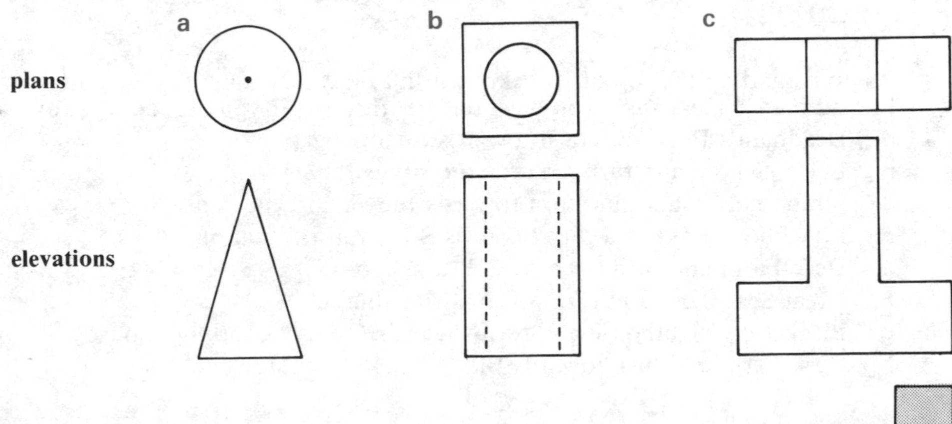

5 Which one of these three solids does *not* have this elevation?

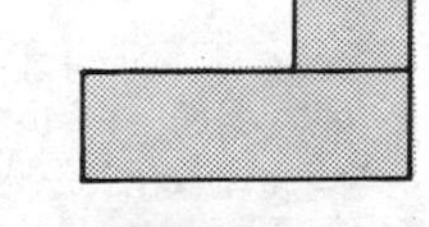

a

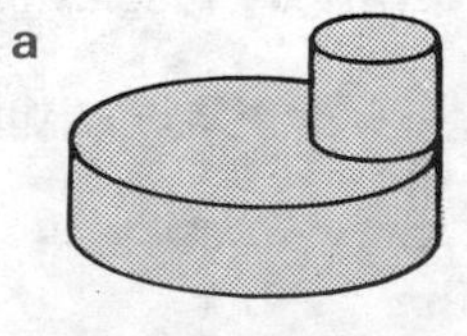

b

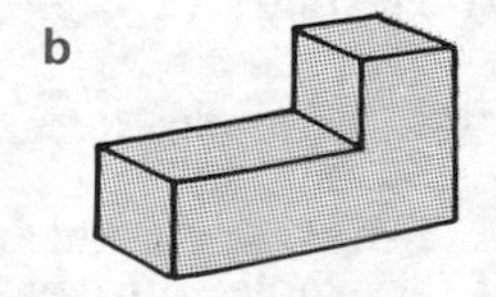

c

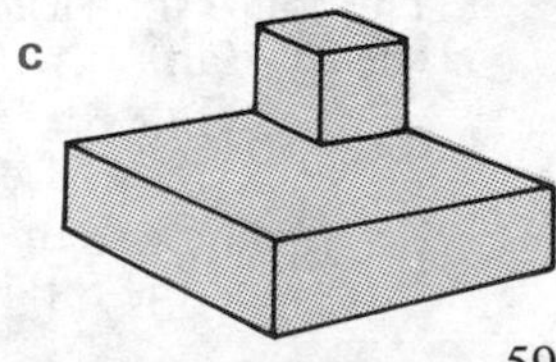

Angles

Part 1 Turns and bearings

1

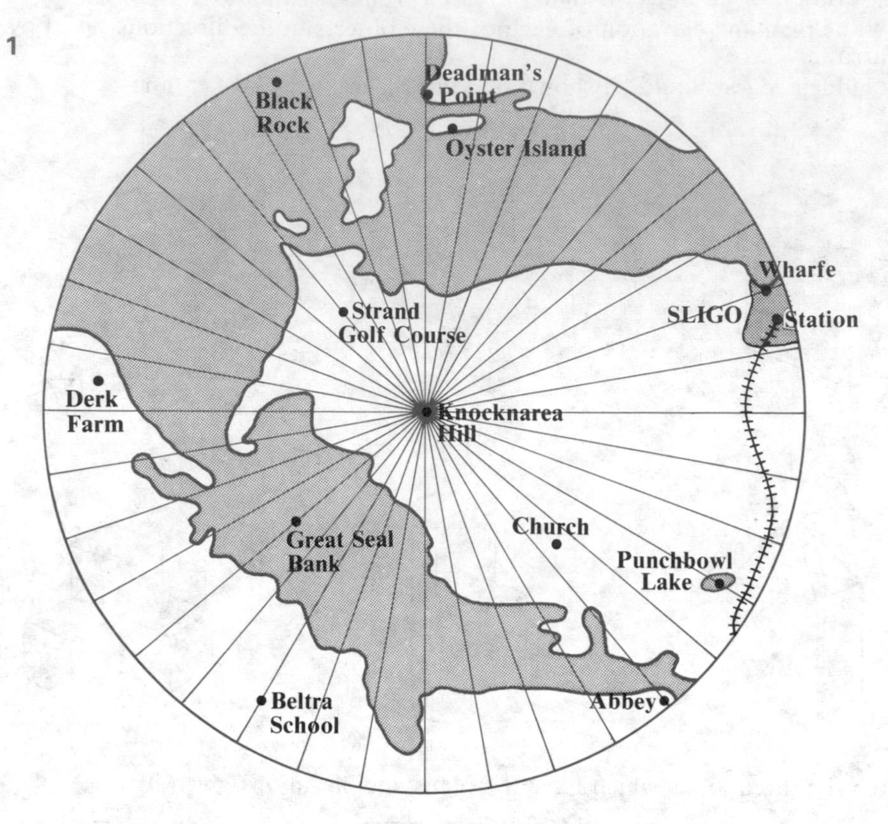

As you stand on the top of Knocknarea hill on a clear day, use your protractor to find the smallest angle you have to turn through if you are looking towards

- **a** Deadman's Point and turn towards Sligo wharf
- **b** Beltra School and turn towards the ruined abbey
- **c** Strand golf course and turn towards the Great Seal Bank
- **d** Punchbowl Lake and turn towards Sligo railway station
- **e** Derk Farm and turn towards Beltra school
- **f** Great Seal Bank and turn towards the church
- **g** Black Rock lighthouse and turn towards Oyster Island lighthouse
- **h** Derk Farm and turn towards Black Rock lighthouse.

2 A boat leaves Maryport in Cumbria to sail off the west coast of England. The route shown here is in seven stages before the boat berths at Douglas in the Isle of Man.

- **a** Use your protractor to find the angles a to f through which the boat turns at the end of each leg of the journey.
- **b** If the scale of the map is 1 mm = 1.5 km, find the length of each leg and hence the total length of the whole journey.

Angles

Stranraer
Carlisle
a
Maryport
b
e
Douglas
Blackpool
f
c
d
Liverpool
Holyhead

3 To find a three-figure bearing, imagine that you face due north and write the angle which you turn through clockwise to reach the required direction.
Use this map to find the bearing from Southampton of

a Basingstoke *Ba*
b Aldershot *Al*
c Petersfield *Pe*
d Ventnor *V*
e The Needles *TN*
f Swanage *Sw*
g Bournemouth *Bo*
h Andover *An*
i Portsmouth *Po*
j Ryde *R*
k Shaftesbury *Sh*
l Salisbury *Sa*
m Devizes *D*.

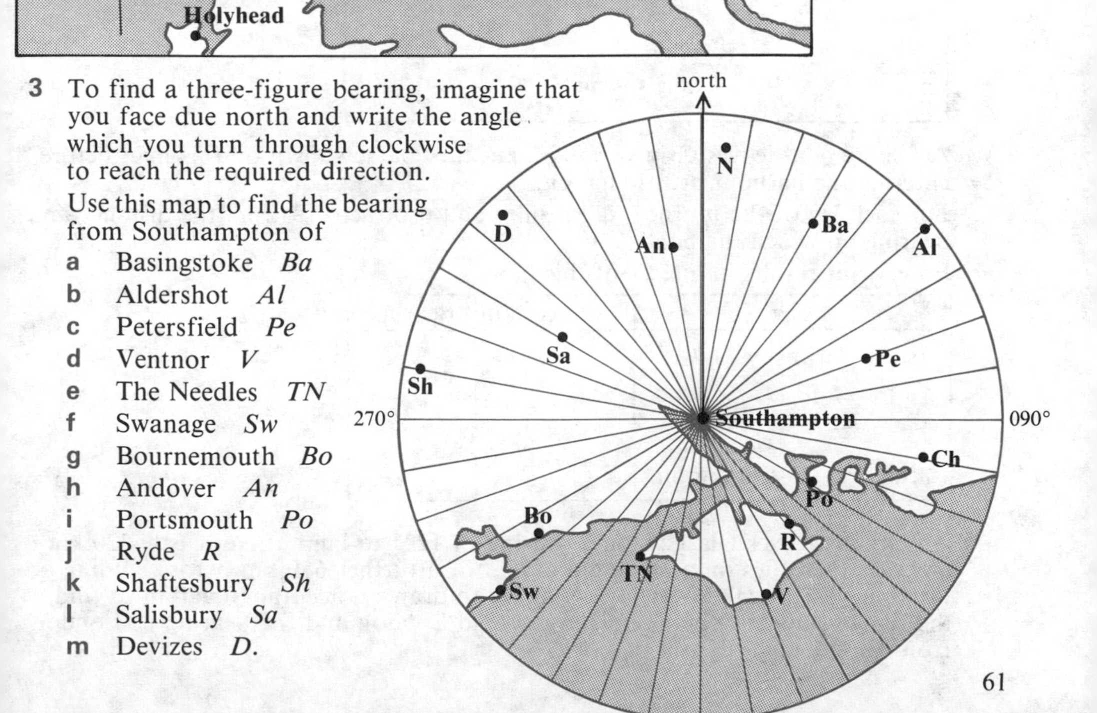

Angles

4

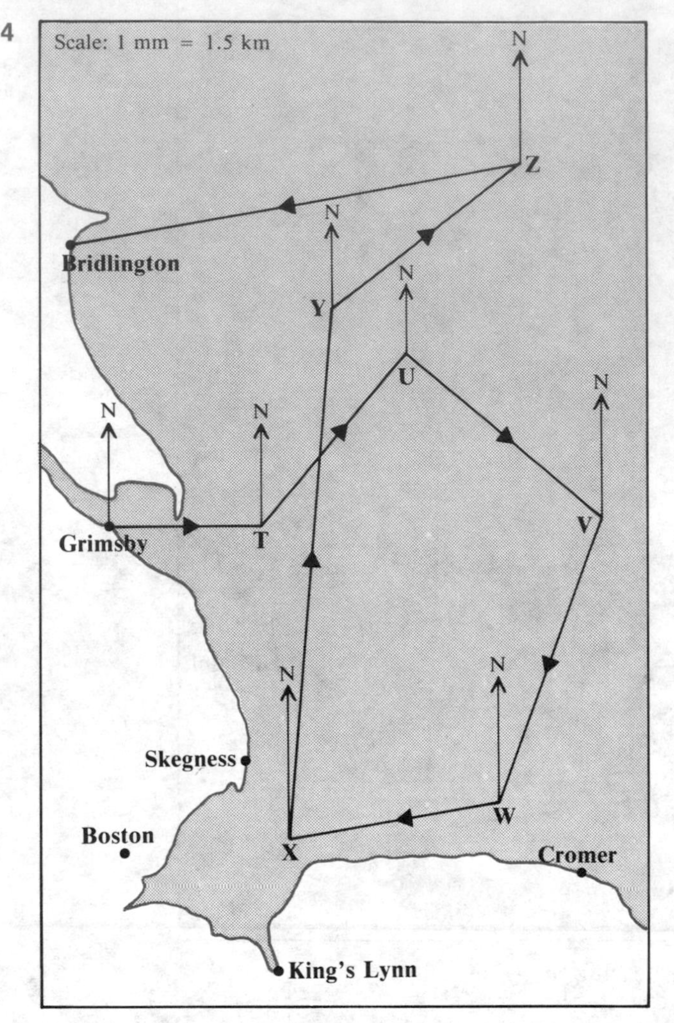

A fishing boat leaves Grimsby and takes the course shown on this map before entering the harbour at Bridlington.

For each leg of the journey, use a ruler and protractor to find the distance and bearing on which the boat sails.

Enter your results in a copy of this table.

Leg		Distance, km	Bearing
1st	Grimsby to T		
2nd	T to U		
8th	Z to Bridlington		

5 An aircraft leaves Glasgow on a bearing of 122° to land at Newcastle 320 km away. It then flies on a bearing of 176° for a further 650 km before landing at London. Use a scale of 1 cm = 100 km to draw an accurate diagram to find the shortest distance between Glasgow and London and the bearing of London from Glasgow.

Angles

6 From a hilltop, a farmer sees a barn 5 km away on a bearing of 072° and his farmhouse 4 km away on a bearing of 220°. On what bearing is the barn from the farmhouse and what is the shortest distance between them?

7 From the centre of Birmingham,
Derby is 57 km away on a bearing of 030°,
Leicester is 54 km away on a bearing of 070°,
Worcester is 38 km away on a bearing of 215°,
and Wolverhampton is 23 km away on a bearing of 310°.
Draw a scale diagram to find the shortest distance between

a Wolverhampton and Leicester **b** Wolverhampton and Worcester

and also the bearing of

c Derby from Wolverhampton **d** Worcester from Wolverhampton.

Part 2 Points, lines and shapes

Find the sizes of the lettered angles in these diagrams.
Angles marked with the same letter are equal.

1 At a point

a

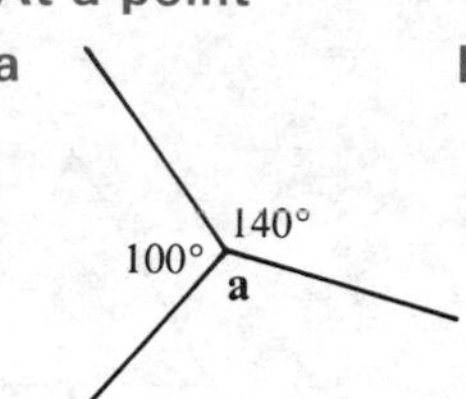

b

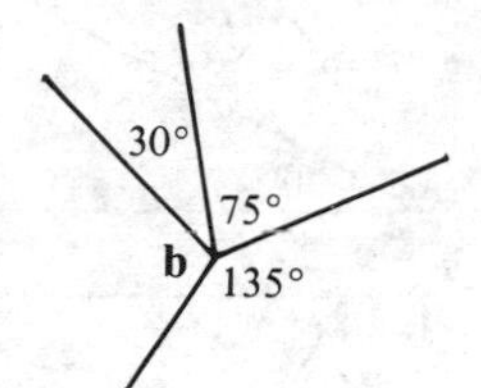

c

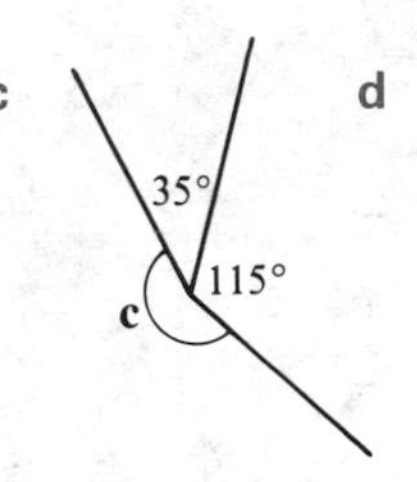

d

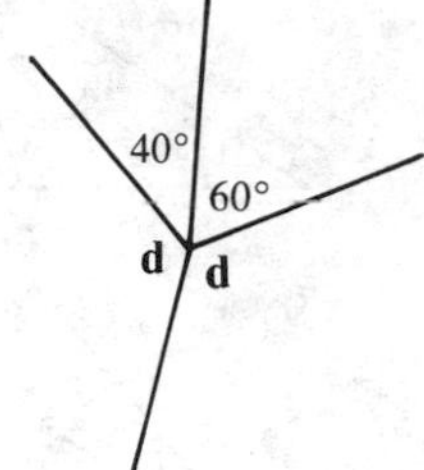

2 On a straight line

a

e
150°

b

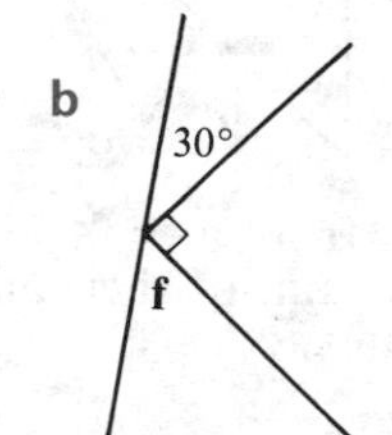

c

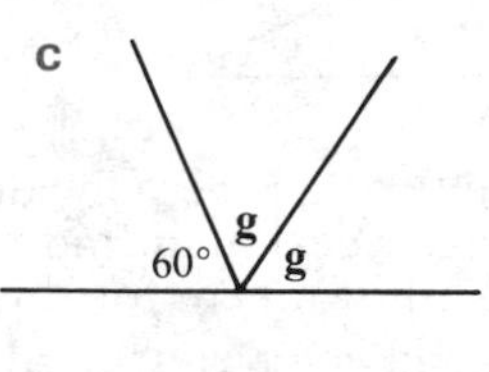

d

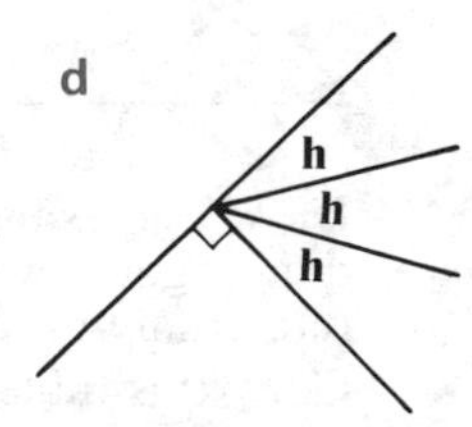

3 Vertically opposite

a

130°
j k
i

b

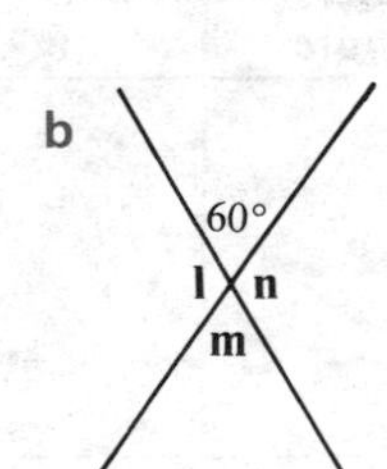

c

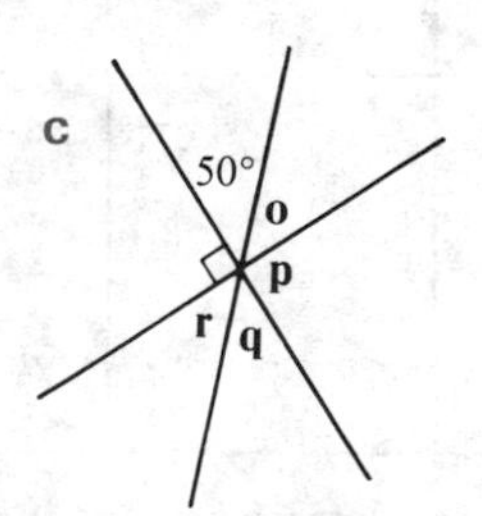

d

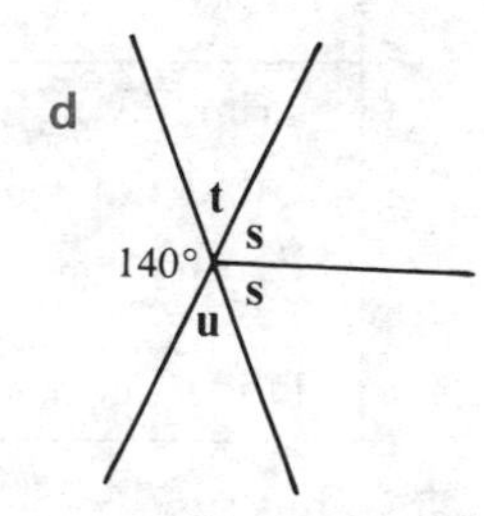

4 Parallel lines

a

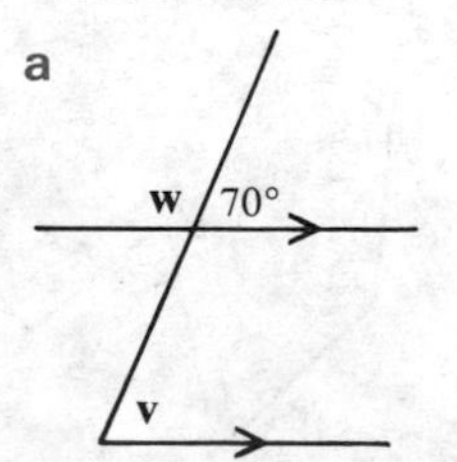

b

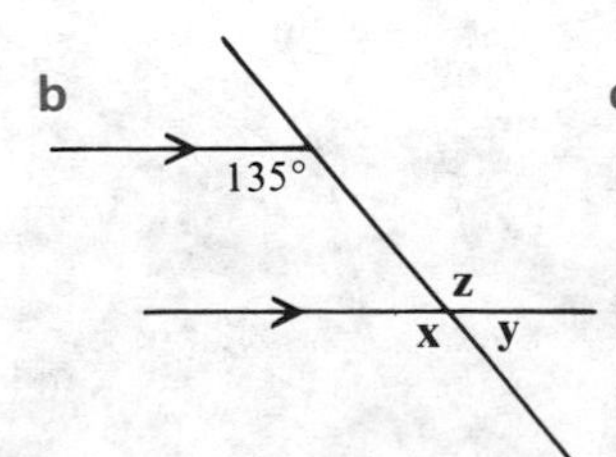

c

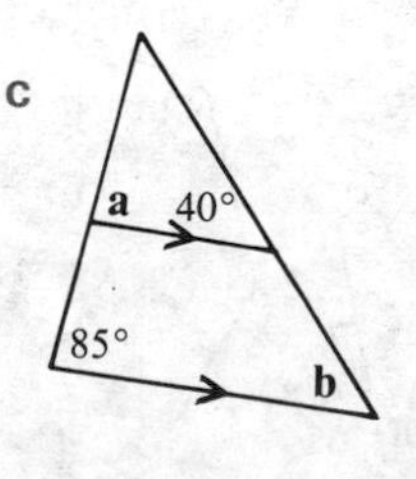

d

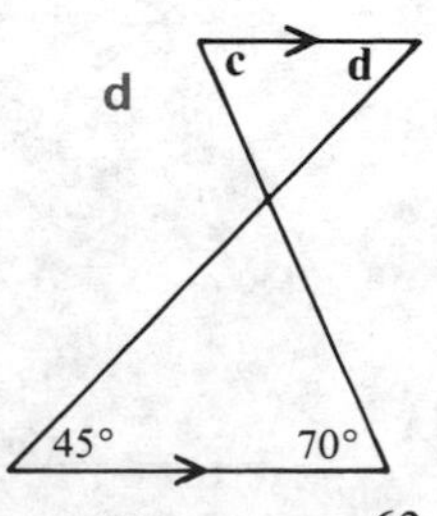

Angles

5 Triangles

a

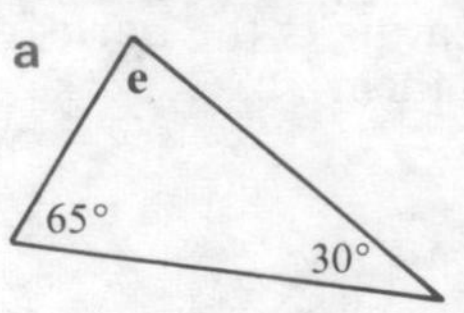

b

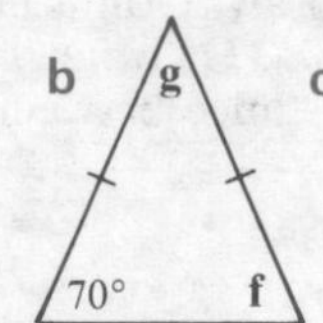

c

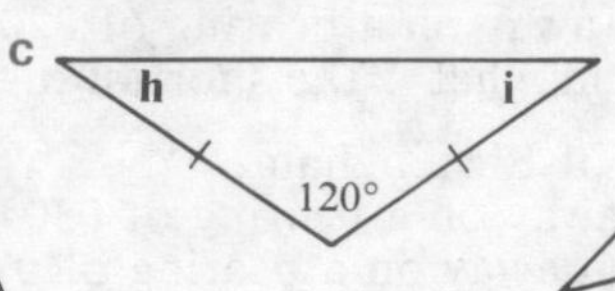

d

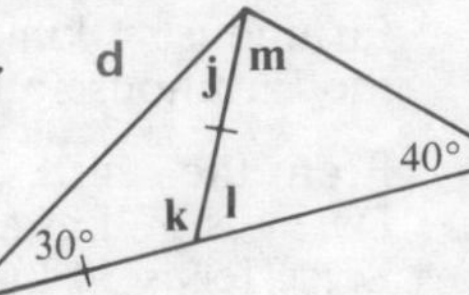

6 A mixture

a

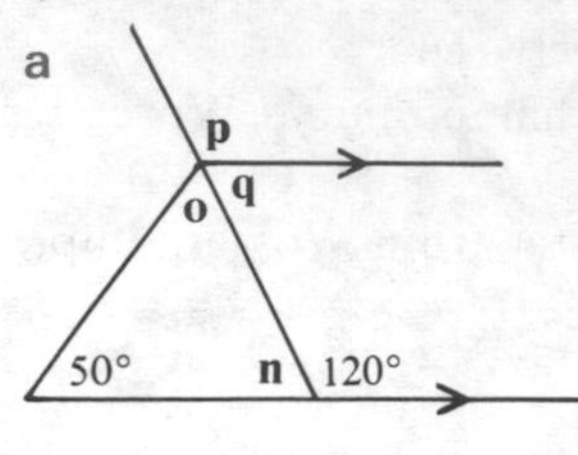

b

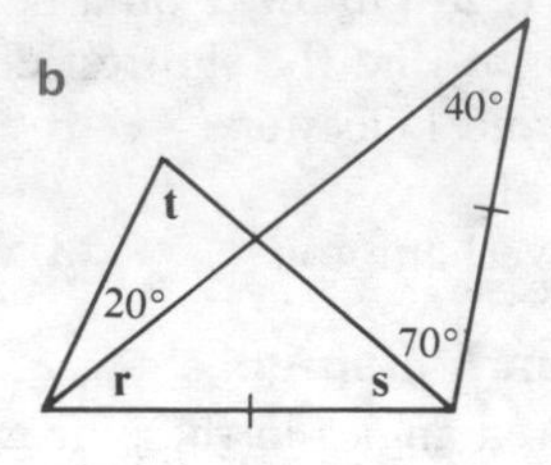

c

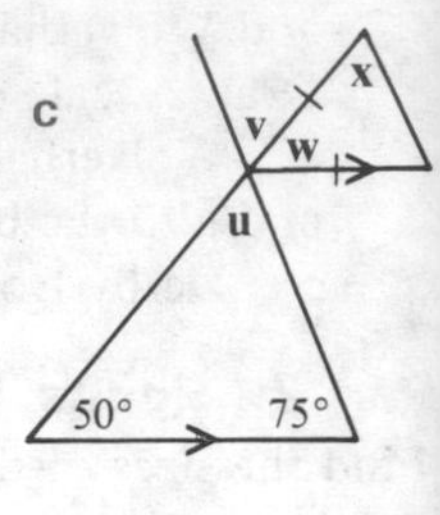

d

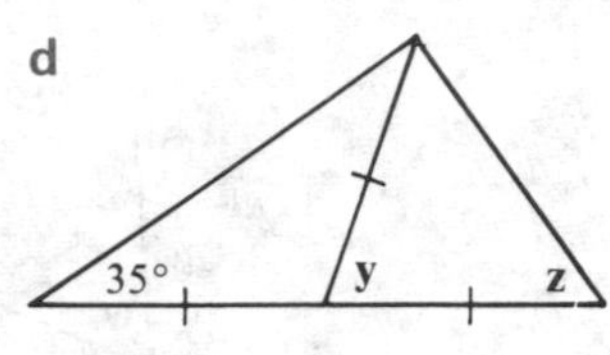

e

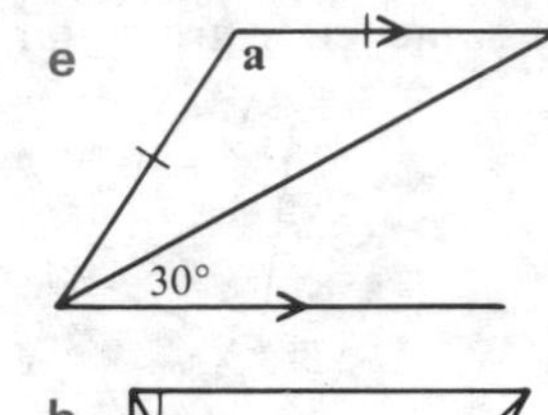

f

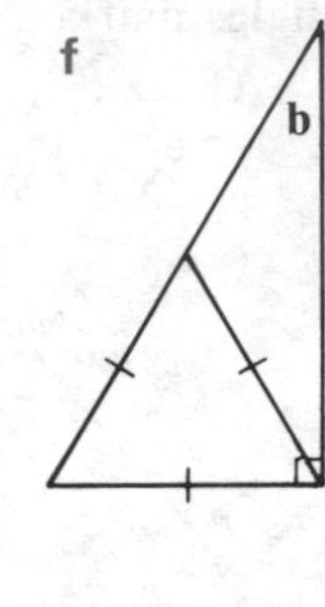

g

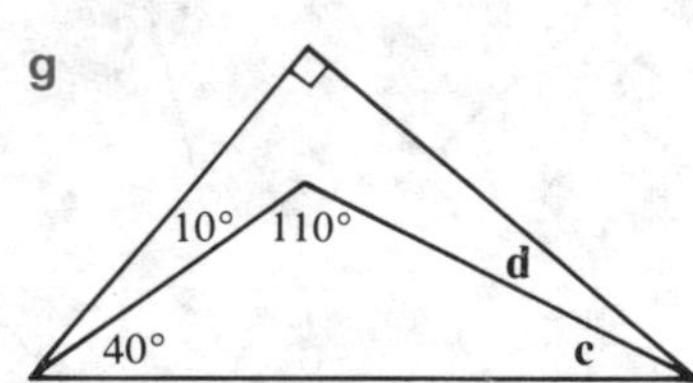

h

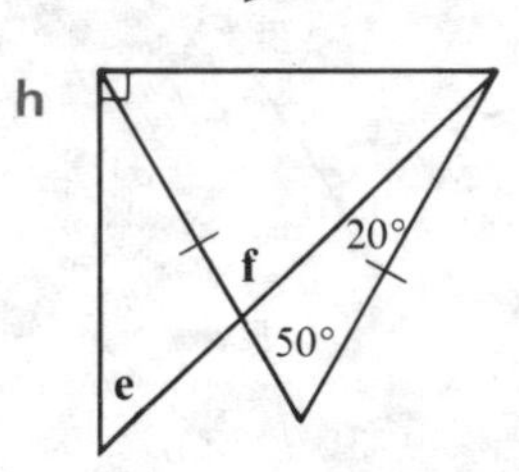

7 Draw one example of all the polygons from a triangle (three sides) to a dodecagon (twelve sides), and find the sum of their angles by splitting them into triangles, as shown here for this hexagon.

Copy this table and enter your results.

Polygon	Number of sides	Number of triangles	Sum of all angles
Triangle			
Quadrilateral			
Pentagon			
Dodecagon			

8 Find the size of the lettered angle in each of these polygons.

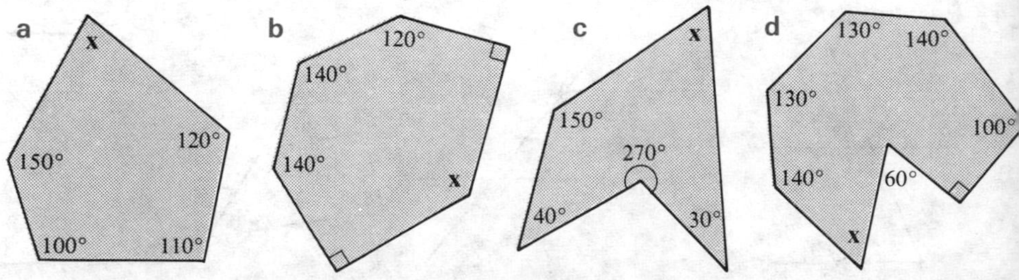

Angles

9 A decagon has four angles of 125° each and four angles of 150° each. If the other angles of the decagon are equal, what is their size?

10 A quadrilateral has one angle $x°$, and one angle twice this size, one angle 10° bigger than $x°$ and one angle of 70°. Find the value of x.

11 A pentagon has one angle $x°$, one angle three times this size, one angle 20° less than x, and two angles of 145° each. Find the value of x.

12 Regular polygons

a Find the internal angles of a regular pentagon by

First method (i) using the table in **7** above to write down the sum of all the angles of any pentagon
(ii) hence calculating one internal angle of a *regular* pentagon.

Second method (i) writing down the sum of all the *external* angles of *any* polygon
(ii) calculating one external, and hence one internal, angle of a *regular* pentagon.

b Repeat these methods to find one internal angle of
(i) a regular octagon (ii) a regular hexagon
(iii) a regular decagon (iv) a regular dodecagon.

Part 3 Angles in circles

Find the sizes of the lettered angles in these diagrams.
Angles marked with the same letter are equal.

1 Between a tangent and radius

a

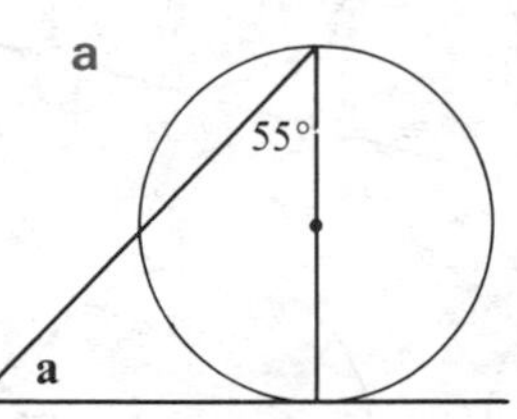

b

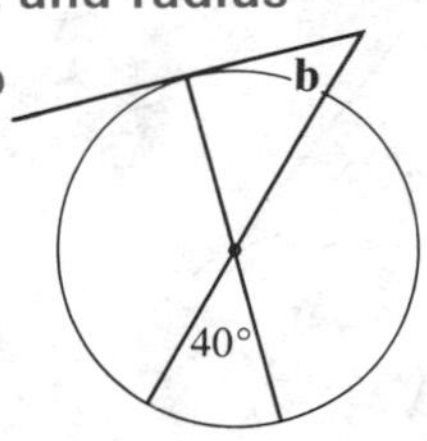

c **d**

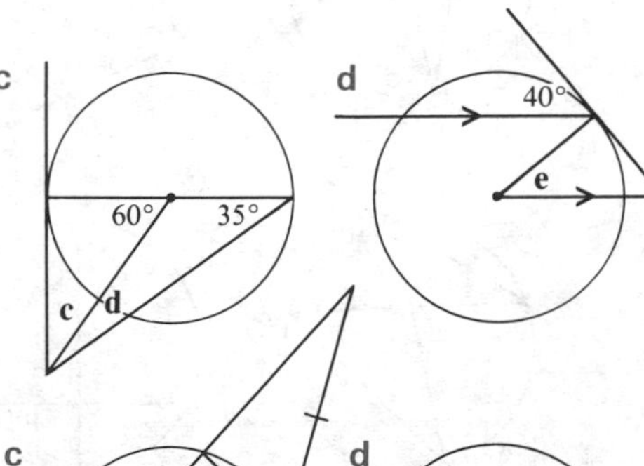

2 In a semicircle

a

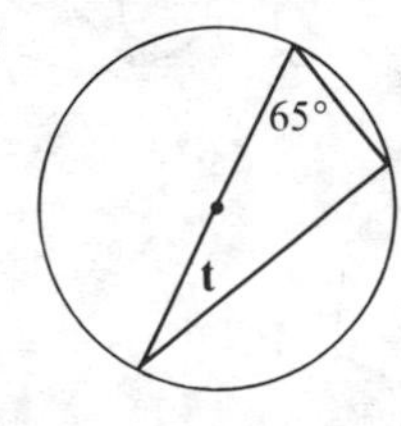

b

v
u
35°

c **d**

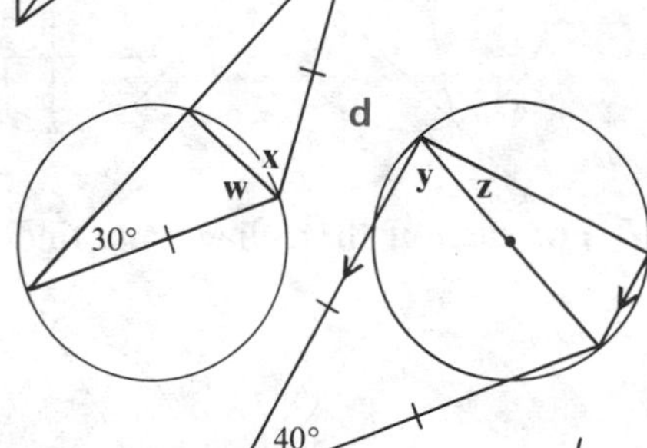

3 On the same arc

a

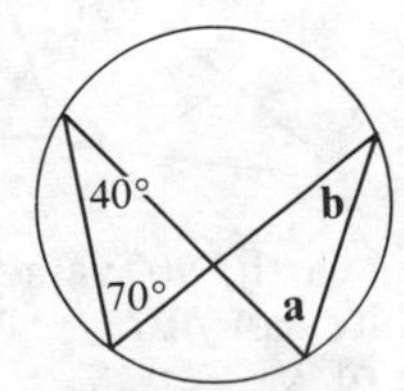

b

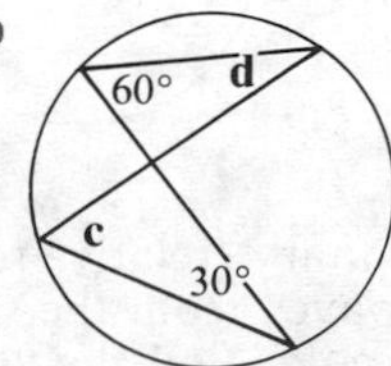

c

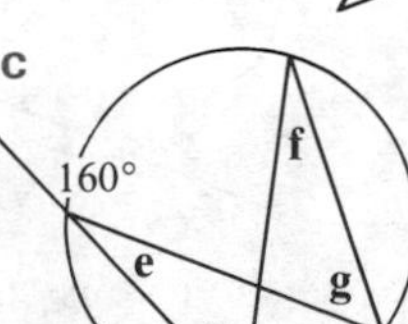

d

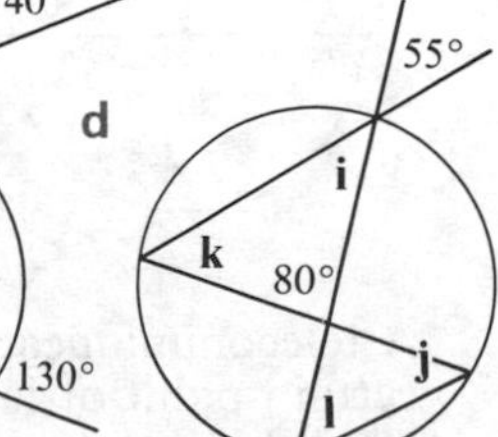

Angles

4 At centre and circumference

a

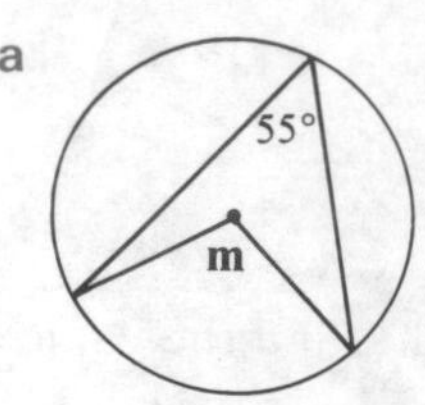

b c

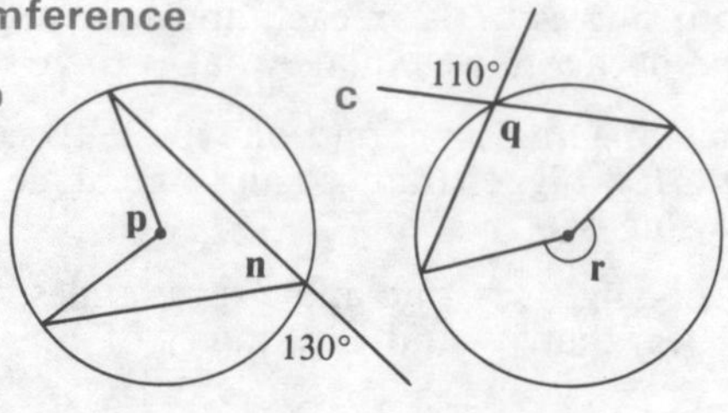

d

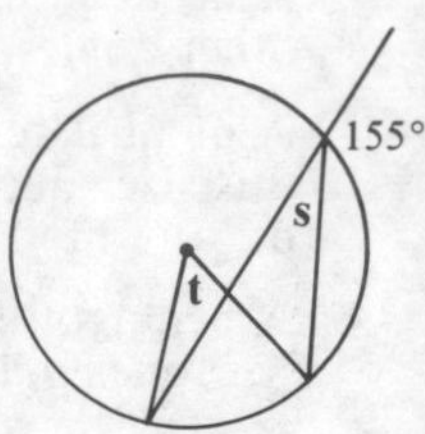

5 In a cyclic quadrilateral

a b

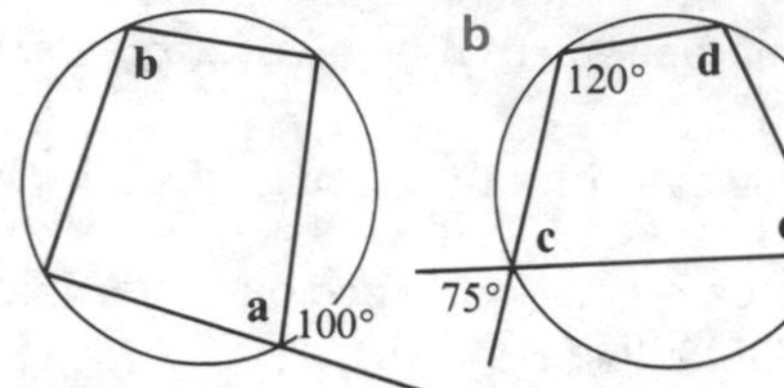

c

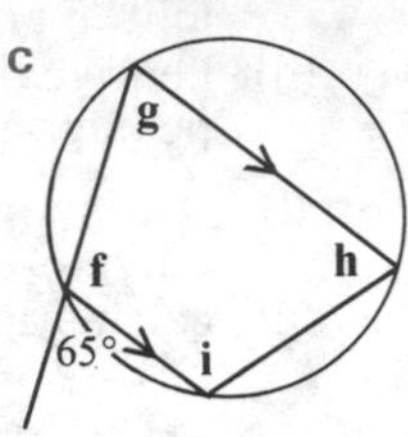

d

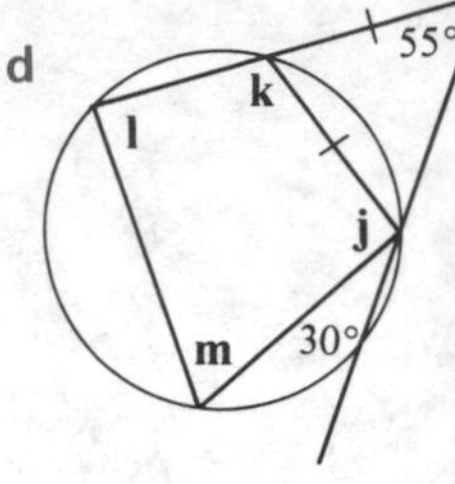

6 A mixture

a

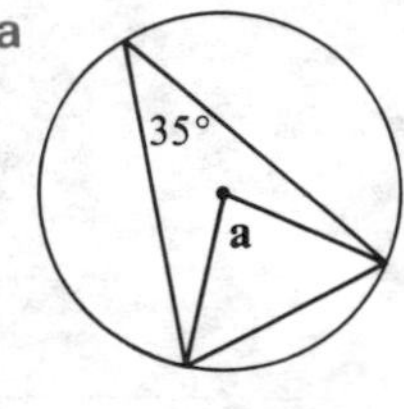

b

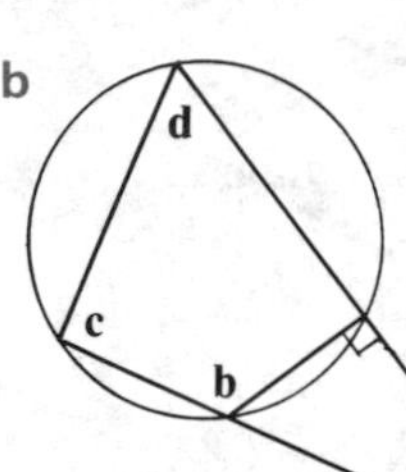

c

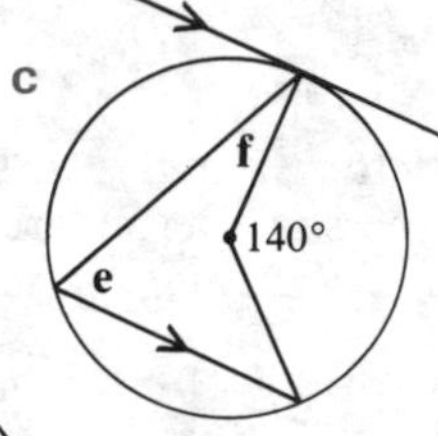

d

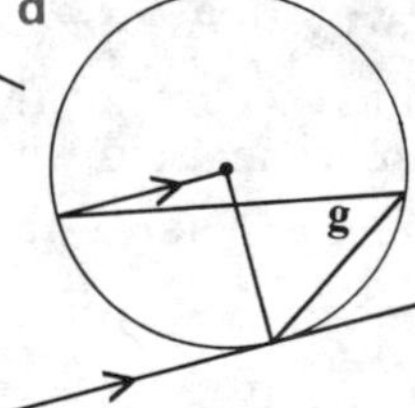

e

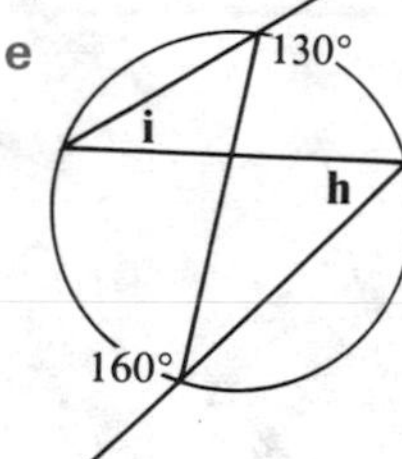

f

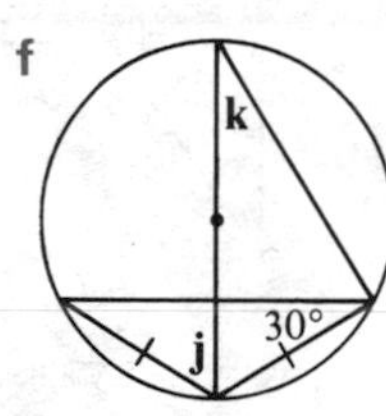

g

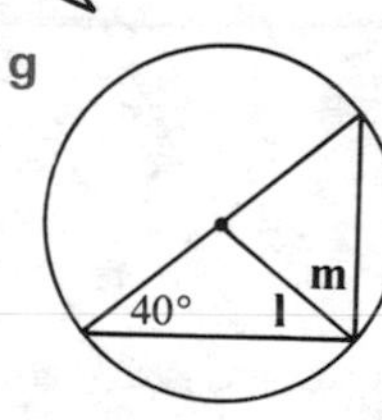

h

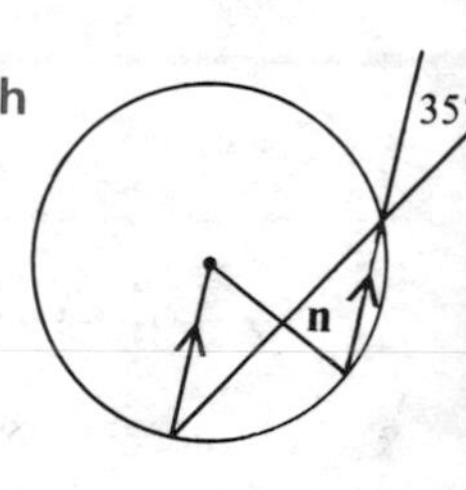

7 For each of these diagrams, find angle y in terms of angle x.

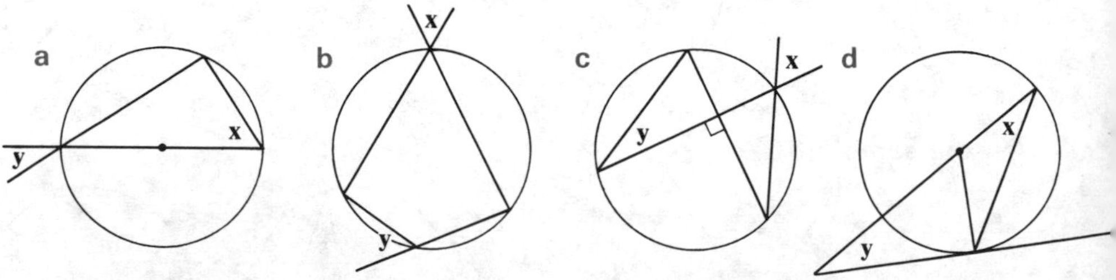

a

b

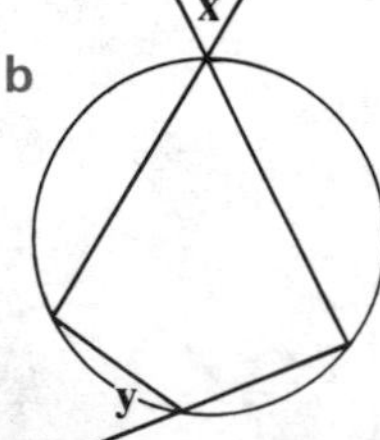

c

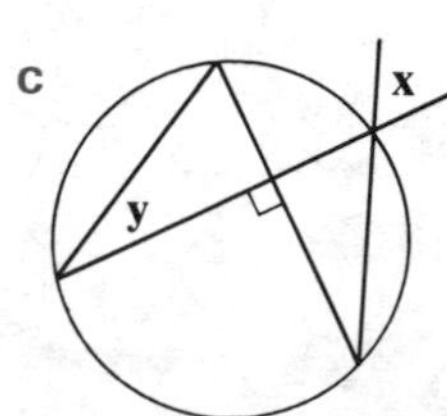

d

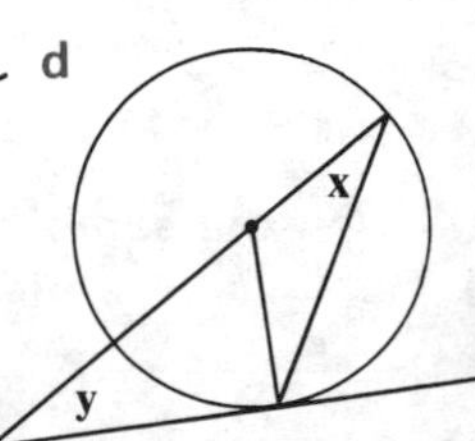

8 A telecommunications satellite is stationary (relative to the earth) above a certain point of latitude $\lambda°$. If radio waves from the satellite can *just* reach both the equator and the north pole, find the value of λ.

9 Stoke on Trent S (53°N, 2°W) is on the same meridian as the point E (0°, 2°W) on the equator. If O is the centre of the earth and P is the south pole, find

a angle SOE b angle SPE.

10 a Find angle E_1E_2N where E_1 and E_2 are the two points (0°, 0°) and (0°, 180°W) on the equator and N is the north pole.

b If London L is at ($51\frac{1}{2}$°N, 0°) find angle $E_1 LN$.

11 Harrogate H (54°N, $1\frac{1}{2}$°W) is on the same meridian as the point E (0°N, $1\frac{1}{2}$°W). If N and S are the north and south poles, find

a angle ENS b angle EHS.

Part 4 Constructions

All problems in this exercise must be solved by using only ruler and compasses. No protractor should be used.

1 a Construct triangle ABC such that AB = 8 cm, BC = 5 cm and CA = 6 cm. Measure its height from C to AB and calculate its area.

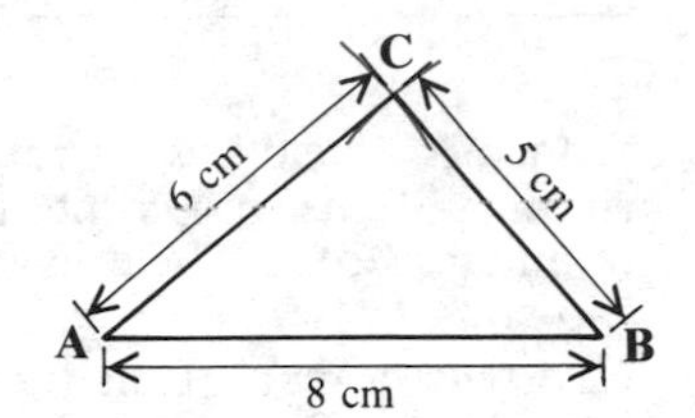

b

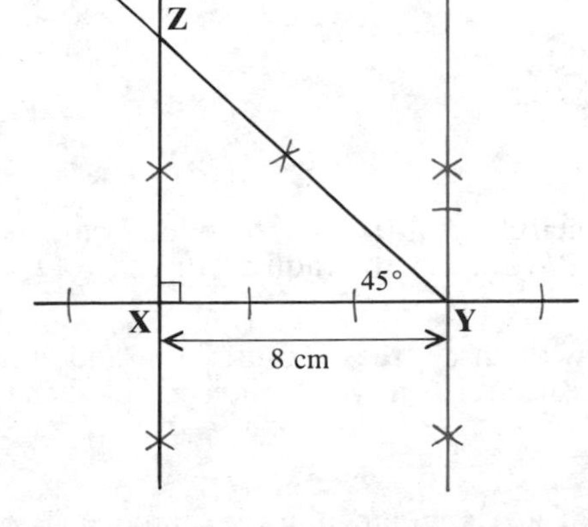

Construct triangle XYZ such that XY = 8 cm, angle X = 90° and angle Y = 45°.
Measure the length XZ and calculate the area of the triangle.

c Construct triangle PQR such that PQ = 8 cm, angle RPQ = 60° and PR = 7 cm.
Measure the height from R to PQ and calculate the area of the triangle.

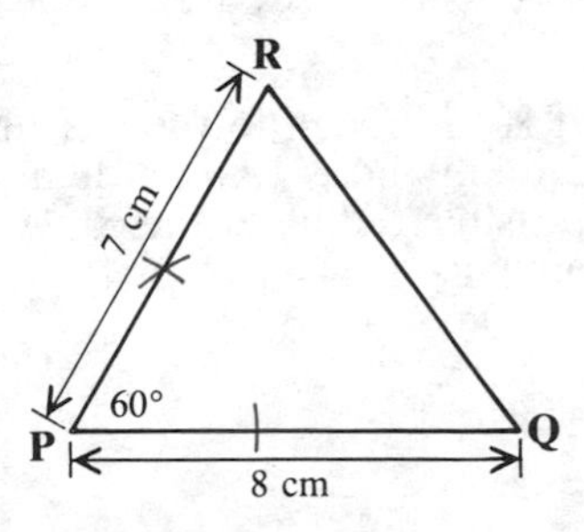

2

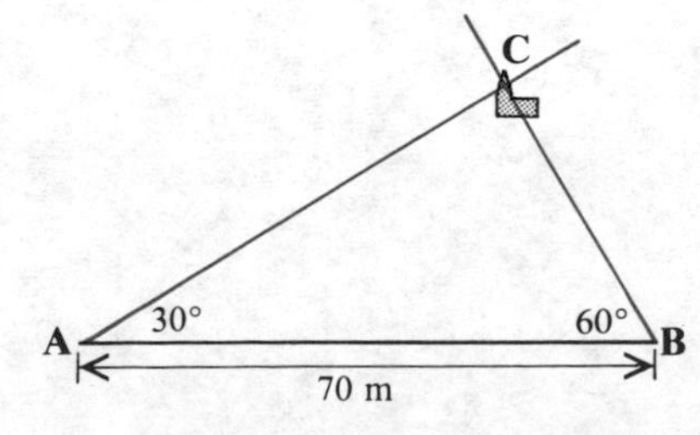

A surveyor marks out a base line AB 70 metres long. A church spire C makes angles of 30° and 60° with AB as shown. Construct triangle ABC and find the distance of C from A and from B.

Angles

3 Thornton T is 5 km due south of Brampton B. Woodford W is on a bearing of 090° from Brampton and 060° from Thornton. Construct the triangle TBW and find the distance between

a Brampton and Woodford
b Thornton and Woodford.

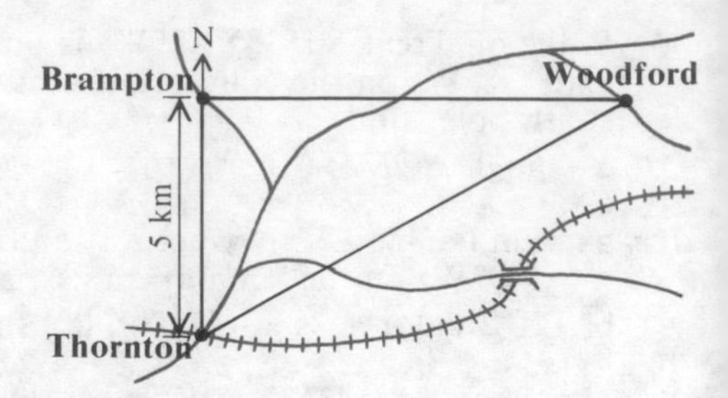

4 A yacht leaves Abermawr on the course shown to Shellisle. Construct the route accurately and find the shortest distance between Abermawr and Shellisle.

5 A farmer's field has a straight track AB running 90 m from corner to corner. The boundaries of the field have lengths as shown.
Construct a map of the track and the boundaries. Take measurements from your map to calculate the areas on either side of the track and hence the total area of the field.

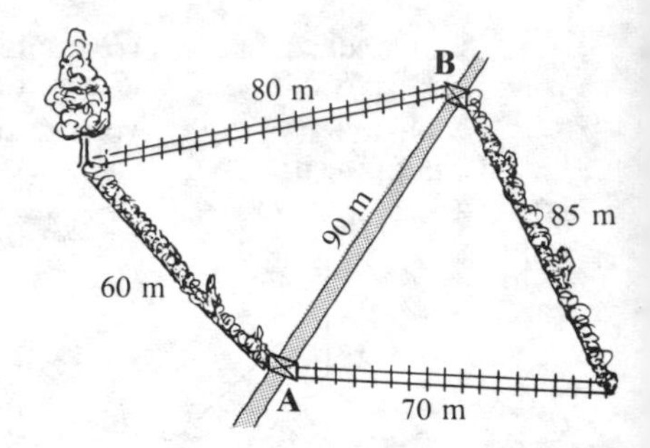

6 Construct the circumcircle and the incircle of a triangle of sides 7 cm, 6 cm and 5 cm, on two separate diagrams. What are the radii of these two circles?

7 A girl rides her horse in a paddock with three trees 100 m, 85 m and 70 m apart. What is the radius of the largest circle on which she can ride her horse so that the trees are outside it?

8 It is forbidden to fly over any part of a triangular military camp of sides 2 km, 2.6 km and 3.4 km. A light aircraft circles the camp. What is the smallest possible radius of the circle on which it flies if it is not to enter forbidden airspace?

Scale drawings and loci

1 This diagram shows a flat metal plate which is fixed by two screws to a kitchen wall to hold a wall can-opener. The scale of the diagram is 1:4. This means that 1 cm on the diagram represents 4 cm on the actual metal plate.

a Find the actual height h and width w of the plate.

b What is the actual distance d between the two screw holes?

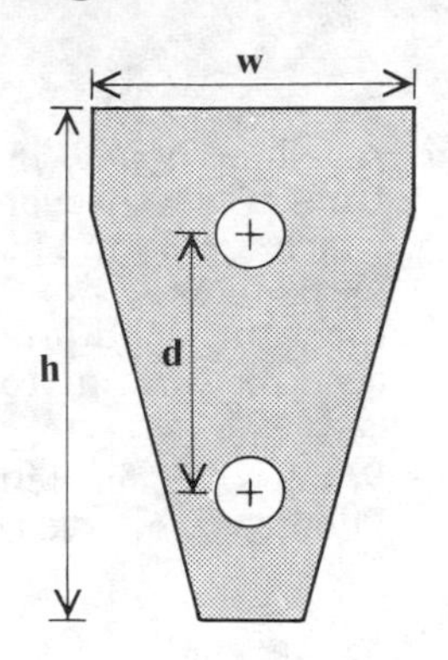

2

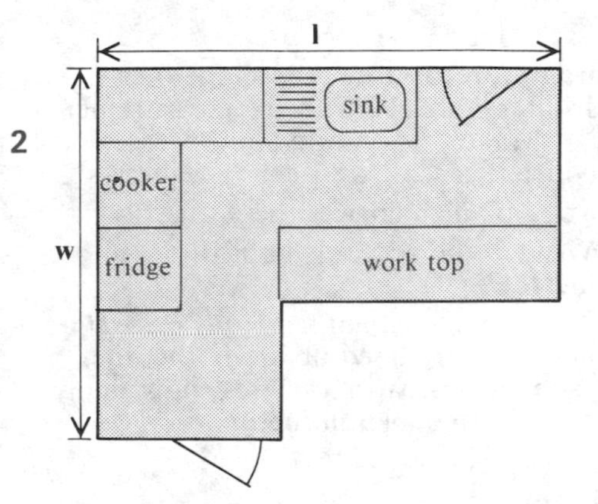

This plan of a kitchen is drawn to a scale of 1:100.

a What is the actual length l and width w of the kitchen in metres?

b Find the perimeter of thc kitchen in metres.

3 A house has a large lawn with a garden hut H, a greenhouse G, a rotating washing line W, a child's swing S and three fir trees F_1, F_2 and F_3. This plan is drawn to a scale where 1 cm represents 10 metres.

a What is the shortest distance from the hut to the greenhouse?

b A young child on the swing sees her mother hanging out the washing, her elder brother in the hut and her father in the greenhouse. What is the shortest distance she has to run to visit each of them and then return to the swing?

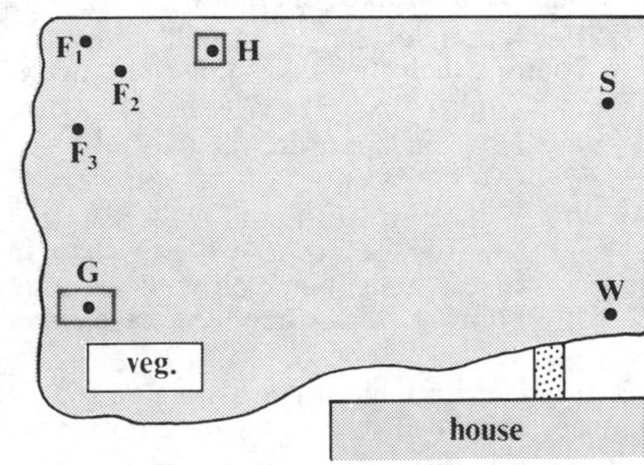

4

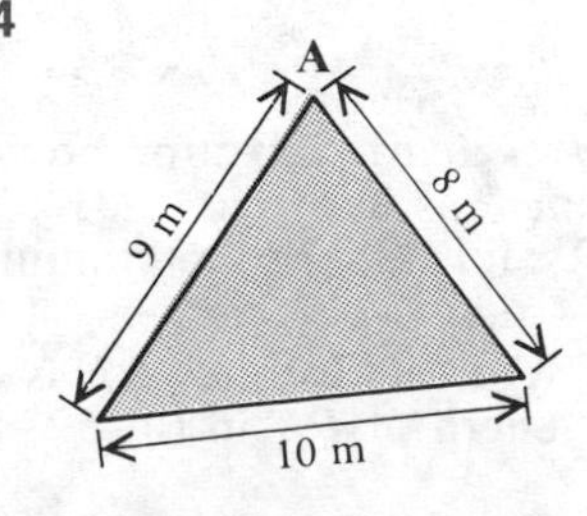

A vegetable plot of triangular shape has fences 10 metres, 9 metres and 8 metres long around its perimeter.

a Use a ruler and a pair of compasses to construct an accurate drawing of the plot using a scale of 1 cm for each metre.

b What is the shortest distance from corner A to the 10-metre fence?

c Calculate the area of the plot to the nearest m^2.

Scale drawings and loci

5 A fishing boat B in distress can be seen from two coastguard stations C and D which are 8 km apart. If the coastguards measure angles of 70° and 40° as shown, use a ruler and protractor to make an accurate drawing to a scale of 1 cm for each kilometre.

What is the shortest distance in kilometres between B and C?

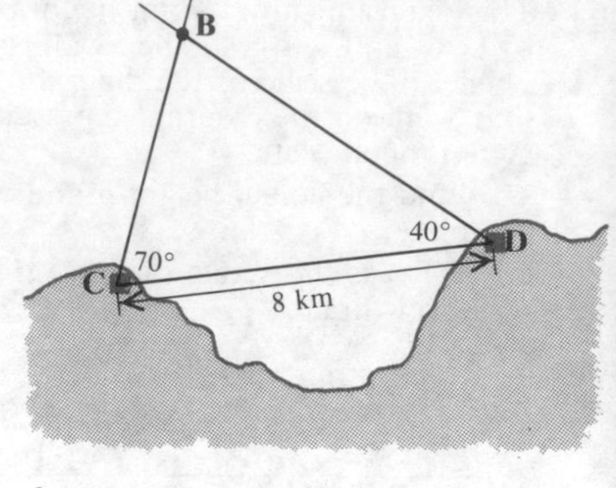

6

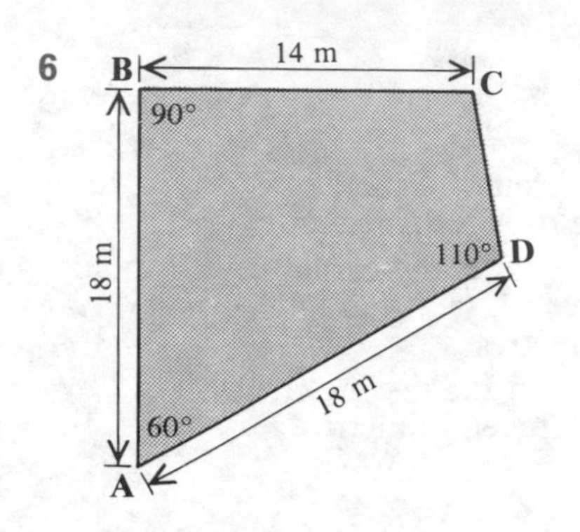

A garage forecourt has the shape of a quadrilateral $ABCD$ with the measurements shown. Make an accurate plan of the forecourt to a scale of 1 cm = 2 metres.

Use your plan to answer these questions.

a What is the actual length of CD in metres?

b A white line is painted from B to D and new cars are lined up along the line. If each car takes up 3 metres, how many cars can be lined up here?

7 Two hikers H and K leave the Youth Hostel Y at the same time to walk straight across flat moorland in two directions 60° apart. After two hours, H has walked four miles and K has walked six miles.

Use a scale of 2 cm = 1 mile and draw the routes which they take. Use only a ruler and compasses (but *not* a protractor).

a How far apart are the two hikers after two hours?

b If they now turn and walk directly towards each other without changing their speeds, mark on your diagram the point at which they meet each other.

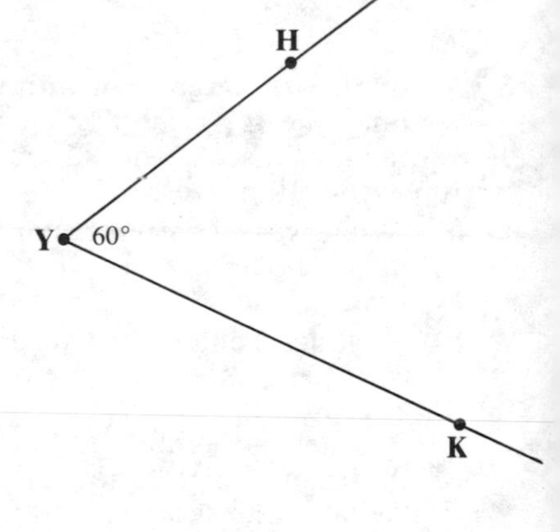

8 PQ is a straight garden path 8 metres long. Two clothes posts X and Y are to be positioned so that they are both 5 metres from P and 6 metres from Q.

a Use a scale of 1 cm for each metre to find by accurate drawing the positions of X and Y.

b What is the distance XY in metres?

9 **a** Two guard dogs are tied to points P and Q 12 metres apart. A young boy X walks so that he is always the same distance from P as he is from Q. Draw a scale diagram to show the path taken by X and state the minimum length of PX.

b Draw a second scale diagram to show the locus of X if the distance PX is always *twice* the distance QX. State the minimum length of PX in this case.

Scale drawings and loci

10 Two straight roads OP and OQ intersect each other at an angle of 60° at the point O. A man X walks so that he is always as far from the road OP as he is from the road OQ.

a Draw accurately the locus of X.

b How far is X from the road OP when he is 60 metres from point O?

11 Two parallel lines L and M are 6 cm apart. Draw the locus of

a the point X which moves so that it is always 2 cm from line L and 4 cm from line M

b the point Y which moves so that it is always 2 cm from line L and 8 cm from line M.

12 A circle has a radius of 3 cm. Draw the locus of a point which moves so that it is always

a 4 cm from the circle and label this locus X

b 3 cm from the circle and label this locus Y

c 2 cm from the circle and label this locus Z.

13 A bathroom tile has a pattern based on a square of side 6 cm. Draw the square accurately full-size.

The next part of the design shows the locus of a point P which moves so that it is always 2 cm from the sides of the square. Draw the locus of P accurately.

14 A garden roller of diameter 40 cm is being pulled towards a step 20 cm high.

Use a scale of 1 cm for 10 cm and draw the locus of the centre C of the roller before and after reaching the step.

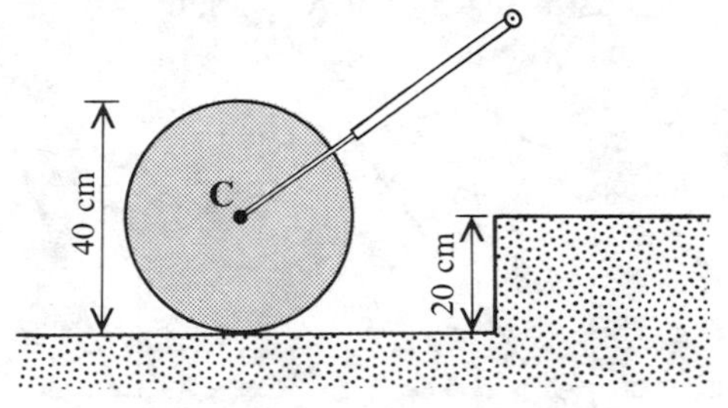

15

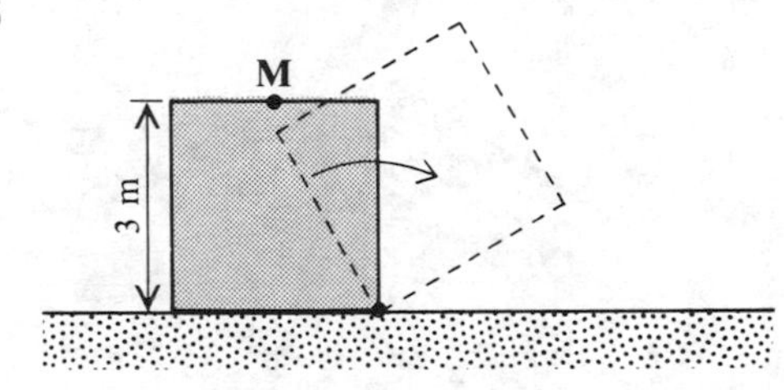

A crate with a vertical square cross-section of side 3 metres has point M as the midpoint of its upper side. It is rolled clockwise on a horizontal floor through four quarter-turns about each corner in succession until M is again on the upper side. Draw an accurate scale diagram and construct the locus of M.

16 Draw the right-angled triangle ABC four times.

Draw the locus of the point which moves inside the triangle so that it is:

a as far from B as it is from C

b nearer to B than it is to C

c as far from C as A is from C

d nearer to C than A is to C.

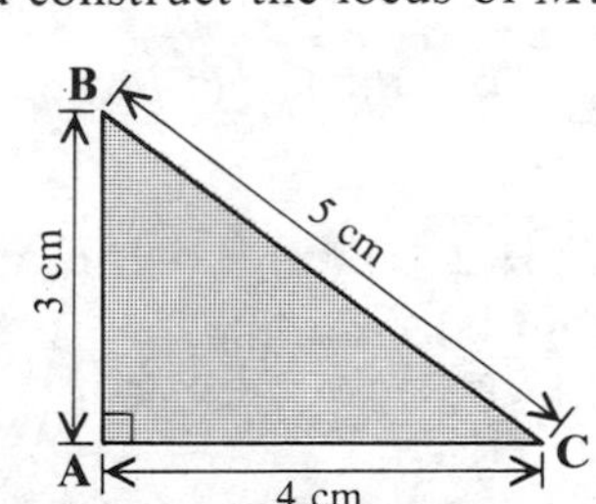

17 Given that $\mathscr{E}$ = {all points inside the square $PQRS$}

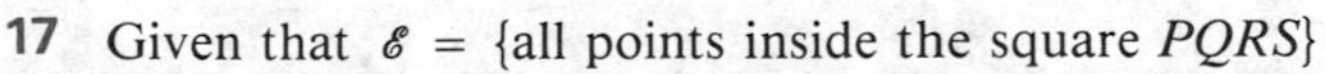

A = {points equidistant from sides PQ and SR}

B = {points nearer side PQ than side SR}

and C = {points nearer side PQ than side QR}

draw three diagrams to indicate the three sets A, B and C.

Trigonometry

Part 1 Pythagoras' Theorem

1 Calculate the lettered lengths in these right-angle triangles to three significant figures.

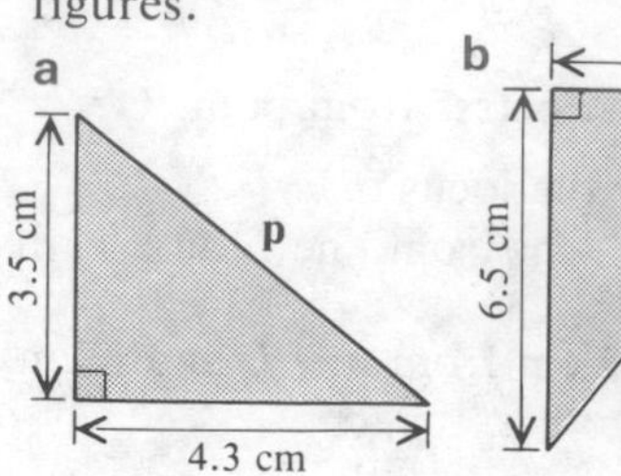

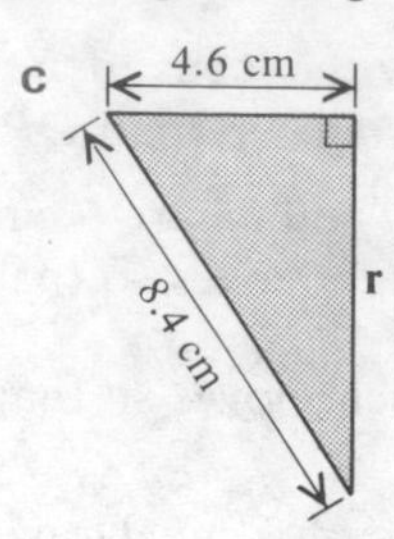

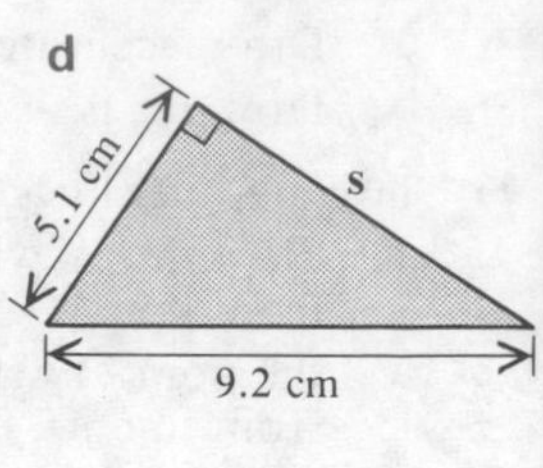

2 A man walks 4.2 km due north and then turns to walk 5.7 km due east. How far will he then be from his starting point?

3 A cone has a slant height of 8.7 cm and a base radius of 4.7 cm. Calculate its vertical height.

4 A telegraph pole of height h is held vertical by two sloping wires, one of length l and the other of length 10.0 m. These wires are fixed at the distances shown. Calculate

a the height h **b** the length l.

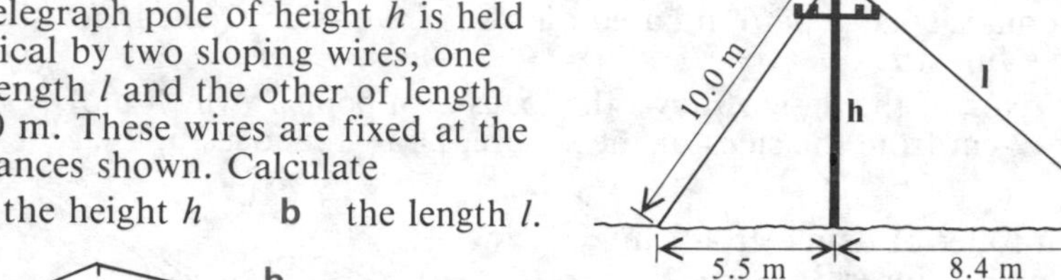

5 **a** **b**

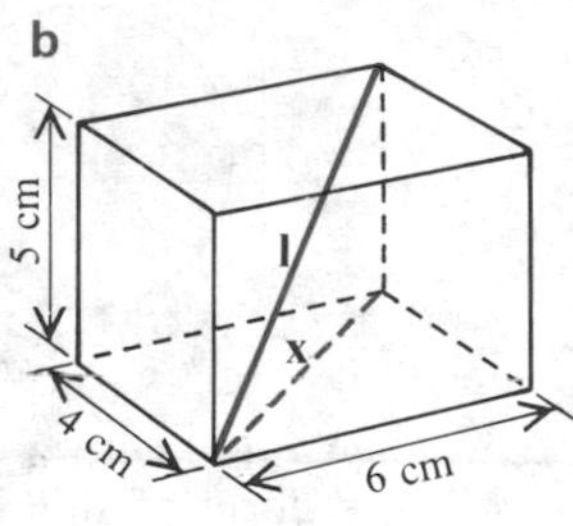

Each of these two boxes is shown holding the longest possible straight rod which can fit inside it. Calculate, for each box,
(i) the length x of the diagonal
(ii) the length l of the rod.

6 A fishing rod is propped in the corner of two walls as shown. Calculate

a the distance x of the end of the rod from the bottom corner of the walls

b the length l of the fishing rod.

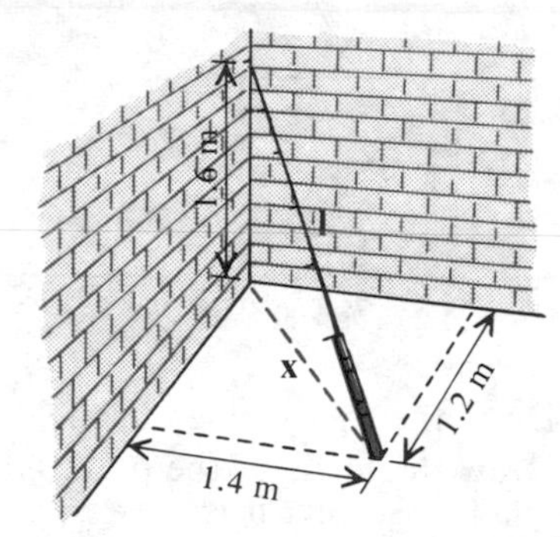

Part 2 Sine, cosine and tangent

Sine of an angle

1 Calculate the angle α in these right-angled triangles.

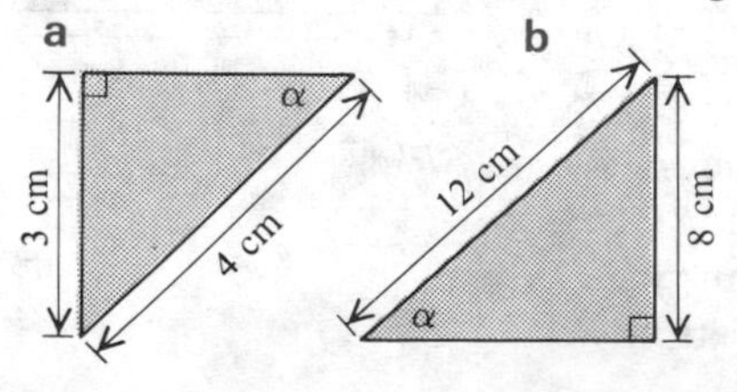

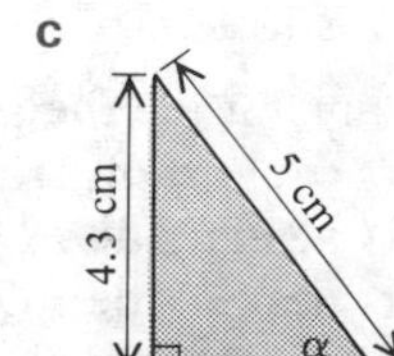

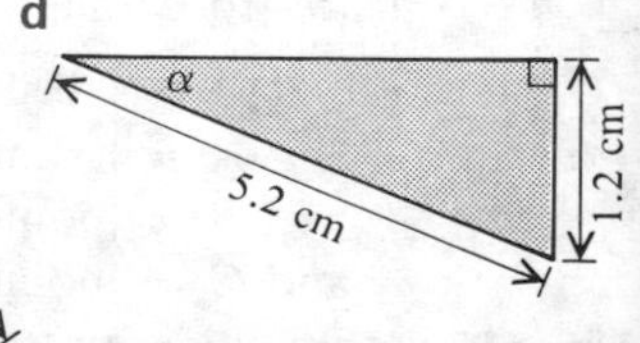

Trigonometry

2 Calculate the lettered lengths in these triangles.

a **b** **c** **d**

3 A funicular railway climbs a mountain on a steady incline of track 4 km long. If the mountain is 1.4 km high, calculate the angle of the incline to the nearest degree.

4 A cone has a slant height of 9.5 cm rising at an angle of 67° to the horizontal table on which the cone stands. Find the vertical height of the cone to three significant figures.

5 King Gorm the Old is buried under a huge mound 15 m high. To get to the top, you climb 24 m up the grassy slope of the mound. What is the angle of incline of this slope?

Cosine of an angle

6 Calculate the angle β in these right-angled triangles.

a

b

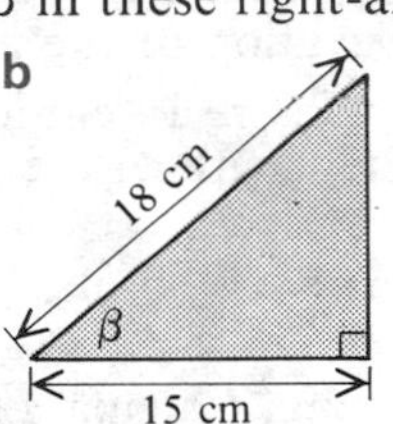

c

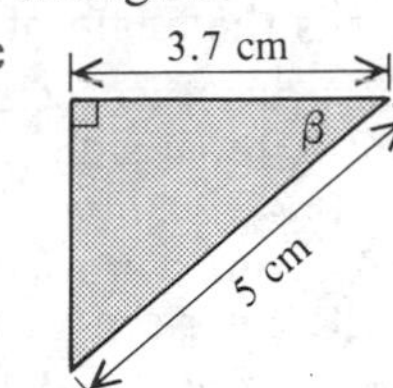

d

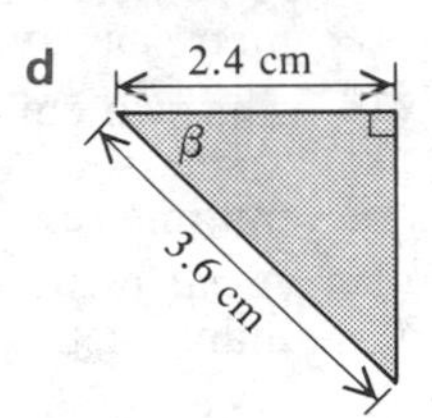

7 Calculate the lettered lengths in these triangles.

a

b

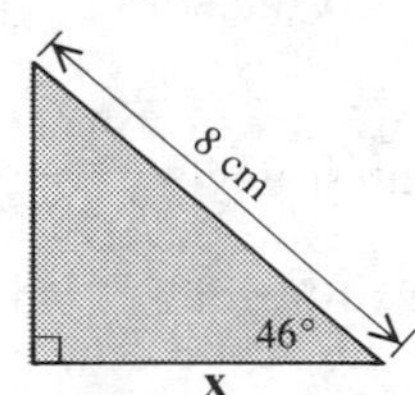

c

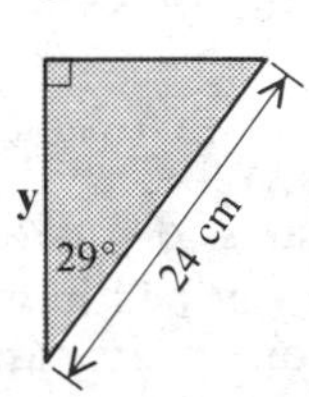

d

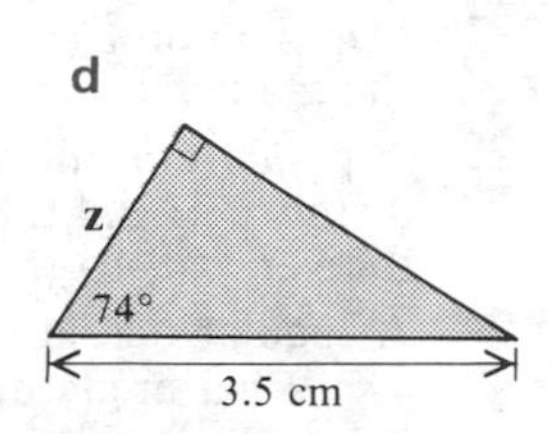

8 A circle, centre O, has a radius OP of 6 cm. PQ is a chord of the circle 11.6 cm long. Calculate the angle OPQ between the radius and the chord.

9 A step-ladder has both legs 2.4 metres long opened and inclined at 70° to the horizontal. Calculate, to the nearest centimetre, the distance between the two feet of the ladder.

10 A car climbs a long hill on a steadily rising road. Its odometer measures 4.8 km whereas a map gives the distance as 4.5 km. At what angle, to the nearest degree, is the road rising?

Tangent of an angle

11 Calculate the angle θ in these right-angled triangles.

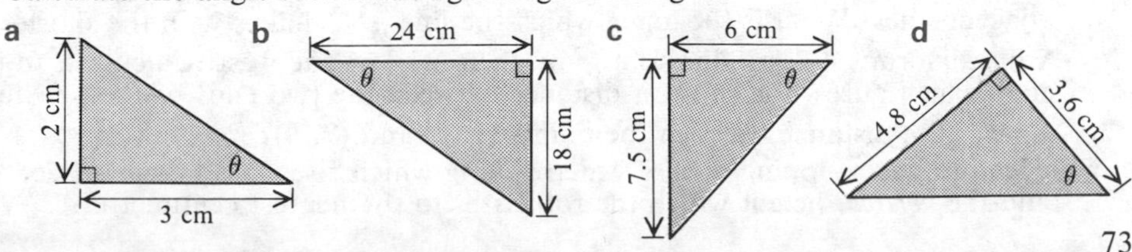

Trigonometry

12 Calculate the lettered lengths in these triangles.

a

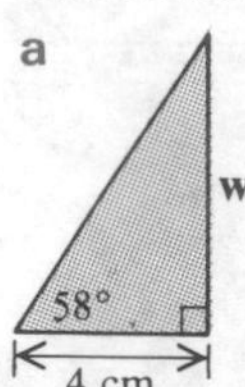

b c

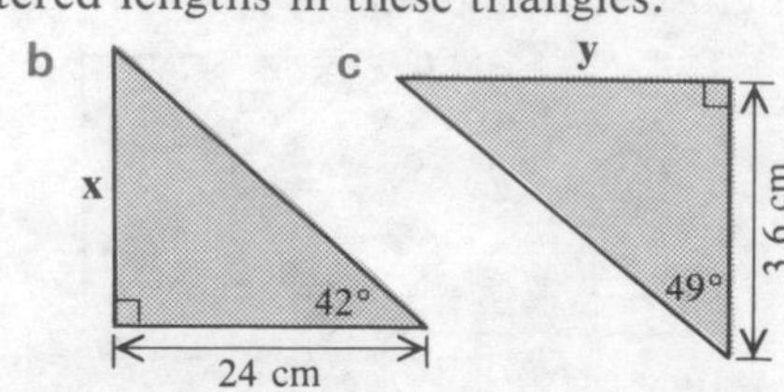

d

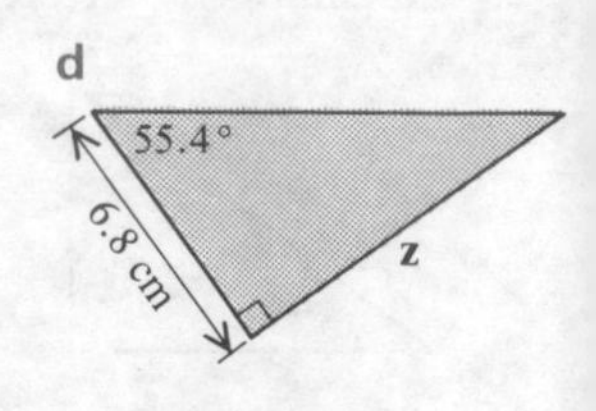

13 A ski-lift climbs steeply so that for every 3.2 metres lifted vertically a skier has moved 2 metres horizontally. At what angle to the horizontal does the lift climb?

14 A buoy is anchored 240 metres out from the foot of a cliff. If the angle of elevation of the cliff top from the buoy is 32°, calculate the height of the cliff.

15 A factory chimney 28 metres tall casts a shadow on level ground 40 metres long. What is the altitude of the sun at this time?

A mixture

16 A boat sails from port P 6 km due north and then 10 km due east to reach harbour H. Calculate the bearing of H from P to the nearest degree.

17 A vertical cliff behind our house rises 35 metres. If the foot of the cliff is 80 metres away, what is the angle of the elevation of the top of it from the house?

18 The two equal sides of an isosceles triangle are 14 cm long and each makes an angle of 65° with the third side. Find the length of the third side to the nearest millimetre.

19 A car is driven 6 km on a motorway slowly rising at 5° to the horizontal. How many metres has the car risen vertically?

20 A rectangle has sides 16 cm and 12 cm long. Calculate the angle between a diagonal and one of the longest sides.

21 P is the point (8, 18) and O is the origin. Calculate the angle between OP and the x-axis.

22 A pendulum bob swings on 1.8 metres of thread so that it rises a maximum distance of 0.4 metres. What is the greatest angle which the thread makes with the vertical, to the nearest degree?

23 A regular octagon is inscribed in a circle of radius 5 cm.
- **a** What angle does one of its sides make at the centre of the circle?
- **b** Calculate the length of the sides of the octagon.

Part 3 The right-angled triangle

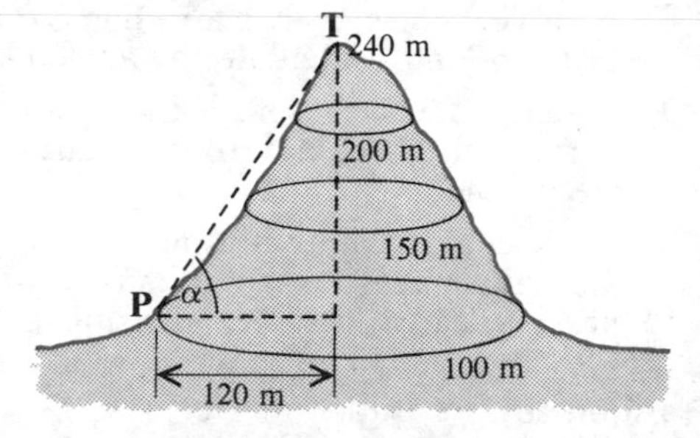

1 A hilltop T is 240 m above sea-level. Point P is 120 m horizontally from T and on the 100 m contour. Calculate the angle of elevation, α of T from P.

2 A plane flies 7 km after take-off to rise to a height of 3 km. Calculate the angle at which it climbs.

3 A rectangle has sides of 24 cm and 14 cm. M and N are the midpoints of two adjacent sides. What is the angle which the line MN makes with the shorter side?

4 A pipeline runs 4.2 km due south and then 3.1 km due west. Calculate, to two significant figures, the shortest distance between the two ends of the pipeline.

5 What is the distance between the points (1, 2) and (7, 9)?

6 A lean-to has a sloping roof 3.8 metres long which rises at an angle of 25°. Find the *vertical* height which the roof rises to the nearest centimetre.

Trigonometry

7 A pipe crosses a canal as shown. Calculate that length of the pipe which is above ground level.

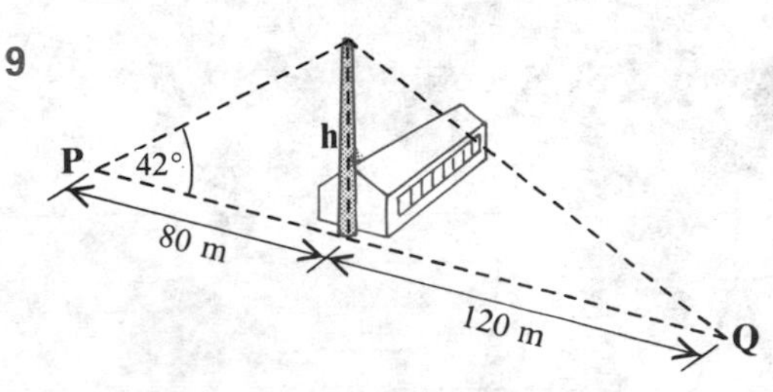

8 The two sides of a roof slope asymmetrically at 35° and 60° as shown. Find, to the nearest metre, the distance x between the two gutters.

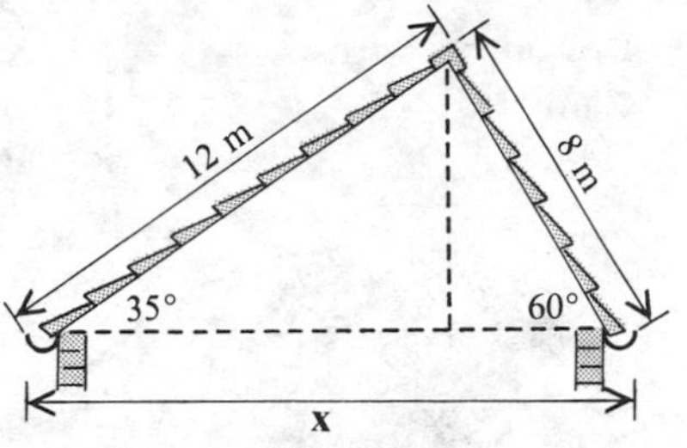

9 Point P is 80 m due west of a factory chimney. The angle of elevation of its top is 42° from P. Point Q is 120 m due east of the chimney. Calculate

- **a** the height h of the chimney
- **b** the angle of elevation of the top from Q.

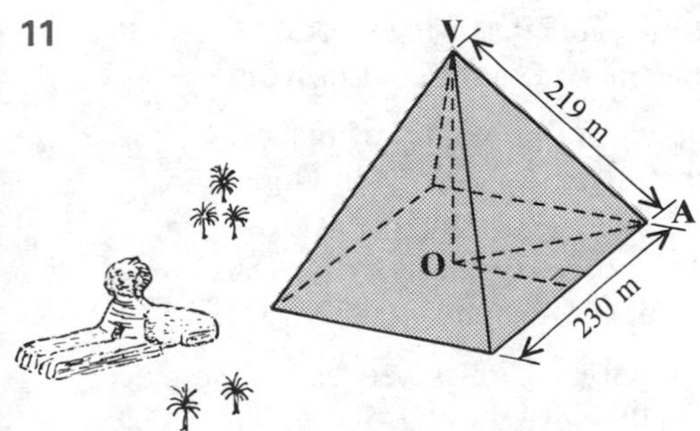

10 I am standing at point B which is 2 km due south of point C and 2.5 km from airport A. C is due west of A and I can see a plane P vertically above C at an elevation of 24°. Calculate

- **a** the distance AC
- **b** the height of the plane above C
- **c** the angle α to the horizontal at which the plane comes in to land at A.

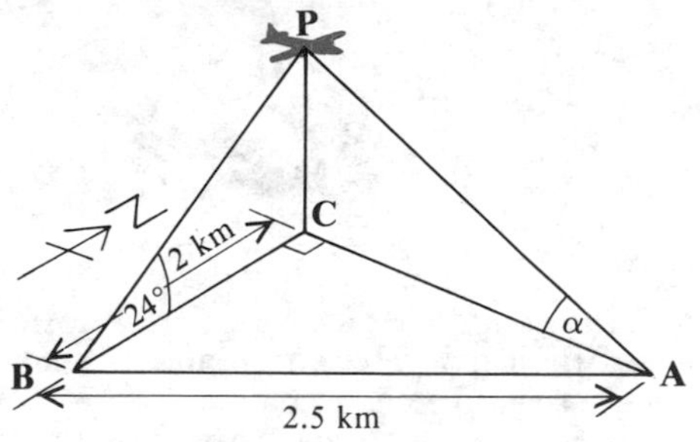

11 One of the Great Pyramids at Giza in Egypt is built on a square base of side 230 metres and centre O, with slanting edges of length 219 metres meeting at the vertex V. What would you expect to be

- **a** the length of OA, to the nearest metre
- **b** the vertical height OV, to the nearest metre
- **c** the inclination of AV to the horizontal?

12 Two rods AY and AZ of equal length are both placed in a box. Rod AY overhangs the box by 9 cm as shown. Calculate, to three significant figures,

- **a** the total length of rod AY
- **b** the length AC
- **c** by how much rod AZ overhangs the box.

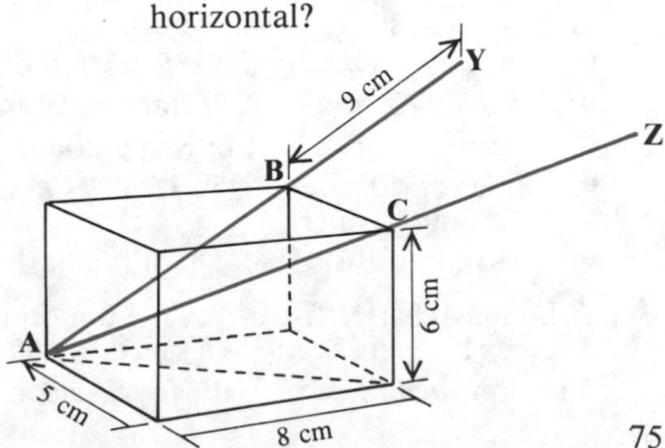

Trigonometry

Part 4 Latitude and longitude

1 Write the positions of these places given by their initials on the map.

a	Oslo *O*	**b**	Grand Cayman *GC*	**c**	Hangchow *H*
d	Ruwenzori Mtns *R*	**e**	Santiago *S*	**f**	Beira *B*
g	Canberra *C*	**h**	Berkner Island *BI*	**i**	Wellington *W*
j	Denver *D*	**k**	Patna *P*	**l**	Fiji Islands *F*

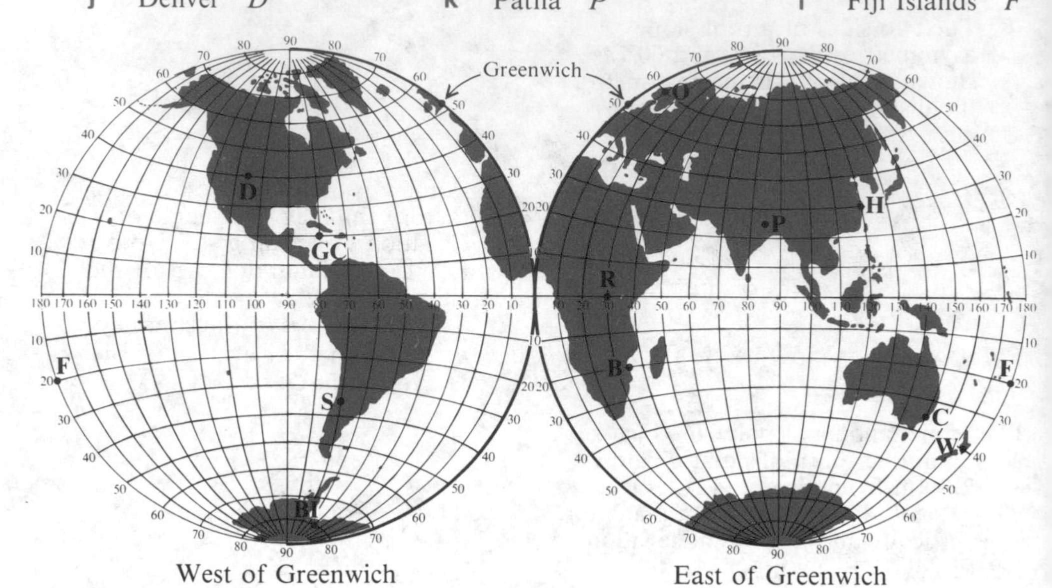

2 *Antipodes* are places at the opposite ends of any diameter through the centre of the earth. Here are pairs of such places. Find the latitude and longitude of the second in each pair.

- **a** Talavera (40°N, 5°W) in Spain and Wanganui in New Zealand
- **b** The Bay of Plenty (37°S, 177°E) in New Zealand and Granada in Spain
- **c** Bermuda (32°N, 65°W) in the Atlantic and Perth in Australia
- **d** Montevideo (36°S, 55°W) in South America and Seoul in Korea

3 Describe in words the shortest routes between these pairs of places.

- **a** Ostend (51°N, 3°E) in Belgium and Lagos (6°N, 3°E) in Nigeria
- **b** Entebbe (0°, 33°E) in Uganda and Macapa (0°, 51°W) in Brazil
- **c** London ($51\frac{1}{2}$°N, 0°) in Britain and Wrangel Island ($71\frac{1}{2}$°N, 180°) in USSR
- **d** Fairbanks (64°N, 147°W) in Alaska and Murmansk (69°N, 33°E) in USSR

4 Calculate the shortest distances along the Great Circles between these places. Give your answers to three significant figures in kilometres.

- **a** Dnepropetrovsk (48°N, 35°E) in the USSR and Tel-Aviv (32°N, 35°E) in Israel
- **b** Rome (42°N, 13°E) in Italy and Luanda (9°S, 13°E) in Angola
- **c** Kismayu (0°, 42°E) in Somalia and Nanyuki (0°, 37°E) in Kenya
- **d** Sulawesi (0°, 120°E) in Indonesia and the Galapagos Islands (0°, 90°W) off South America
- **e** Anchorage (61°N, 150°W) in Alaska and Kiev (50°N, 30°E) in the Ukraine

5 A holiday charter flight leaves London ($51\frac{1}{2}$°N, 0°) for Alicante ($38\frac{1}{2}$°N, 0°) in Spain. The plane averages a speed of 600 km/h and leaves London at 1430 GMT. How long in kilometres is the flight and at what time (GMT) does it land in Alicante?

Symmetry

Part 1 Two dimensions

1 Copy these shapes and draw all their lines of symmetry.

a  **b** 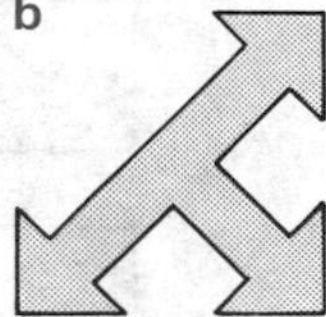 **c** 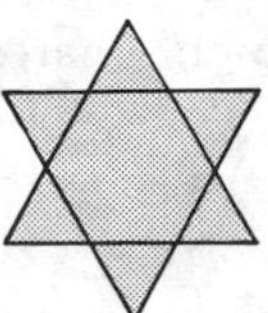 **d** 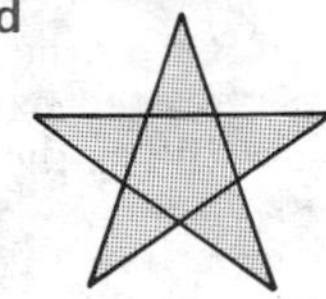

2 How many lines of symmetry has

a a rectangle **b** a square **c** a parallelogram
d a rhombus **e** a kite **f** a trapezium.

3 These diagrams are incomplete.
Copy each one and complete it using the dotted lines as lines of symmetry.

a 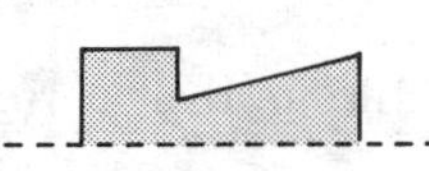 **b** 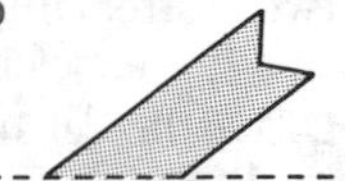 **c** 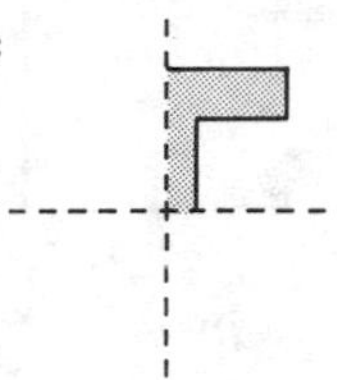 **d** 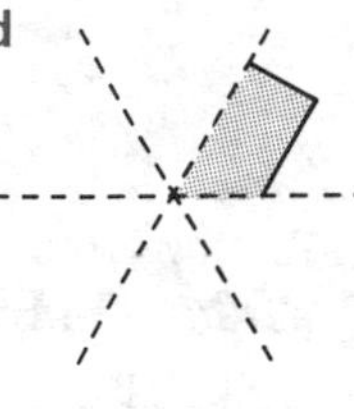

4 Say which of these shapes have rotational (or point) symmetry.
Give the *order* of any rotational symmetry.

a 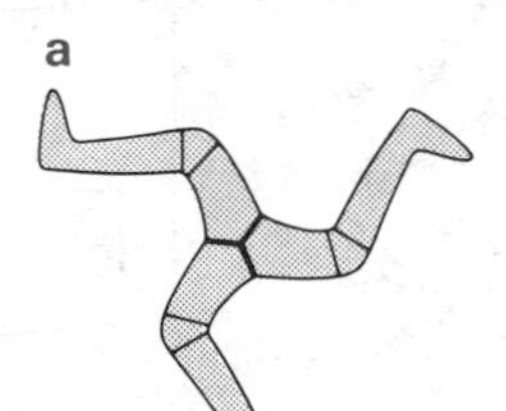 **b** 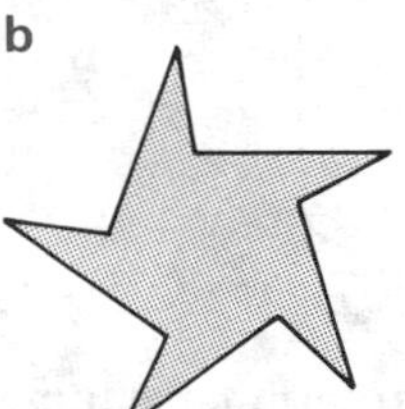 **c** 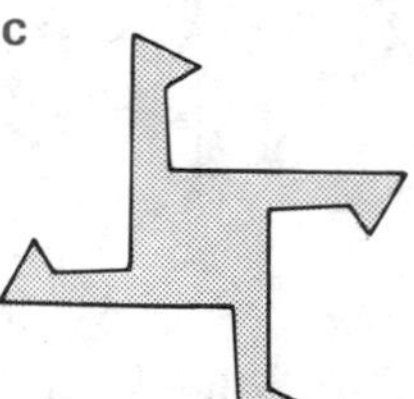 **d** 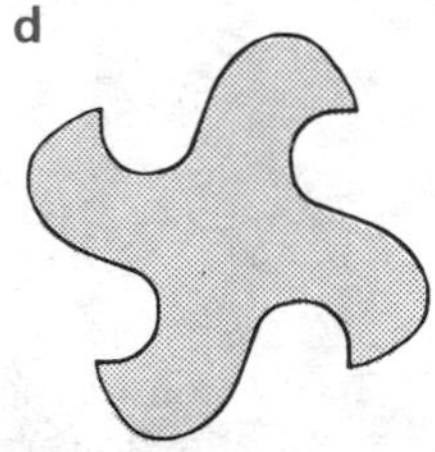

5 Which of the shapes in **1** above have rotational symmetry, and of what order?

6 For each set of points, draw and label both axes from 0 to 10.
Join the points in order to give *part* of a shape which has rotational symmetry.
Complete each shape using the given order and centre of the symmetry.

	Points	Order	Centre
a	(4, 6), (4, 8), (6, 10), (8, 8), (6, 8), (6, 6)	4	(5, 5)
b	(6, 8), (10, 8), (8, 6) (8, 4), (10, 4), (8, 2), (6, 4)	2	(6, 6)
c	(4, 6), (6, 8), (5, 8), (9, 10), (9, 8), (8, 8), (6, 6)	2	(5, 6)

7 Here are three playing cards from three different packs.
For each card, describe

a any lines of symmetry
b the order of any rotational symmetry.

OXO card:
OXO OXO
OXO
OXO OXO

Symmetry

8 This is part of a pattern found on a certain wallpaper.
Copy the pattern onto squared or dotted paper and complete it by filling the gap.

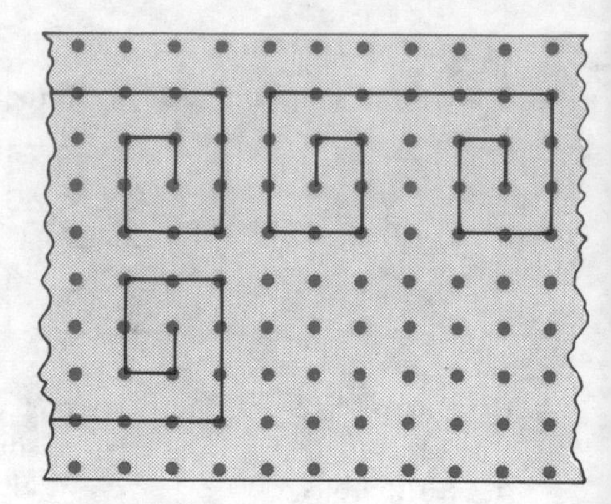

9

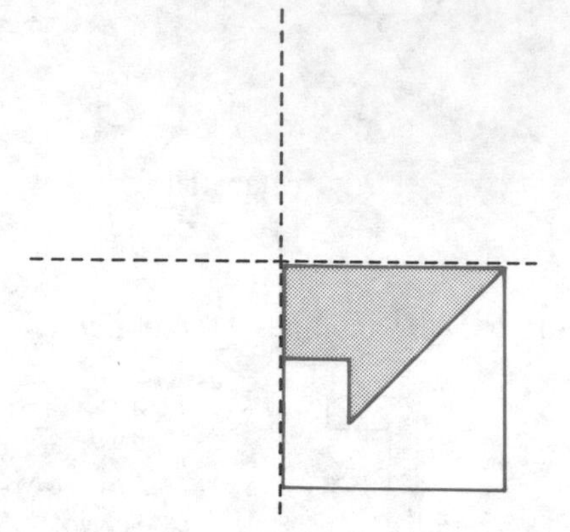

Tiles are sold in packs of four, all with an identical pattern. One tile is shown here. The other three tiles are placed so that the two dotted lines are lines of symmetry for the whole set of four.
Draw the set of four tiles in their correct positions.

Part 2 Three dimensions

1 Sketch these solid objects, draw their axes of symmetry and state the order of the rotational symmetry.

a

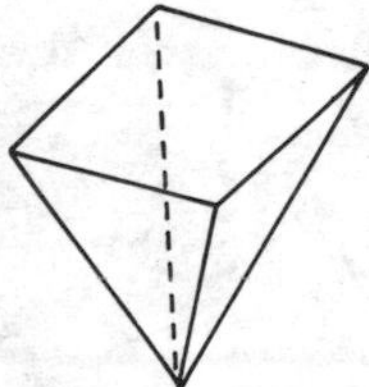

b

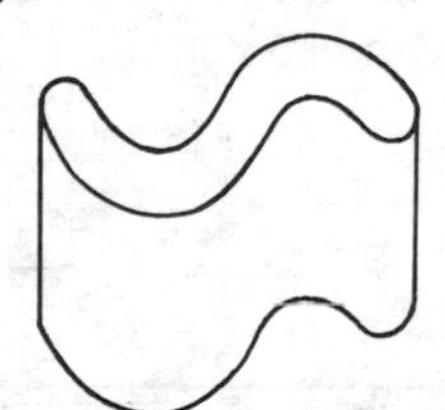

c

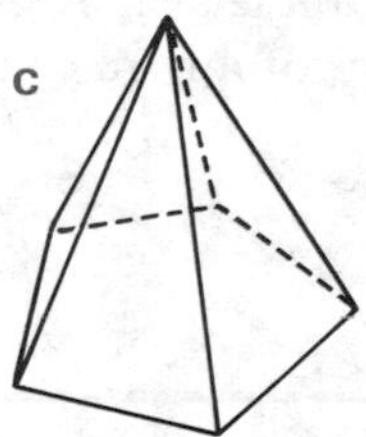

d

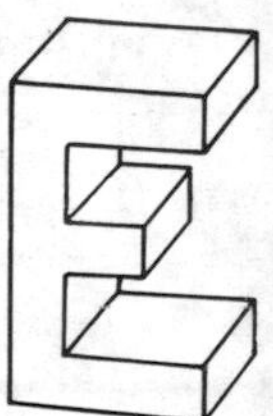

2 Which of these objects have axes of rotational symmetry?

a a flowerpot **b** a rocking chair **c** a light bulb
d a squash racquet **e** a cup **f** a cricket bat

3 Which of the shaded planes in these diagrams are planes of symmetry?

a

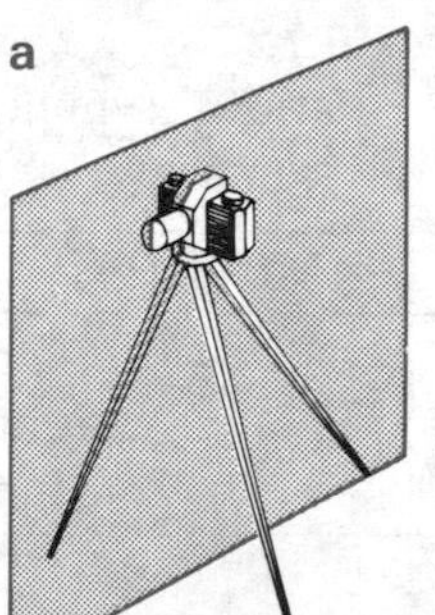

b

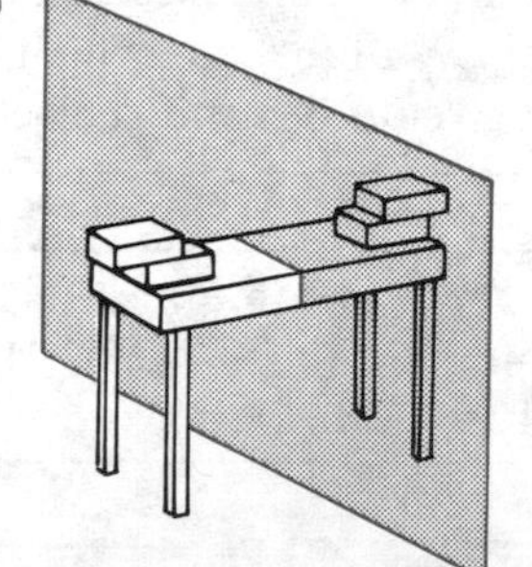

c

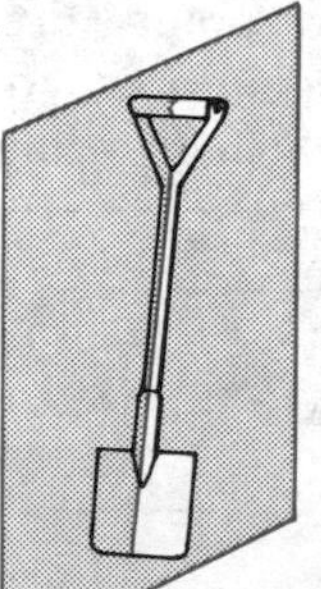

d

4 Which of the objects in **3** above have a plane of symmetry not shown in the diagrams?

5 How many planes of symmetry has

a a teaspoon **b** a pair of binoculars **c** a hexagonal washer
d a screwdriver **e** a lidless shoe box **f** a cube?

Transformations

Part 1 Reflections

1 If you saw this sign in a car mirror, which of the letters would appear unchanged?

RED LINE BUS

2 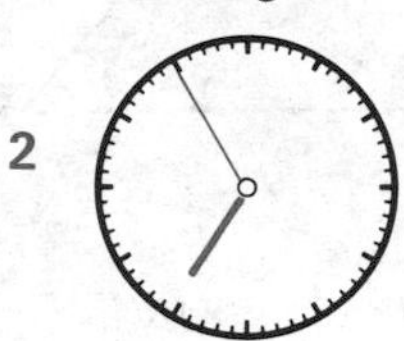

It is five to seven in the morning. If this clock is seen in a mirror, what time does it appear to show?

3

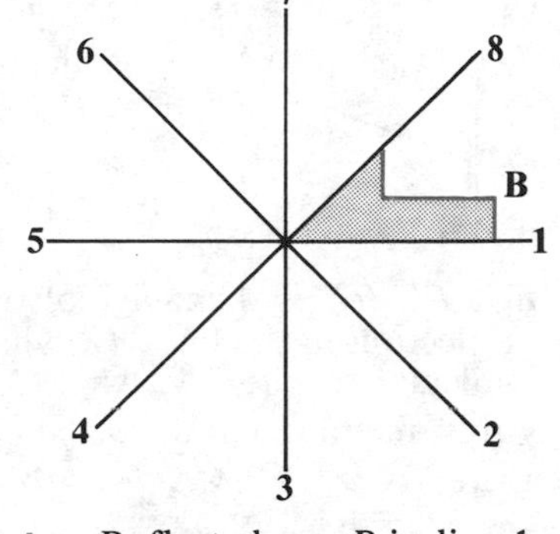

a Reflect triangle A in line 1.
Reflect the image in line 2, and then reflect the new image in line 3. You now have a complete diagram with triangle A and three images.

b Reflect shape B in line 1.
Reflect the image in line 2, and then the new image in line 3, and so on until you have a complete diagram of shape B and seven images.

4 On this snooker table, the white ball W has to hit the black ball B without hitting the coloured balls which are in the way.
At which point on the cushion PQ should ball W be aimed?
Draw the diagram accurately and show all your lines of construction.

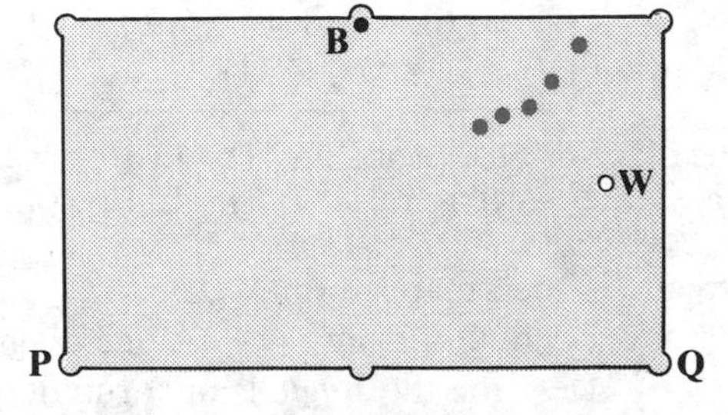

5 The right-angled triangle ABC is reflected in line PQ onto $AB'C'$ so that CAC' is a straight line.
If $\angle ACB = 30°$, find

a $\angle AC'B'$ **b** $\angle B'AC'$
c $\angle BAP$ **d** $\angle BAB'$.

6 The square $KLMN$ is reflected in the line YZ such that $\angle NKZ = 20°$.

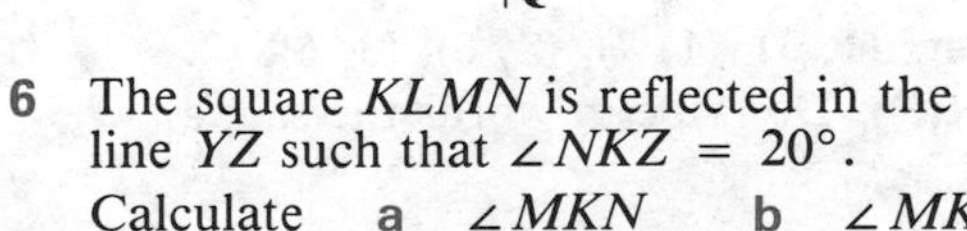

Calculate **a** $\angle MKN$ **b** $\angle MKZ$

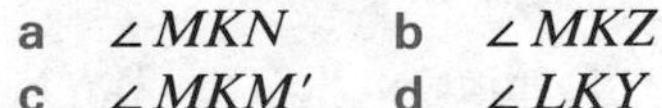

c $\angle MKM'$ **d** $\angle LKY$
e $\angle LKL'$.

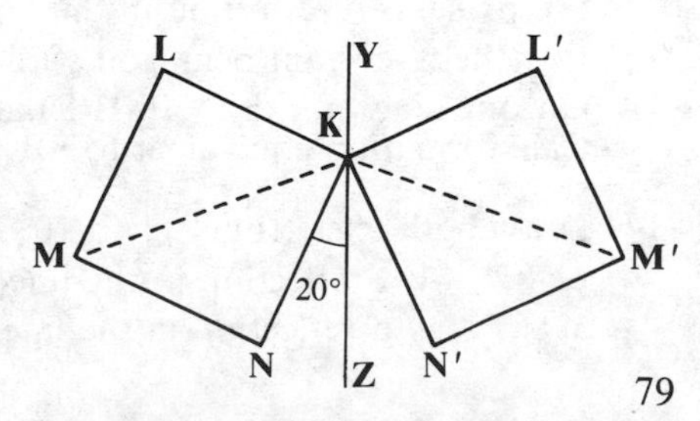

Transformations

7 The right-angled triangle PQR is reflected in the line AB onto $PQ'R'$. The side QR is parallel to AB.
Sketch this diagram and draw the image triangle $PQ'R'$.
Calculate **a** $\angle QPA$ **b** $\angle QPQ'$ **c** $\angle RPB$ **d** $\angle RPR'$.

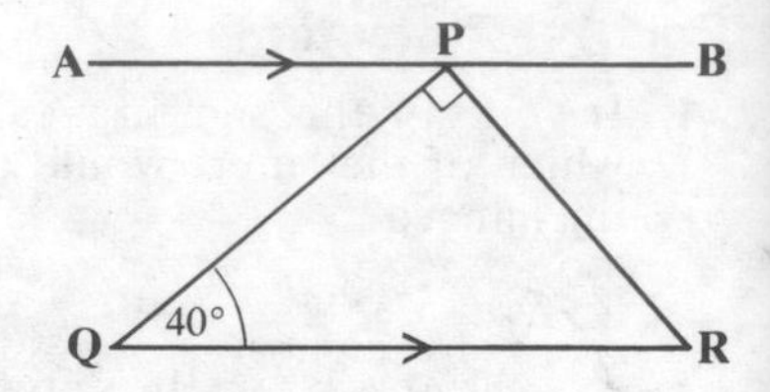

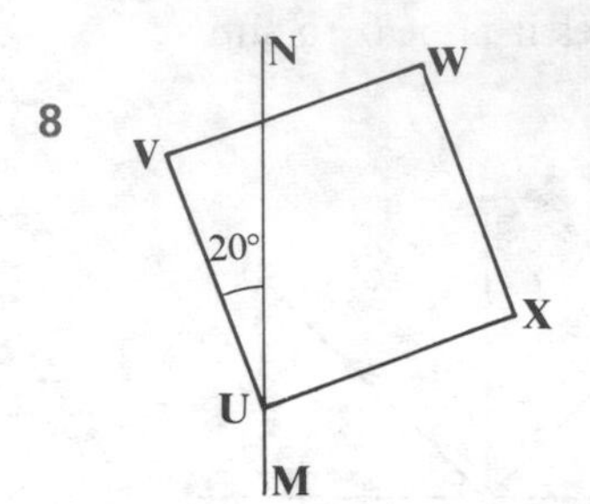

8 The square $UVWX$ is reflected in the line MN shown in the diagram onto its image $UV'W'X'$.
Copy this diagram and draw its image.
Calculate **a** $\angle NUX$ **b** $\angle NUV'$ **c** $\angle V'UX$ **d** $\angle XUW'$.

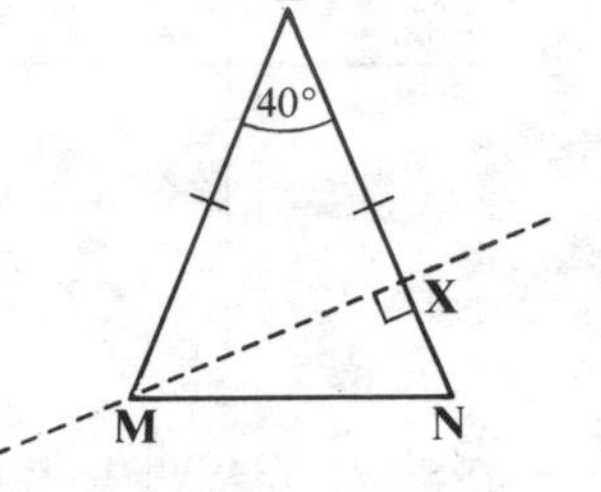

9 Triangle $L'MN'$ is the image of triangle LMN after reflection in the line MX which is perpendicular to side LN. Angle MLN is 40°.
Copy this diagram and draw the image triangle.
Calculate **a** $\angle LMN$ **b** $\angle XMN$ **c** $\angle XML'$ **d** $\angle NMN'$.

10 Draw each pair of object and image shapes on axes labelled from 0 to 10.
Construct the line of reflection which maps the object onto the image.

	Object shape	Image of shape
a	(3, 1), (0, 5), (2, 7), (3, 6), (2, 5)	(5, 1), (8, 5), (6, 7), (5, 6), (6, 5)
b	(2, 3), (6, 2), (8, 6), (6, 6)	(1, 4), (0, 8), (4, 10), (4, 8)
c	(10, 3), (8, 7), (7, 1)	(1, 6), (5, 8), (2.2, 2.6)

The diagrams for problems **11** to **13** are to be drawn with both axes labelled from 0 to 10.

11 The quadrilateral $Q(1, 7), (1, 10), (4, 7), (2, 8)$ is transformed onto Q' by a reflection in the line $y = 6$.
Q' is then mapped onto Q'' by a reflection in the line $y = x$.
a Draw Q, Q' and Q'' on one diagram.
b Describe the *single* transformation which maps Q directly onto Q''.

12 The hexagon $H(1, 3), (1, 2), (2, 1), (4, 1), (2, 2), (2, 3)$ is reflected in the line $x + y = 6$ onto its image H'. H' is then reflected in the line $x + y = 10$ onto H''.
Draw H, H' and H'' on one diagram and describe the single transformation which maps H directly onto H''.

13 **R** represents a reflection in the line $y = x$.
S represents a reflection in the line $x + y = 10$.
P is the re-entrant pentagon with corners (1, 3), (1, 7), (2, 6), (3, 6), (2, 3).
On one diagram, draw P, **R**(P) and **SR**(P). Describe in words the *single* transformation equivalent to **SR**.

14 Draw both axes from −1 to 10. Transform triangle $T(3, 2), (3, 5), (7, 5)$ onto
a T' by a reflection in the line $y = \frac{1}{2}x$
b T'' by a reflection in the line $y = 2x$.

Transformations

Part 2 Rotations

1 Estimate the angle through which

a the revolving door has to turn to allow a person to enter

b the tin-opener has cut in the top of the tin

c the windscreen wiper has turned.

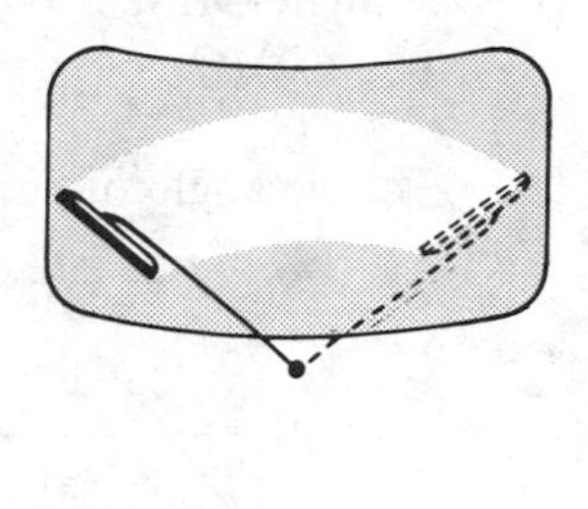

2 Through what angle does blade W of this fan have to rotate to be in the same position as

a blade X

b blade Y

c blade Z?

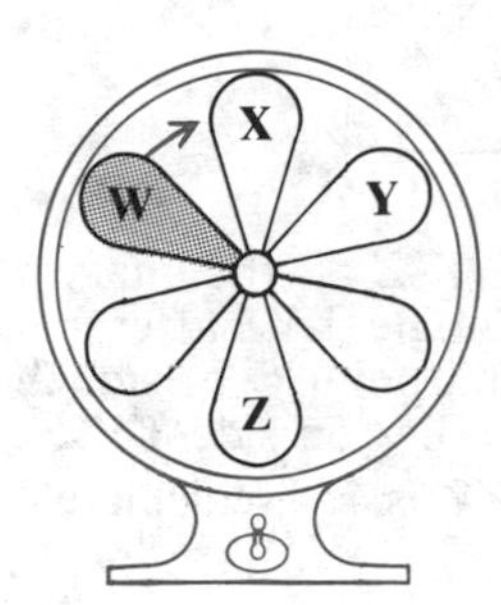

3 Which *one* of these uncoloured shapes could be rotated onto the coloured shape?

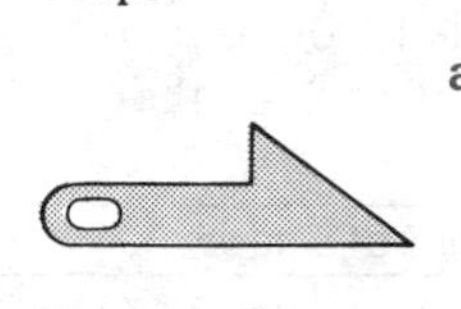

a 

b 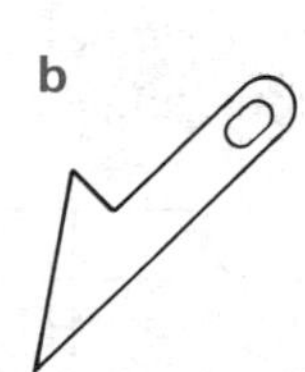

c 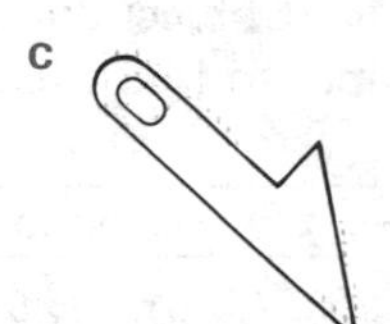

d 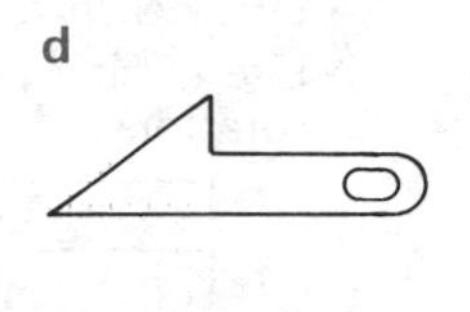

4 The triangle OAB and rectangle $OXYZ$ are rotated clockwise about their corners O onto images $OA'B'$ and $OX'Y'Z'$. Draw both diagrams accurately using ruler and protractor.

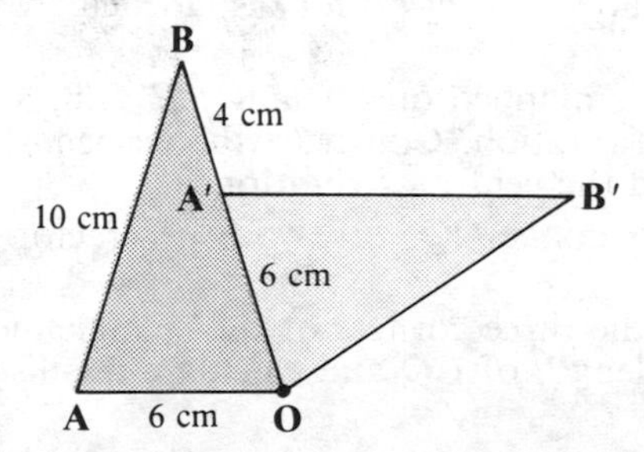

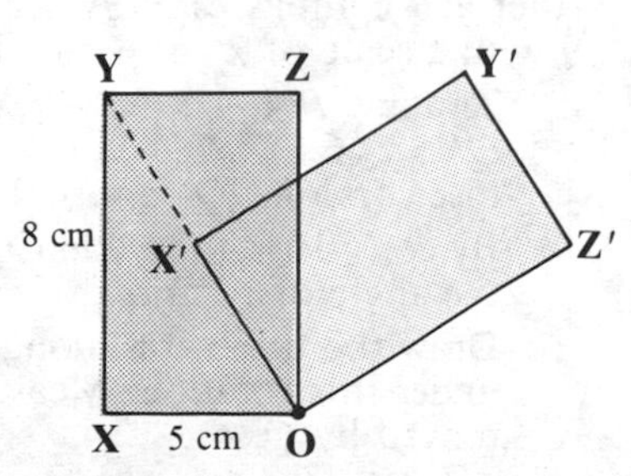

a Use compasses to draw the paths along which the corners A, B, X, Y and Z move.

b Use a protractor to find the angle of each rotation.

c Calculate the distance moved by point B and by point X.

Transformations

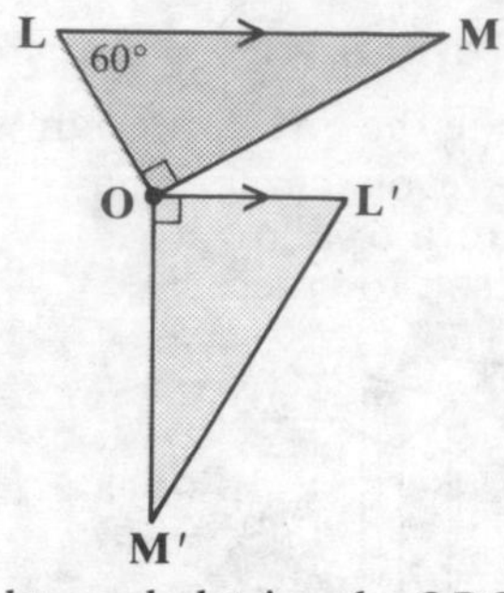

5 Triangle OLM is rotated about O onto its image $OL'M'$ such that LM is parallel to OL' as shown.
If angle OLM is 60°, calculate

a $\angle LMO$ b $\angle MOL'$

c $\angle OM'L'$

d the angle of rotation.

6

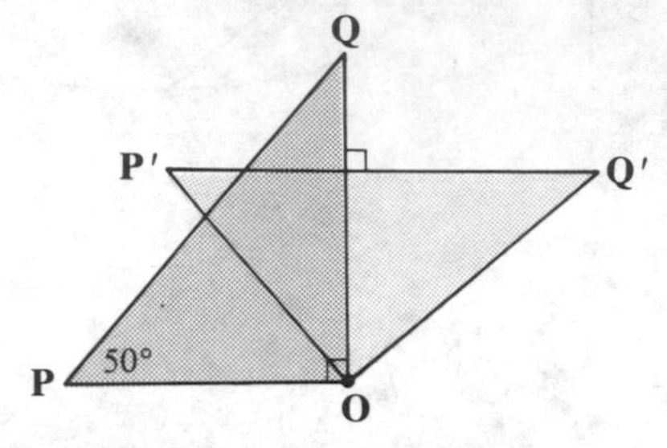

The right-angled triangle OPQ is rotated about O onto $OP'Q'$ such that OQ is at right angles to $P'Q'$.
If angle OPQ is 50°, calculate

a $\angle OP'Q'$ b $\angle P'OQ$

c the angle of rotation.

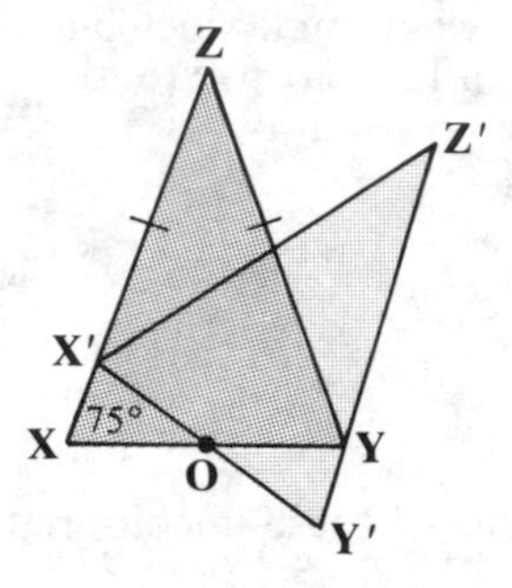

7 Triangle $X'Y'Z'$ is the image of the isosceles triangle XYZ after a rotation about the midpoint O of side XY, such that point Y lies on side $Y'Z'$.
If angle YXZ is 75°, calculate

a $\angle XYZ$ b $\angle OY'Y$

c $\angle OYY'$

d the angle of rotation.

8 Draw each pair of object and image shapes on axes labelled from 0 to 10. Use tracing paper to find the centre, angle and direction of each of the rotations.

	Object shape	Image of shape
a	(3, 9), (1, 7), (3, 3), (3, 5), (2, 7)	(4, 4), (6, 2), (10, 4), (8, 4), (6, 3)
b	(10, 10), (10, 4), (8, 7), (3, 8), (3, 9)	(2, 2), (2, 8), (4, 5), (9, 4), (9, 3)
c	(1, 3), (3, 8), (8, 2), (6, 1)	(4, 8), (9, 6), (3, 1), (2, 3)

9 The centres of rotation in these problems are to be found by construction using ruler and compasses on A4 squared paper having both axes labelled from 0 to 9 with a scale of 2 cm per unit.

a Triangle $P(4, 1)$, $Q(7, 2)$, $R(7, 5)$ is mapped onto triangle $P'(3.8, 3.4)$, $Q'(5.2, 6.3)$, $R'(3.1, 8.4)$ under a rotation. Construct the perpendicular bisectors of QQ' and RR' to find the centre of rotation, C.
By considering the line CQ and its image CQ', use your protractor to find the angle of rotation.
Draw the three arcs along which the three corners of the triangle move under this rotation. Measure the length of CQ and calculate the distance moved by Q to Q'.

b Triangle $X(1, 3)$, $Y(2, 5)$, $Z(4, 2)$ rotates onto its image $X'(7, 5)$, $Y'(5, 6)$, $Z'(8, 8)$. Construct the centre of rotation C, draw the line CZ and its image CZ' and measure the angle of rotation.
Draw the three arcs along which the corners of the triangle move and calculate the distance travelled by Z.

c The quadrilateral $J(1, 4)$, $K(4, 4)$, $L(4, 5)$, $M(2, 6)$ maps onto its image $J'(7, 8)$, $K'(7, 5)$, $L'(8, 5)$, $M'(9, 7)$ under a rotation. Construct the centre of the rotation, C and measure the angle of rotation.
Measure the length CM, draw the path of M as it maps onto M' and calculate the distance it travels.

10 **R** represents a clockwise rotation of 90° about the point (3, 6).
S represents an anticlockwise rotation of 90° about the point (7, 3).
a Draw the quadrilateral Q(1, 7), (1, 10), (3, 10), (3, 9) and its images **R**(Q) and **SR**(Q) on one diagram.
b Describe the single transformation which maps Q directly onto **SR**(Q).

11 $\mathbf{R}_1$ represents a half-turn about the point (5, 3).
$\mathbf{R}_2$ represents a quarter turn clockwise about the point (5, 8).
a Draw the pentagon P(1, 3), (2, 1), (4, 1), (4, 2), (2, 2) and its images $\mathbf{R}_1$(P) and $\mathbf{R}_2\mathbf{R}_1$(P) on one diagram.
b Describe the single transformation equivalent to $\mathbf{R}_2\mathbf{R}_1$.

Part 3 Vectors and translations

1 Say whether a *vector* or a *scalar* quantity is involved in each of these statements.
a A ship is sailing due east at a speed of 15 knots.
b A bottle of lemonade has a volume of $\frac{3}{4}$ of a litre.
c The minimum overnight temperature was -4°C.
d A stone is thrown down a cliff with a speed of 15 m/s.

2 This map, drawn to a scale of 1 cm = 1 km, shows six vectors which describe the route taken by a yacht from Caregddu to Traethbach.
Use a ruler and a protractor to find the magnitude and direction of each vector.

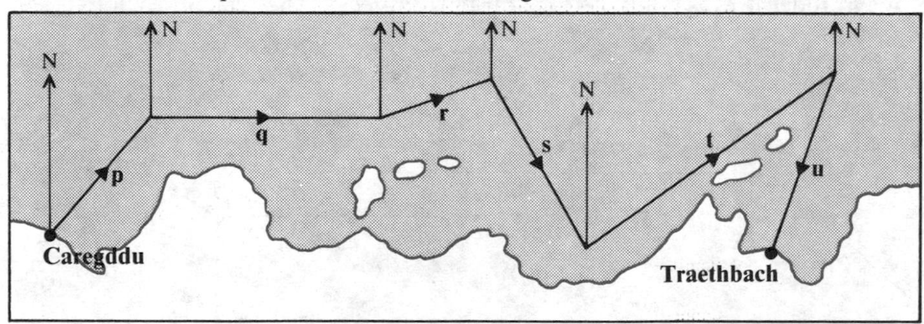

3 The route of a power line across country from a power-station P to a substation S is shown on this map where each square of the grid has sides of 10 km.
Write the components of each of the vectors **a** to **g**. For example, $\mathbf{a} = \begin{pmatrix} 1 \\ 3 \end{pmatrix}$

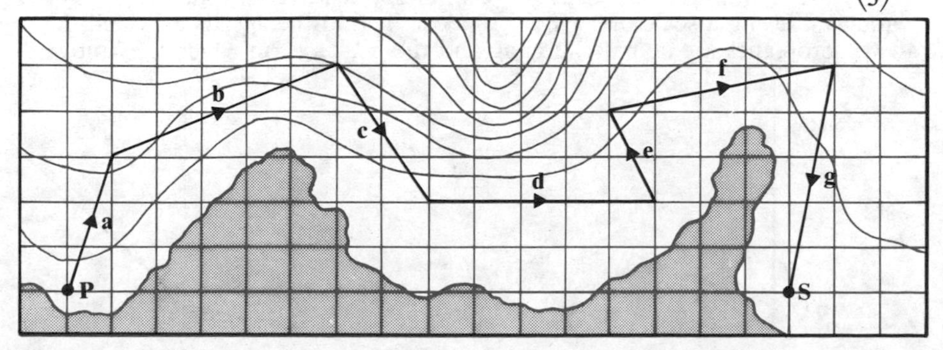

Transformations

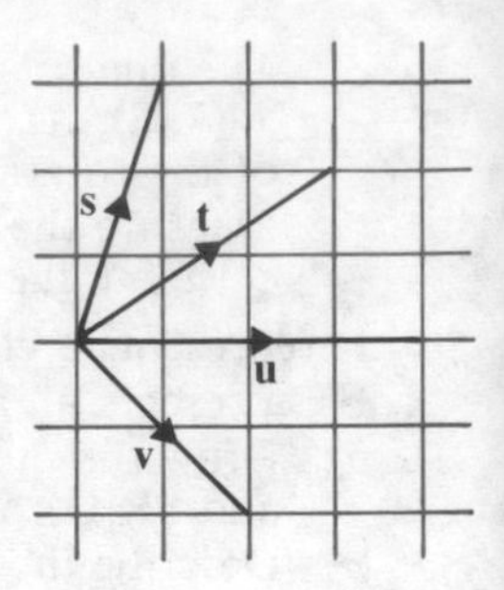

4 a By drawing these four vectors tip-to-tail on squared paper, find the resultant vector $\mathbf{r} = \mathbf{s} + \mathbf{t} + \mathbf{u} + \mathbf{v}$. Write the components of **r**.

b Does the value of **r** depend on the order in which you draw the four given vectors?

5 If $\mathbf{u} = \begin{pmatrix} 3 \\ 4 \end{pmatrix}$ $\mathbf{v} = \begin{pmatrix} 5 \\ -2 \end{pmatrix}$ and $\mathbf{w} = \begin{pmatrix} -3 \\ -5 \end{pmatrix}$

draw the vector $\mathbf{r} = \mathbf{u} + \mathbf{v} + \mathbf{w}$ on squared paper and write the components of **r**.

6 A ship is being pulled by three tugs which exert forces $\mathbf{T}_1$, $\mathbf{T}_2$ and $\mathbf{T}_3$. The arrows on this diagram give the directions of the forces and their magnitudes to a scale of 1 cm = 1 unit of force.
Draw the arrows tip-to-tail using tracing paper and find the magnitude of the resultant force on the ship.

7 A car *C* and a telegraph pole *T* are acted upon by the forces shown in these diagrams. Draw the arrows tip-to-tail using tracing paper, and hence find the magnitude of the resultant force on both objects. The scale in each case is 1 cm = 1 unit of force.

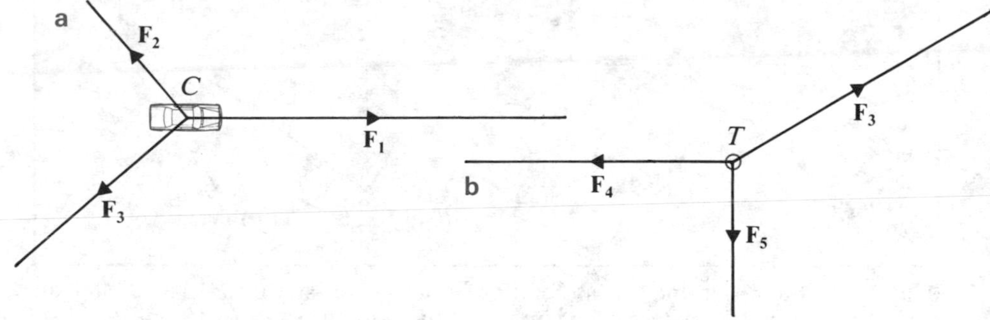

8 Two aeroplanes P_1 and P_2 have *air speeds* given by the vectors $\mathbf{A}_1$ and $\mathbf{A}_2$. They are flying in winds which have *wind speeds* given by the vectors $\mathbf{W}_1$ and $\mathbf{W}_2$. Using a scale of 1 cm = 100 km/h, find the speeds at which the two aeroplanes are actually moving and the bearings on which they move.

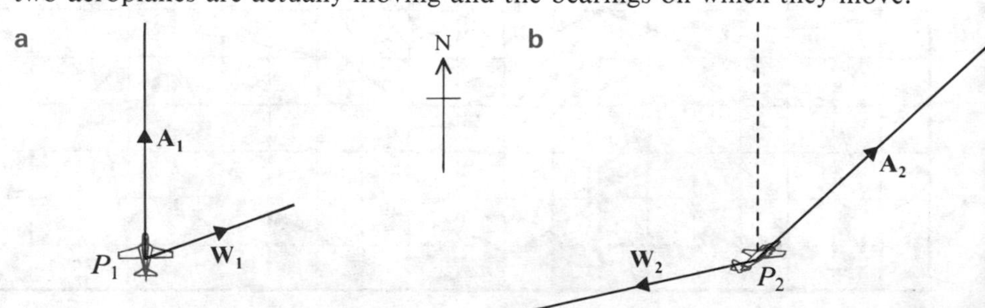

Transformations

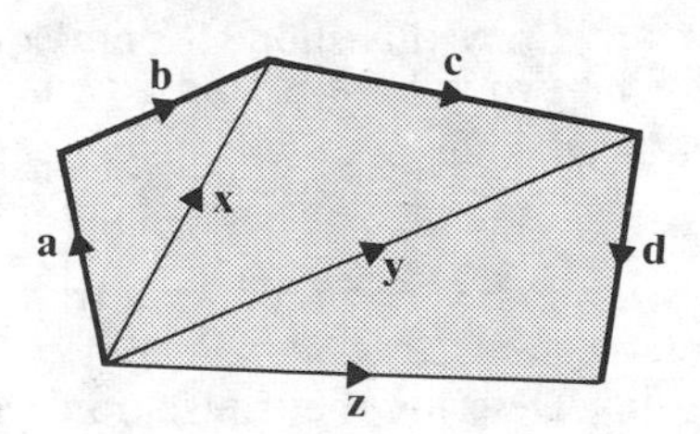

9 Find the vectors **x**, **y** and **z** in terms of **a**, **b**, **c** and **d**.

10

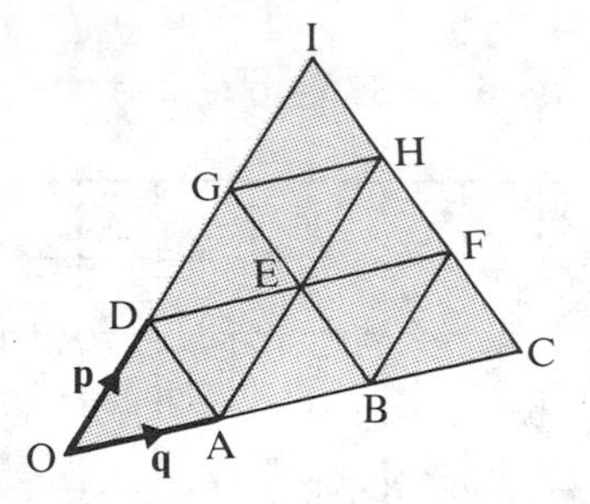

The triangle OCI has its sides divided into thirds as shown. If **OD** = **p** and **OA** = **q**, find the following vectors in terms of **p** and **q**.

a **OI** b **OB** c **AO** d **GO**
e **EA** f **FD** g **HG** h **AD**
i **BG** j **CI**

11 Two equilateral triangles overlap to form a star so that all sides are cut in thirds. If **OQ** = **a** and **OY** = **b**, find the following vectors in terms of **a** and **b**.

a **QR** b **YX** c **XW** d **XU**
e **XT** f **PQ** g **PS** h **PT**
i **ZY** j **ZV**

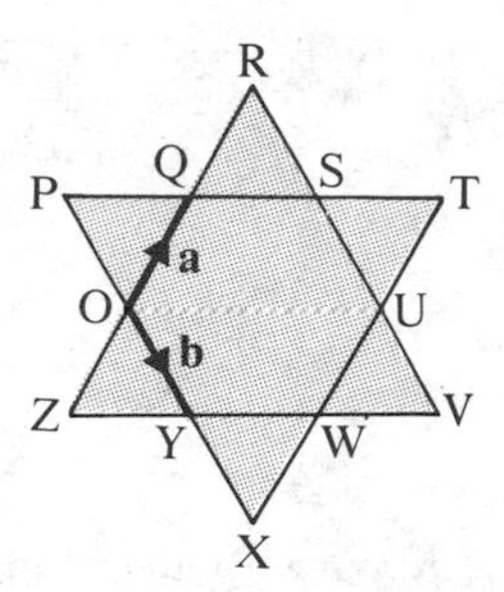

12

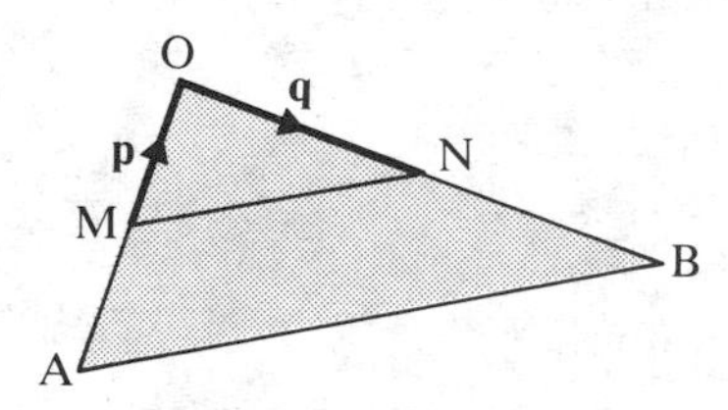

M and N are the midpoints of two sides of triangle OAB. If **MO** = **p** and **ON** = **q**, find the vectors **MN** and **AB** in terms of **p** and **q**.

What conclusions can you draw from your answers about the lines MN and AB?

13 The diagonal BD of the parallelogram ABCD is divided into thirds by the points P and Q. If **AB** = **u** and **BP** = **v**, find the following vectors in terms of **u** and **v**.

a **AP** b **QD** c **DC** d **QC**

What conclusion can you draw about the lines AP and QC?

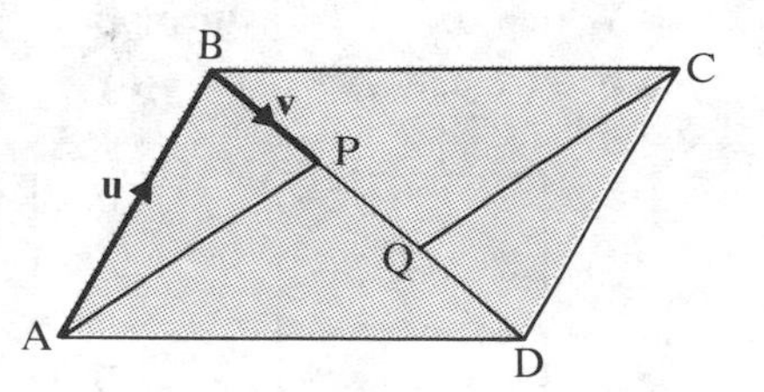

14 On axes labelled from 0 to 10, draw each of these object shapes and their images.

Write the vector of the translation which maps each object onto its image.

	Object shape	Image of shape
a	(1, 1), (1, 3), (2, 3), (3, 2)	(4, 3), (4, 5), (5, 5), (6, 4)
b	(1, 10), (5, 10), (1, 6), (2, 9)	(3, 9), (7, 9), (3, 5), (4, 8)
c	(9, 7), (10, 7), (10, 4), (8, 4), (9, 6)	(8, 10), (9, 10), (9, 7), (7, 7), (8, 9)
d	(8, 3), (10, 3), (9, 2), (10, 2), (8, 1)	(5, 2), (7, 2), (6, 1), (7, 1), (5, 0)

Transformations

15 a Copy this shape S onto squared paper and draw two images S′ and S″ where

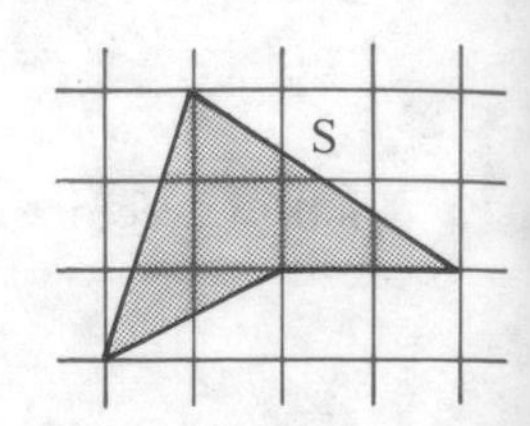

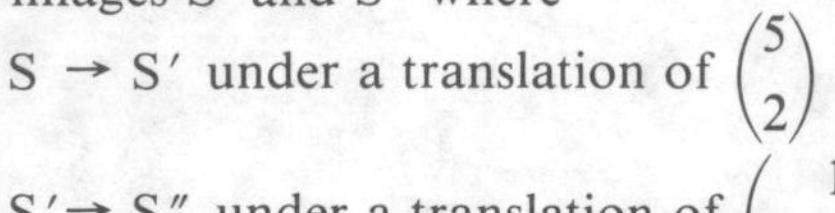

$S \rightarrow S'$ under a translation of $\begin{pmatrix} 5 \\ 2 \end{pmatrix}$

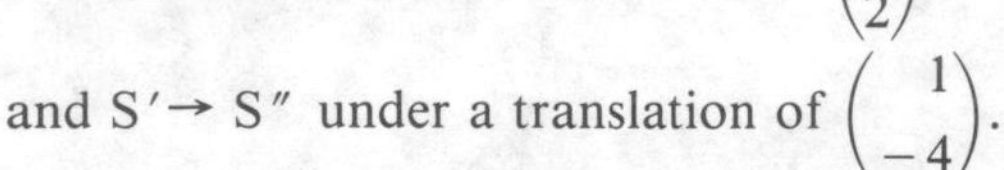

and $S' \rightarrow S''$ under a translation of $\begin{pmatrix} 1 \\ -4 \end{pmatrix}$.

b Describe the translation which maps S directly onto S″ in one step.

16 a Translation **A** is given by the vector $\begin{pmatrix} 3 \\ -2 \end{pmatrix}$ and translation **B** by $\begin{pmatrix} 1 \\ 3 \end{pmatrix}$

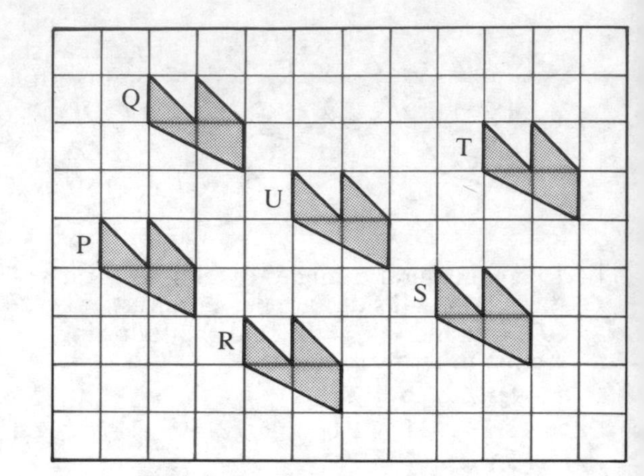

Say whether these statements are *true* or *false*.

(i) U = **A**(Q)
(ii) Q = **B**(P)
(iii) T = **A**(U)
(iv) S = **AB**(R)
(v) T = $\mathbf{A}^2$(P)
(vi) S = $\mathbf{A}^2$(Q)

b On axes labelled from 0 to 12, draw the triangle T(4, 1), (4, 3), (7, 3) and its images **B**(T), **AB**(T) and $\mathbf{B}^2$(T).

Part 4 Stretches

1 A stretch lengthens an object in *one* direction only.

These diagrams show coloured objects transformed into images in black.

Which of the transformations are

(i) stretches (ii) enlargements (iii) neither?

a

b

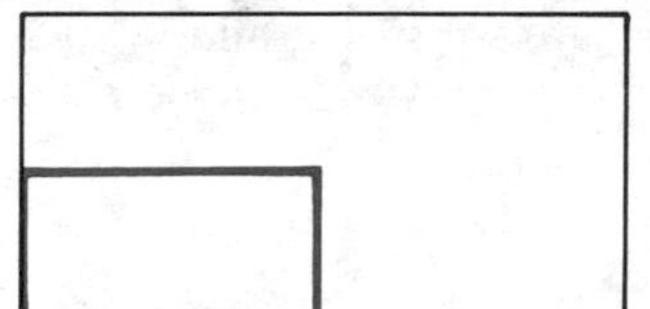

c

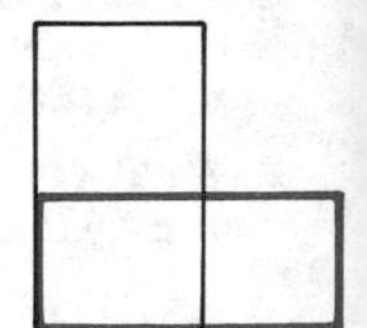

d

e

f

Transformations

2 These two coloured objects are both stretched so that line AB is invariant.

Find the *scale factor* for each stretch.

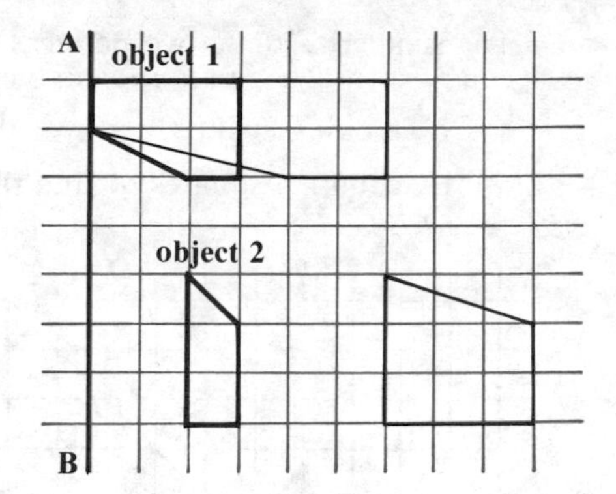

3 On axes labelled from 0 to 10, draw each object and image pair. Find the scale factor for each stretch.

	Object shape	Image shape
a	(0, 8), (2, 10), (2, 9), (1, 8), (1, 7), (2, 6), (2, 5), (0, 7)	(0, 8), (8, 10), (8, 9), (4, 8), (4, 7) (8, 6), (8, 5), (0, 7)
b	(0, 2), (2, 4), (2, 3), (3, 3), (3, 1), (2, 1), (2, 0)	(0, 2), (4, 4), (4, 3), (6, 3), (6, 1), (4, 1), (4, 0)

4 On axes labelled from 0 to 10, draw each object. Take the x-axis as invariant and use the given scale factor to draw each image.

	Object shape	Length scale factor
a	A square (3, 0), (1, 1), (2, 3), (4, 2)	3
b	An arrow (6, 2), (6, 5), (7, 4), (8, 5), (9, 4), (8, 3), (9, 2)	2

5 On axes labelled from 0 to 14, draw the shape S(4, 4), (4, 6), (5, 6), (5, 5), (7, 6), (7, 4).

Draw the image S' of S under a shear of scale factor 2 which keeps the y-axis invariant.

Draw the image S'' of S' under a second shear of scale factor 2 which keeps the x-axis invariant.

Describe and give details of the single transformation which maps S directly onto S'' in one step.

Part 5 Enlargements

1 A transparency or slide 3 cm tall is projected onto a screen so that its image is 60 cm high. What is the *length scale factor* of this enlargement?

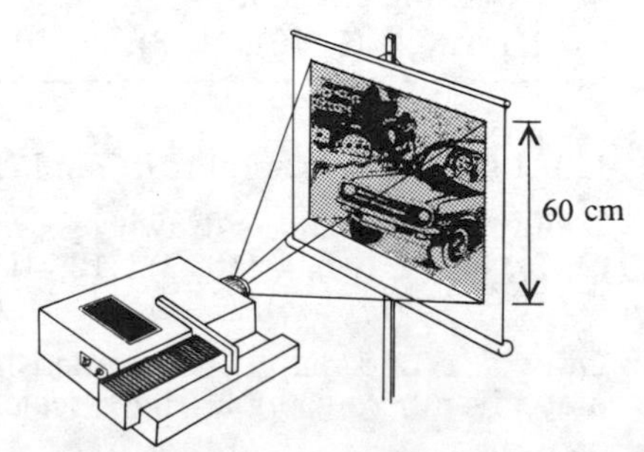

Transformations

2 The smaller of these two patterns would help make a doll's T-shirt. It is enlarged to give a pattern suitable for a young child.

a By looking at the lengths of the edges, write down the *length scale factor*.

b By counting squares to find the two areas, write down the *area scale factor*.

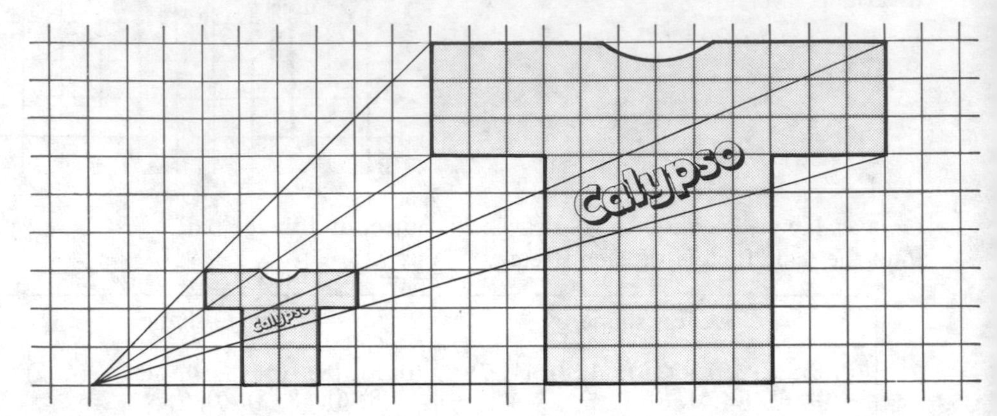

3 Each of these object and image pairs are to be drawn on axes labelled from 0 to 10. For each pair find

(i) the centre of the enlargement (ii) the length scale factor

(iii) the area scale factor.

	Object shape	Image shape
a	(3, 9), (4, 7), (2, 6), (2, 7)	(7, 9), (10, 3), (4, 0), (4, 3)
b	(3, 3), (6, 7), (4, 6), (2, 7)	(3, 1), (9, 9), (5, 7), (1, 9)
c	(1, 2), (1, 6), (5, 10), (9, 6)	(7, 2), (7, 3), (8, 4), (9, 3)
d	(0, 5), (4, 1), (10, 7)	(2, 4), (4, 2) (7, 5)
e	(1, 9), (5, 9), (2, 8), (2, 6), (1, 6)	(10, 3), (2, 3), (8, 5), (8, 9), (10, 9)

4 Draw each of these object shapes on axes labelled from 0 to 10. Enlarge each one using the given centre and length scale factor.

	Object shape	Centre of enlargement	Length scale factor
a	Trapezium (2, 4), (2, 5), (4, 6), (4, 3)	(1, 2)	2
b	Rectangle (1, 5), (3, 9), (9, 6), (7, 2)	(5, 5)	$\frac{1}{2}$
c	Triangle (1, 10), (1, 8), (4, 10)	(4, 7)	-2

5 Label the x-axis from 0 to 16 and the y-axis from 0 to 10.

A map of an island is drawn by joining these points in this order: (1, 6), (1, 4), (2, 2), (4, 1), (6, 2), (10, 2), (12, 4), (14, 8), (12, 10), (11, 9), (9, 8), (9, 10), (7, 10), (4, 9), (2, 8), (2, 7), (4, 5), (1, 6)

Draw a second map of the same island on the same diagram by reducing the map given here using a length scale factor of $\frac{1}{2}$ and the point (8, 4) as centre.

Transformations

6 Map B is an enlargement of map A.

a Copy the table below and use a ruler to help you complete it.

b What is the length scale factor of the enlargement?

c What is the area scale factor of the enlargement?

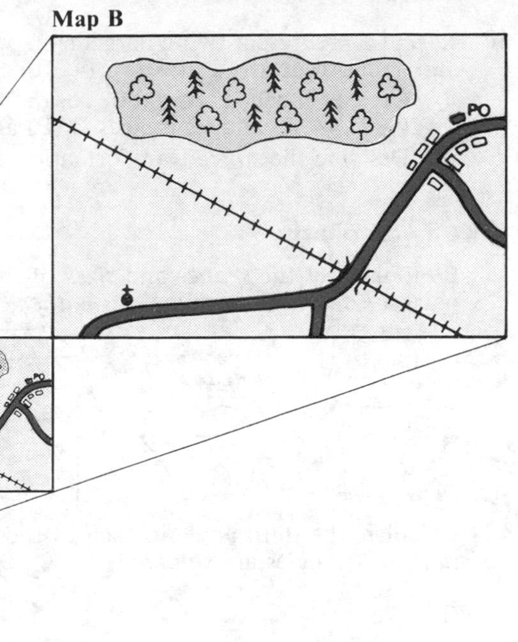

	Map A	Map B
Length of rail track, cm Shortest distance from Post Office to church, cm Approximate area of the wood, cm^2		

7 An enlargement of length scale factor 3 maps trapezium T of area 6 cm^2 onto trapezium T'. Find

a the length x

b the area scale factor

c the area of T'

d the area of the shaded region.

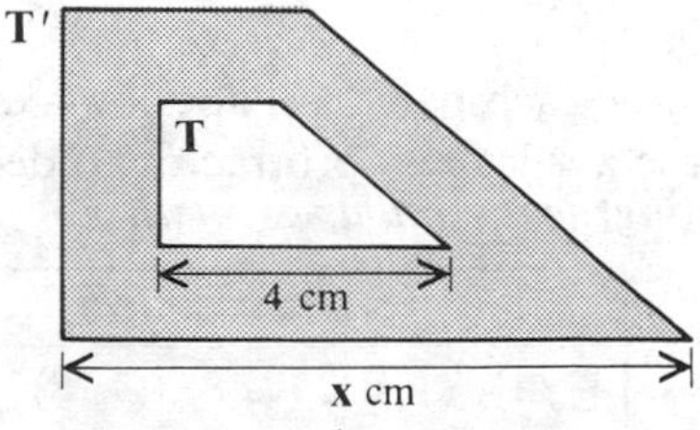

8

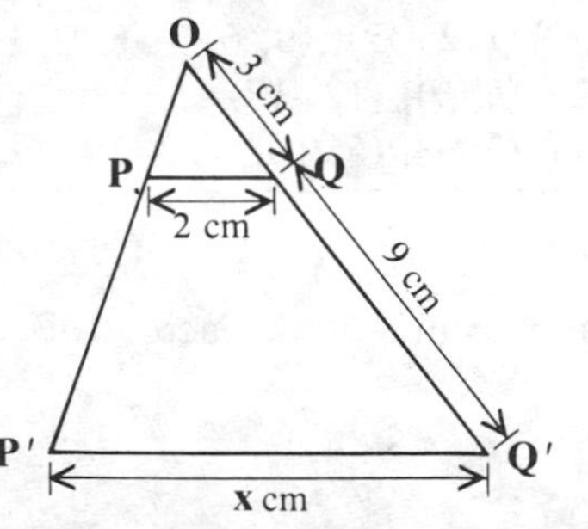

Triangle OPQ of area $2\frac{1}{2}$ cm^2 is enlarged onto triangle $OP'Q'$ as shown. Find

a the length scale factor of the enlargement

b the area scale factor of the enlargement

c the value of x

d the area of triangle $OP'Q'$

e the area of trapezium $PQQ'P'$.

9 **E** represents an enlargement of length scale factor 3 and centre (3, 10).
F represents an enlargement of length scale factor 2 and centre (11, 8).

a On axes labelled from 0 to 12, draw the triangle T(4, 10), (4, 8), (5, 8) and its images **E**(T) and **FE**(T).

b Describe the *single* transformation which maps T directly onto **FE**(T).

Transformations

10 $\mathbf{E}_1$ and $\mathbf{E}_2$ represent two enlargements with respective length scale factors of 3 and $\frac{1}{3}$ and respective centres of (4, 10) and (10, 4).

a On axes labelled from 0 to 10, draw the pentagon, P(3, 9), (4, 8), (5, 9), (5, 7), (4, 7) and its images $\mathbf{E}_1(\text{P})$ and $\mathbf{E}_2\mathbf{E}_1(\text{P})$.

b Describe the *single* transformation equivalent to $\mathbf{E}_2\mathbf{E}_1$.

Part 6 Similarity

1 By looking at the *shapes* and *sizes* of these pairs of figures, decide whether the pairs are *similar*, *congruent* or *neither*.

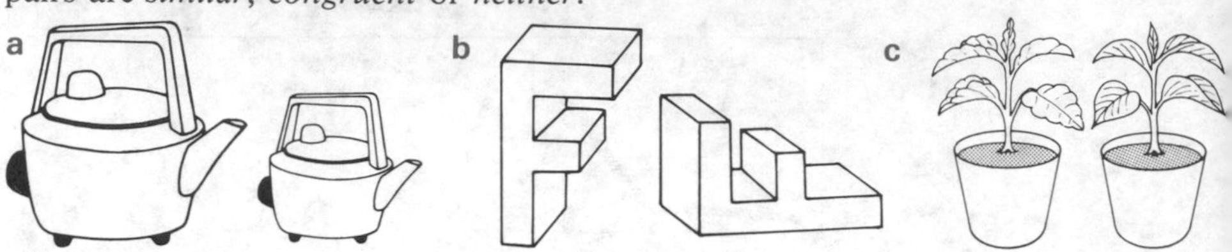

2 Calculate the third angle in each of these triangles and hence decide which pairs of triangles are *similar*.

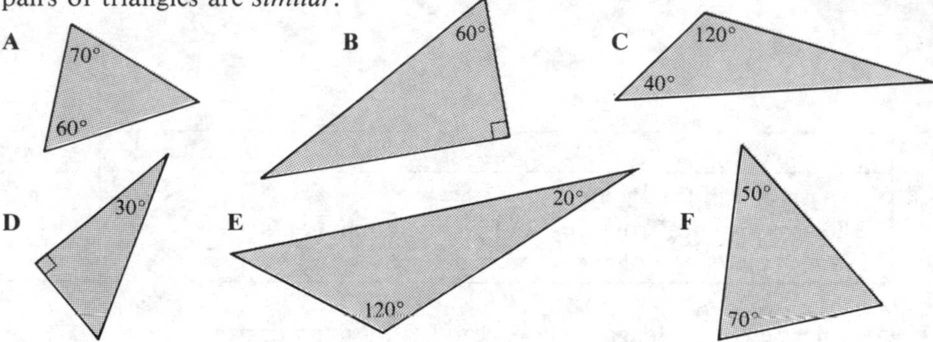

3 Draw each pair of shapes on axes labelled from 0 to 12.
Use a ruler and protractor to decide whether the shapes in each pair are *congruent*, *similar* or *neither*.

	First shape	Second shape
a	(7, 1), (1, 1), (4, 8), (7, 8)	(9, 11), (9, 5), (2, 8), (2, 11)
b	(2, 1), (7, 1), (4, 7)	(12, 2), (12, 12), (0, 6)
c	(0, 8), (10, 0), (7, 7)	(4, 2), (5, 9), (12, 12)
d	(9, 1), (9, 4), (1, 12), (1, 5)	(6, 1), (3, 1), (5, 10), (12, 10)

4 These diagrams show pairs of *similar* solids.
Find the length scale factor, area scale factor and volume scale factor for each pair.

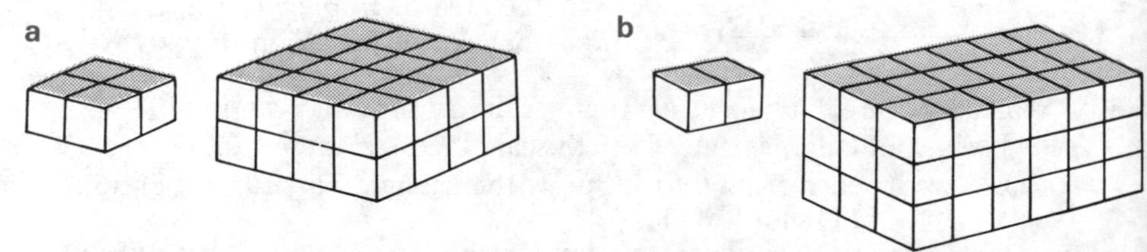

Transformations

Copy this table, enter your results and complete the table by following the pattern.

Length scale factor	2	3	4	5	6	7	8	9	10	m
Area scale factor										
Volume scale factor										

5 Washing-up liquid is sold in two similar containers 20 cm and 40 cm high. The smaller one holds $\frac{1}{2}$ litre. Find

a the length scale factor **b** the volume scale factor

c the capacity of the larger container.

6 A wall poster 24 cm high is similar to a postcard of height 8 cm and area 100 cm^2. Find

a the length scale factor **b** the area scale factor

c the area of the poster.

7 Baked beans are sold in two similar tins of radii 3 cm and 6 cm. The small tin has a label of area 40 cm^2 and holds 300 cm^3 of beans.
Calculate the area of the label and the capacity of the larger tin.

8 A young girl makes a cardboard model of her own home to a scale of 1:10 for length. If the actual house has five rooms, stands 10 metres high and has a ground floor area of 150 m^2, find the number of rooms, the height and ground floor area in her model.

9 A telegraph pole 8 m high is held upright by a sloping wire which just touches the top of a 2 m high fence. If the wire is fixed to the ground 1 m from the fence, find the distance x between the fence and pole.

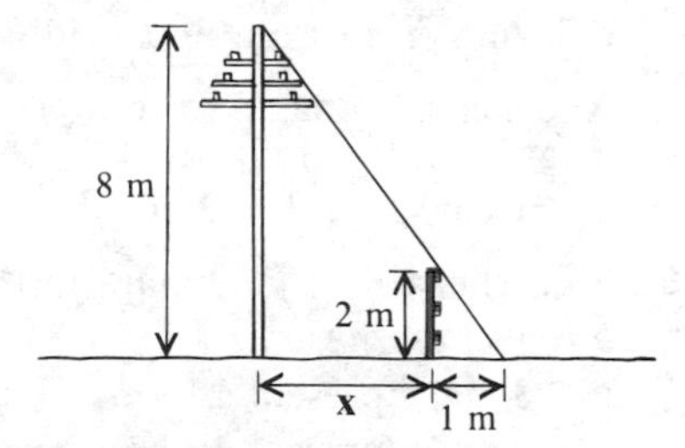

10 A builder carrying a plank on his shoulder, rests one end 2 m behind him on the ground and the other end on a wall 4 m in front of him. If his shoulder is $1\frac{1}{2}$ m high, calculate the height h of the wall.

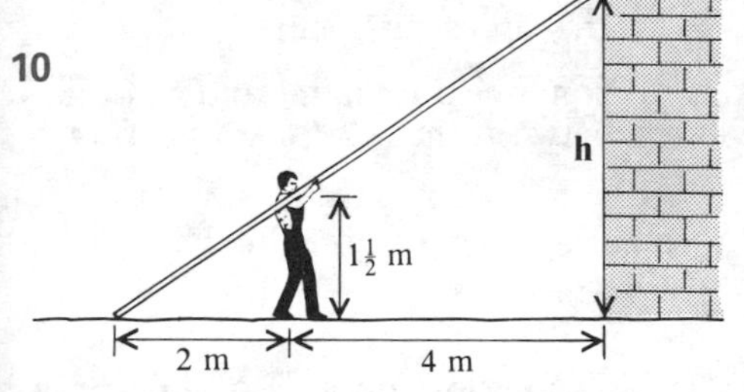

11 A street light 8 m tall stands between two walls, both 2 m high, which cast shadows 2 m and 3 m long as shown. Calculate the lengths x and y and hence the distance between the two walls.

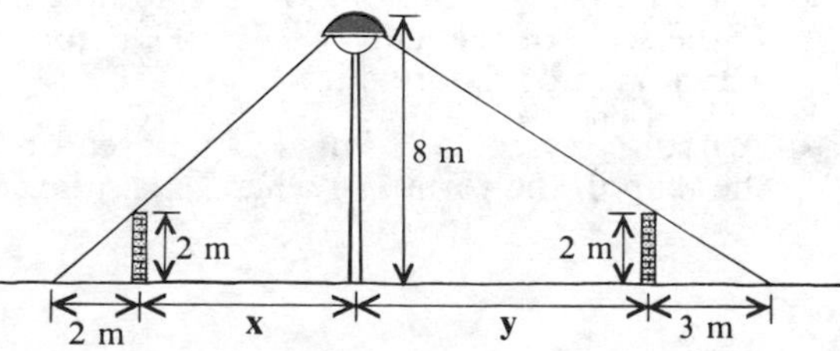

Transformations

12 This is the ground plan of a hotel drawn to a scale of 1:400 for length.

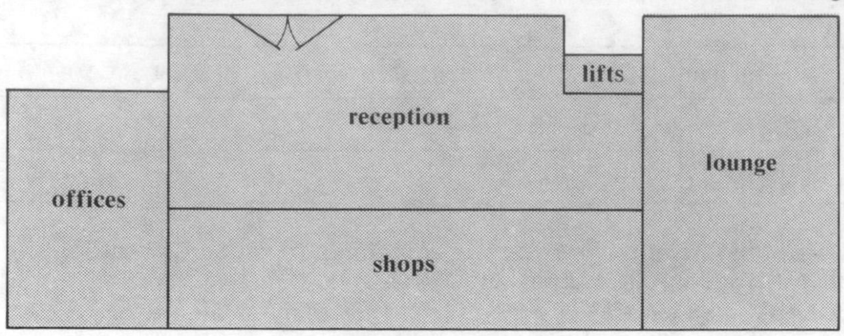

Take measurements with a ruler to find, in metres,

a the length and width of the offices
b the length and width of the shopping area
c the perimeter of the lounge.

13 Find the distance (in kilometres) which 1 cm stands for on a map with a length scale factor or *representative fraction* of:

a 1:300 000 **b** 1:1 200 000 **c** 1:50 000 **d** 1:75 000.

14 A map of Britain has a scale of 1:2 000 000. Find the actual distances (in kilometres) between these places whose distances on the map are given here in centimetres.

a 4.3 cm from Cardiff to Exeter
b 6.6 cm from Shrewsbury to Dewsbury
c 14.1 cm from Belfast to Newcastle
d 26.8 from London to Edinburgh

15 Another map has a scale of 1:250 000. What distance in kilometres do these lengths on the map stand for?

a 1 cm **b** 2 cm **c** 5 cm **d** $\frac{1}{2}$ cm **e** 1.6 cm

16 Find the scale of the map (in the form 1:n where n is a large number) on which

a 1 cm represents 4 km **b** 2 cm represent 10 km
c 3 cm represent 24 km **d** 5 cm represent 40 km
e 4 cm represent 10 km **f** 2 cm represent 1 km.

17 On a map of Europe, the two capital cities of London in England and Sofia in Bulgaria are 8 cm apart. If the actual distance between them is 2000 km, find the scale of the map.

18 A map shows a road 4 cm long running through a forest of area 3 cm^2. If its scale is 1:500 000, find the actual length of the road and the actual area of the forest.

19 A powerline is 8 cm long on a map and it crosses a moorland area of 25 cm^2. If the scale of the map is 1:400 000, find the actual length of the powerline and actual area of moorland.

20 A nature reserve of 8 km^2 has an area on a map of 32 cm^2. Find the scale of the map in the form 1: n where n is a large number.

Topology

1 Here are eight drawings labelled *A* to *H*. Arrange them into pairs which are topologically equivalent.

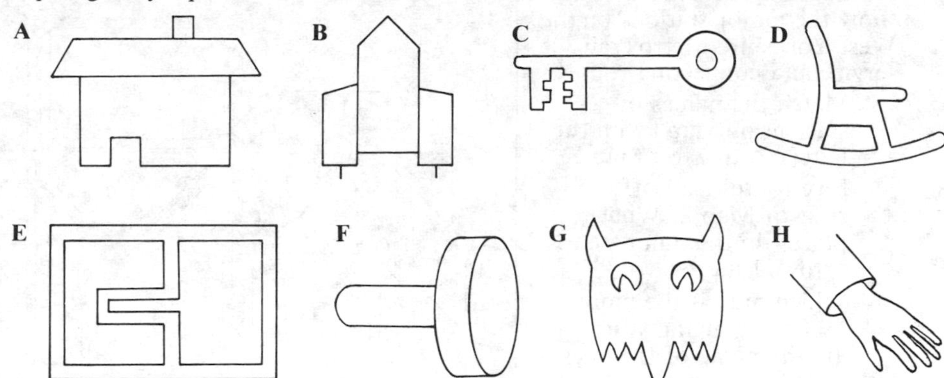

2 Two of these electrical circuits are topologically equivalent. Which is the odd one out?

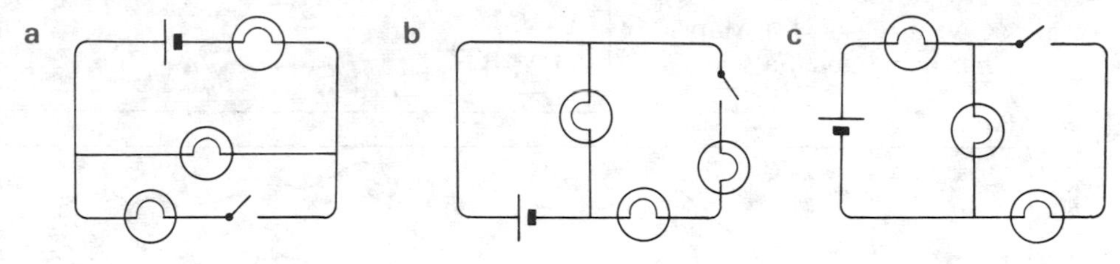

3 Here is a street map of part of a town centre on which eight important buildings are marked.

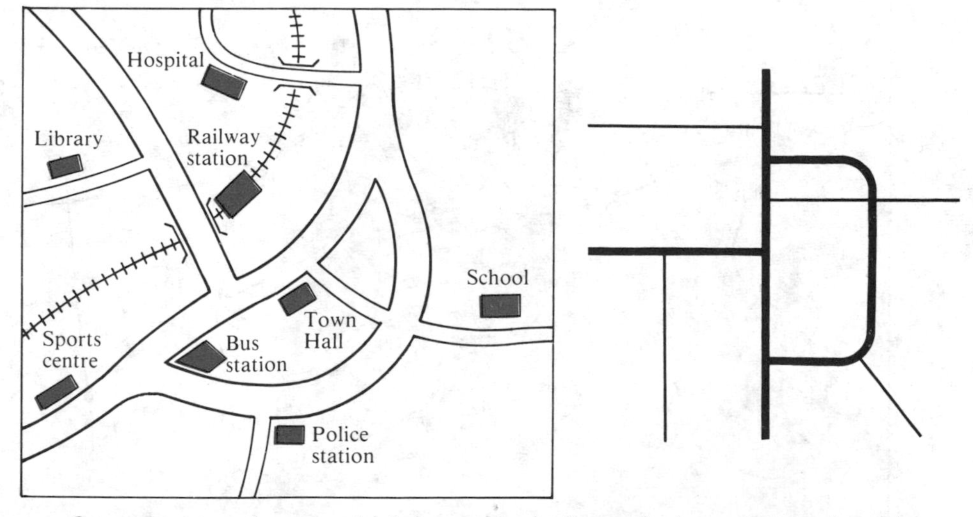

a Copy the second map, which is topologically equivalent, and mark on it the positions of the eight buildings. Also draw a straight line to represent the railway line.

b A new classmate asks you how she might get to the hospital when she leaves school. What directions would you give her?

c Write down the directions which would take a stranger from the railway station to the police station.

Topology

4 These two maps are topologically equivalent. They show the major stations on the West Yorkshire MetroTrain service and connecting routes.

a Match the names of stations on Map 1 with the letters A to Z on Map 2.

b Five routes lead off the edge of Map 2. What places do the numbered arrows lead to?

c Which map is the more useful for giving you
(i) the name of the lines in the network
(ii) the true distance between any two stations?

d Give a reason for Map 1 *not* being given a scale.

Map 1

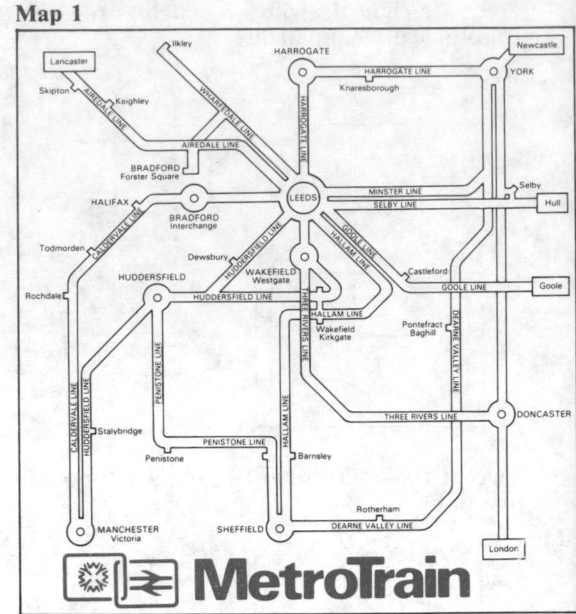

Map 2

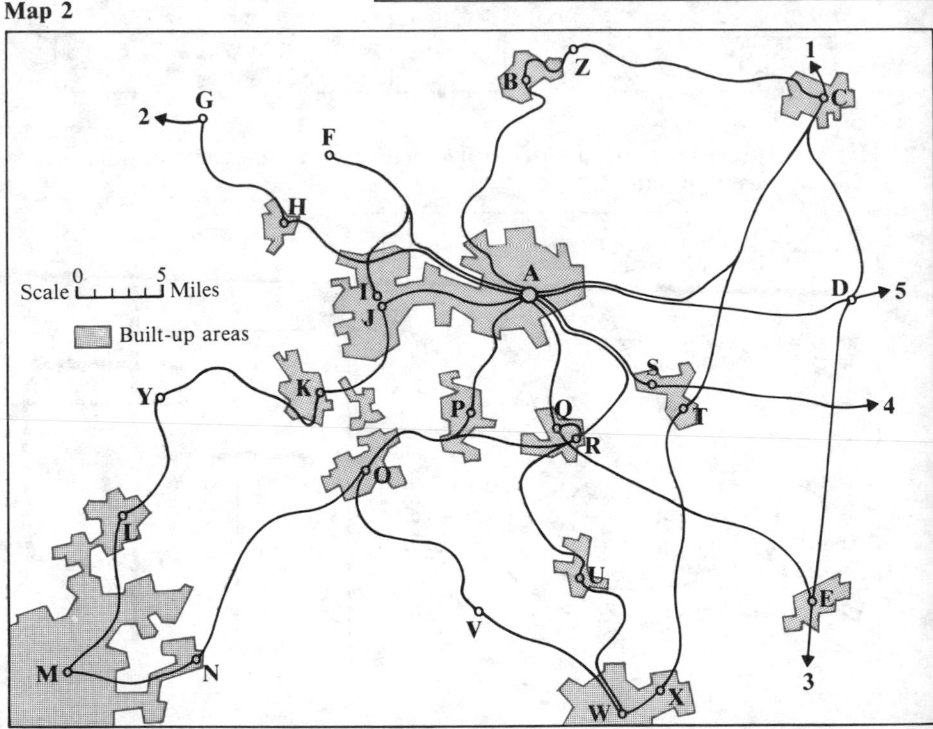

5 Label three columns *sphere*, *torus* and *double-torus*.
Write each of these objects in that column with which it is topologically equivalent.

a golf ball	swimming trunks	a pair of tweezers	a hat
a wedding ring	a screw	a potato	a saw
a cup	a cardigan	a pair of scissors	a kettle

Topology

6 How many odd nodes have these networks?
Which of the networks are *traversable*?

a

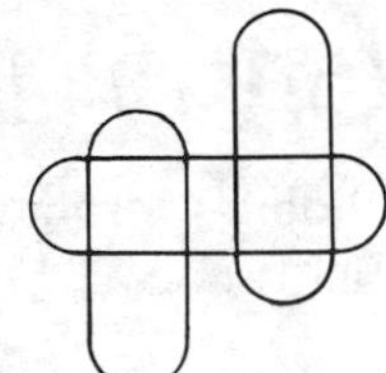

b

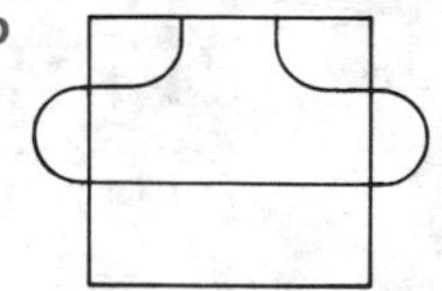

c

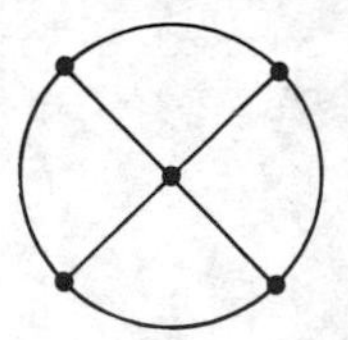

d

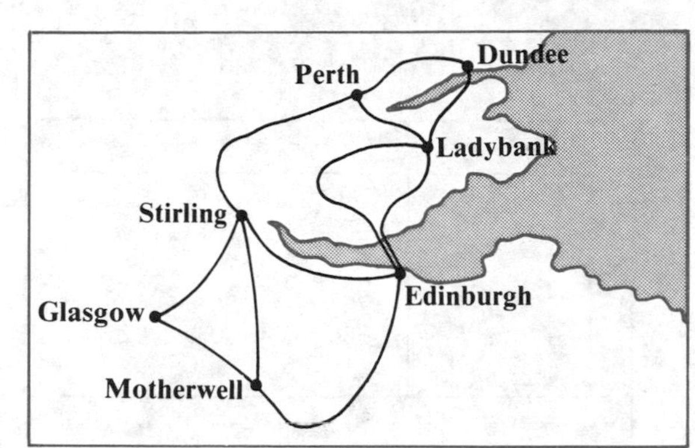

7 This map shows part of the British Rail network. Is it possible for a traveller to arrange a journey using each of these lines once and only once? If so, where should the start and finish be?

8

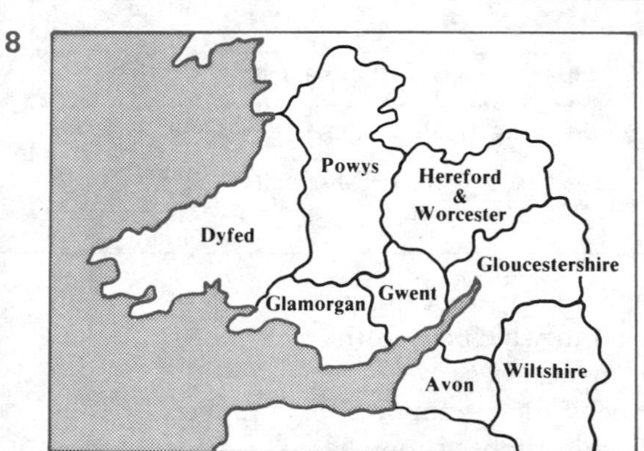

a Can a tourist organise a route through these counties of Wales and the West Country to visit each county just once and arrive back at the start?

b Does the Severn Bridge which connects Avon and Gwent make the problem easier?

9 The main roads between four towns are shown here. Copy and complete this matrix to give the number of *direct* routes.

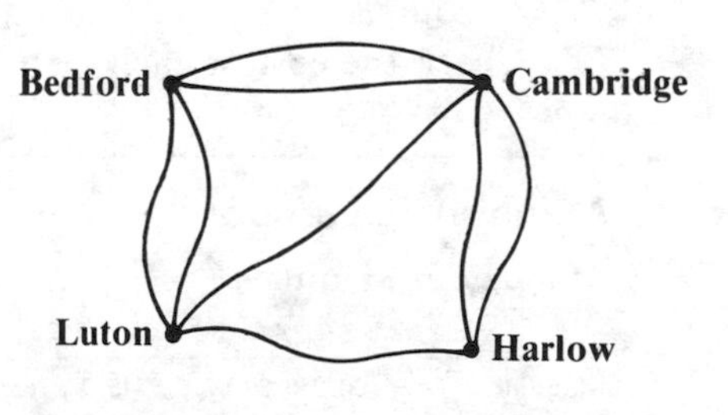

to

$$\text{from}\;\begin{matrix} B \\ C \\ H \\ L \end{matrix}\overset{\begin{matrix} B & C & H & L \end{matrix}}{\begin{pmatrix} & & & \\ & & & \\ & & & \\ & & & \end{pmatrix}}$$

10 Draw a network of routes between the points given in each of these route matrices.

a

$$\text{from}\;\begin{matrix} A \\ B \\ C \end{matrix}\overset{\text{to}\; \begin{matrix} A & B & C \end{matrix}}{\begin{pmatrix} 2 & 3 & 1 \\ 3 & 0 & 2 \\ 1 & 2 & 0 \end{pmatrix}}$$

b

$$\text{from}\;\begin{matrix} P \\ Q \\ R \end{matrix}\overset{\text{to}\; \begin{matrix} P & Q & R \end{matrix}}{\begin{pmatrix} 0 & 2 & 1 \\ 2 & 2 & 0 \\ 1 & 0 & 1 \end{pmatrix}}$$

c

$$\text{from}\;\begin{matrix} X \\ Y \\ Z \end{matrix}\overset{\text{to}\; \begin{matrix} X & Y & Z \end{matrix}}{\begin{pmatrix} 0 & 2 & 1 \\ 3 & 0 & 1 \\ 2 & 1 & 2 \end{pmatrix}}$$

Tables, charts and matrices

Part 1 Tabulation of information

1

To: G & H Gomersal & Son
Canal Wharfe
RUGBY

Menston plc
Timber and Builders Merchants
Harriers Industrial Estate
Tile Hill
Coventry CV4 7PZ
Telephone Coventry (0203) 56144

Reference: DRL/GL/45
Order No: 10472
Date: 3rd March 1986

Quantity	Description	Unit cost	Cost, £
8 tonne	Sand	£1·75 per tonne	14·00
$1\frac{1}{2}$ tonne	Cement	£35 per tonne	52·50
4000	Bricks	£110 per 1000	440·00
200 m	75 mm x 50 mm timber	72p per metre	144·00
150 m	100 mm x 50 mm timber	103p per metre	154·50
		Total before VAT	
		VAT at 15%	
		TOTAL AMOUNT DUE	
		5% discount if paid within 31 days	
		Net cost after discount	

This invoice shows the goods purchased by a local builder.

- **a** Who is selling the materials and in which city is the firm based?
- **b** How much does 1 tonne of cement cost?
- **c** How much does the builder spend on 75 mm × 50 mm timber?
- **d** What length of 100 mm × 50 mm timber is purchased?
- **e** What is the last date on which payment must be made to qualify for discount?
- **f** Check the cost of each material bought and complete the bill.

2

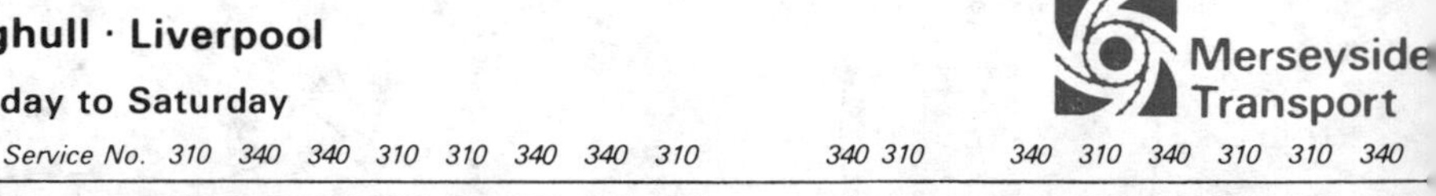

Maghull · Liverpool

Monday to Saturday

Service No.	310	340	340	310	310	340	340	310		340	310		340	310	340	310	310	340
Maghull	0629	0650	0716	0745	0750	0801	0845	0912		45	12		2145	2212	2245	2312	2345	0009
Dodds Lane	0634	0655	0721	0750	0755	0806	0850	0917		50	17		2150	2217	2250	2317	2350	0014
Eastway		0658	0724			0809	0853		then at	53			2153		2253			0017
Highfield Park	0638			0754	0759			0921	these		21					2321	2354	
Hall Lane	0642	0701	0727	0758	0803	0812	0856	0925	minutes	56	25	until	2156	2221	2256	2325	2358	0020
Old Roan	0649	0707	0733	0805	0810	0818	0902	0932	past	02	32		2202	2225	2302	2332	0005	0026
Black Bull	0654	0712	0738	0810	0815	0823	0907	0937	each	07	37		2207	2232	2307			
Walton Church	0702	0720	0746	0818		0831	0915	0945	hour	15	45		2215	2237	2315			
Scotland Road	0713	0731	0757	0829		0842	0926	0956		26	56		2224	2245	2326			
Liverpool centre	0721	0739	0805	0837		0850	0934	1004		34	04		2234	2256	2334			

Tables, charts and matrices

a Mrs Kelly catches the 06.50 from Maghull. When does she reach Old Roan?

b You start work in Liverpool at 08.45. When does the last bus leave Maghull which would get you to work in time?

c You catch the bus in Dodds Lane for your school in Hall Lane. If school starts at 9 a.m., what is the latest time you must be at the bus-stop in Dodds Lane?

d You *just* miss the 09.12 from Maghull to Liverpool. When does the next bus leave and how long have you to wait?

e Which service does *not* stop at Highfield Park? If Mrs Parbold has a dental appointment near Walton Church at 11 a.m., when should she catch the bus in Highfield Park?

f Annette Roberts comes home from school for her midday meal. She lives near the Black Bull and her school is near Walton Church. If afternoon lessons start at 1.50 p.m., when should she catch the bus at the Black Bull?

g You have been visiting your grandmother in Hall Lane and want to catch the 23.15 train from the centre of Liverpool to get home. When does the last possible bus leave Hall Lane for you to make the connection?

h When does the last *310* bus leave Maghull? When does it arrive at Old Roan and how long does the journey take?

3 AIR MAIL RATES (outside Europe)

Blue air mail labels essential See key for zones in which countries are located.	Air mail zone	Not over 10g	Each additional 10g or part thereof
Letters	A	29p	11p
	B	31p	14p
	C	34p	15p
Printed Papers and Small Packets	A	21p	5p
	B	23p	7p
	C	24p	8p
Newspapers and Periodicals Weight limit 2000 g	A	16p	3p
	B	18p	4p
	C	19p	5p
Aerogrammes	To all destinations: 26p Packs of 5 aerogrammes: £1·20		
Postcards	To all zones: 26p each		

Key to Air mail zones

C Australia	B Malaysia
B Bangladesh	C New Zealand
B Barbados	B Nigeria
B Belize	B Pakistan
B Canada	B St Vincent
C Fiji	A Saudi Arabia
B Ghana	B Singapore
B Guyana	B South Africa
B Hongkong	B Sri Lanka
B India	B Tanzania
B Indonesia	B Trinidad & Tobago
A Israel	B Uganda
B Jamaica	B United States
B Kenya	B Zambia
B Malawi	B Zimbabwe

a How much does it cost to send
(i) a letter of 8 grams to Australia
(ii) a letter of 12 grams to Bangladesh
(iii) a small packet of 18 grams to Jamaica
(iv) a newspaper of 32 grams to India?

b Mrs Crummack's two daughters and her son all live overseas. One daughter lives in Canada and the other in New Zealand. Her son is working in Saudi Arabia. She sends them three identical letters at Christmas, each weighing 14 grams. What is the total cost of the postage?

c Shenaz Hussain's young daughter has got her photograph in the local newspaper after taking part in her school's concert. Her mother sends a copy of the newspaper to her uncle in the USA and two copies to the rest of the family in Pakistan. If the newspaper weighs 85 grams, find the total cost of the postage.

Tables, charts and matrices

4 Ulsbrit Ferries charge the prices given in this table for the sea crossing between Britain and Northern Ireland.
All prices are for single journeys. For return journeys the prices are doubled.

ULSBRIT FERRIES	Single fares	
	Standard Sept-May	Summer June-August
Each adult	£10	£12
Each child (aged 5 to 15)	£5	£6
Each car, van or minibus:		
Length: up to 4 metres	£45	£48
up to $4\frac{1}{2}$ metres	£55	£58
up to 5 metres	£65	£68
Each extra metre or part of a metre	£10	£10
NOTES children under 5: free	children 16 and over: adult fare	

a Stewart Anderson (aged 9) and his mother go as foot passengers on this ferry service during October. How much does the single journey cost the two of them?

b (i) Mr and Mrs McCann and their daughter Avril (aged 12) go on holiday by car (of length 3.7 metres) on June 5th and return on July 4th. How much does the total journey there and back cost them?

(ii) How much cheaper would it have been for them if they had started their holiday on May 30th instead of June 5th (but still returned on July 4th)?

c A school trip is organised for September. Two teachers and twelve third-formers will make a return journey in the school minibus which is 5.4 metres long. How much will the total return fare be?

5 This table gives the costs of dialled telephone calls of different lengths over different distances. The second table is an extract from the STD dialling codes booklet which gives the dialling codes and the charge letter for various telephone exchanges.

DIALLED CALLS

		Approximate **Cost** to the customer including VAT					
		1 min	2 mins	3 mins	4 mins	5 mins	Time for one unit
Local							
Local *L*	Cheap	6p	6p	6p	6p	6p	8 mins
	Standard	6p	6p	12p	12p	17p	2 mins
	Peak	6p	12p	12p	17p	23p	1 min 30 secs
National and Irish Republic							
Calls up to 56 km (35 miles) *a*	Cheap	5p	6p	12p	12p	17p	2 mins
	Standard	12p	17p	29p	35p	46p	40 secs
	Peak	12p	23p	35p	46p	58p	30 secs
Calls over 56 km (35 miles) *b*	Cheap	12p	17p	23p	29p	40p	48 secs
	Standard	17p	35p	46p	63p	81p	22.5 secs
	Peak	23p	40p	63p	81p	£1·04	17.1 secs
Calls to the Irish Republic from Great Britain and the Isle of Man	Cheap	23p	46p	69p	92p	£1·15	15 secs
	Standard	46p	86p	£1·32	£1·73	£2·19	8 secs
	Peak	46p	86p	£1·32	£1·73	£2·19	8 secs

Tables, charts and matrices

NATIONAL DIALLING CODES Charges: *L* – Local, *a* – up to 56 km, *b* – over 56 km

Charge letter		Code	Charge letter		Code
b	Abbeytown	096 56	*b*	Abingdon	0235
b	Abbotsbury	030 587	*b*	Accrington	0254
b	Abercrave	063 977	*b*	Achnasheen	044 588
b	Abercynon	0443	*a*	Addingham	0943
b	Aberdare	0685	*b*	Aghalee	0846
b	Aberdeen	0224	*b*	Ahoghill	0266
b	Aberdovey	0654	*b*	Airdrie	023 64
L	Aberford	928 49	*b*	Airton	072 93
b	Abergavenny	0873	*b*	Alcester	0789
b	Aberystwyth	0970	*b*	Aldershot	0252

a How much does a 3-minute local call cost at the standard rate?

b Canterbury is over 56 km away. How much does a 4-minute call cost at the peak rate?

c Mr McIver rings up his mother in Dublin from Liverpool. How much does a 5-minute call cost at the cheap rate?

d How much does a 4-minute call cost to Accrington at the peak rate?

e How much does a 2-minute call cost to Aberford at the standard rate?

f You make a 5-minute call at the cheap rate to your friend whose 'phone number is Addingham 4308. How much does it cost and what number do you dial?

g Susan Fisher's boy-friend lives locally and he rings her up for 24 minutes. How many units does he use at the standard rate? If each unit costs 5.75 pence, how much does the phone call cost?

h A business man makes a call to a company 20 miles away. The call lasts 18 minutes at the peak rate. How many units does he use? If each unit costs 5.75 pence, what does the call cost?

6 This is a price list for pairs of ready-made curtains. They can be bought in four different widths A, B, C and D and in many different lengths varying from 120 cm to 400 cm.

a How much does a pair of curtains 155 cm wide and 160 cm long cost?

b Find the cost of a pair of curtains 205 cm wide and 220 cm long.

c Mrs Woodward wants a pair of curtains for her kitchen and a pair for her dining room. Both pairs are to be 180 cm long. If the kitchen curtains are to be 155 cm wide and the dining room ones 255 cm wide, what will be the total cost?

d What is the cost of a pair of curtains
(i) 155 cm wide and 200 cm long
(ii) 160 cm long and 205 cm wide?
What is the area of each pair and which pair is the 'better' buy?

e Mrs Askwith wants curtains 105 cm wide and 310 cm long. These are not a standard length but the shop is willing to make them for her. What price do you think the shop might charge her?

COTTON CURTAINS PRICE LIST — All prices are per pair

	Width of plain curtains			
Finished lengths	A 105 cm	B 155 cm	C 205 cm	D 255 cm
120 cm	£32·07	£47·38	£62·64	£78·45
140 cm	£35·35	£52·37	£69·26	£86·77
160 cm	£38·41	£57·33	£75·98	£95·09
180cm	£42·25	£62·35	£83·04	£103·94
200 cm	£45·56	£67·68	£89·70	£112·27
220 cm	£48·86	£72·61	£96·37	£120·55
240 cm	£52·19	£77·65	£102·99	£122·88
260 cm	£55·74	£82·98	£110·08	£137·78
280 cm	£59·09	£87·94	£116·77	£146·11
300 cm	£62·41	£92·93	£123·42	£154·39
320 cm	£65·73	£97·96	£130·05	£162·73
340 cm	£69·26	£103·27	£137·16	£171·16
360 cm	£72·59	£108·22	£143·83	£179·92
380 cm	£75·93	£113·26	£150·43	£188·23
400 cm	£79·25	£118·15	£157·12	£196·54

Tables, charts and matrices

7 This advertisement shows the prices of ready-cut carpet pieces. They can be bought for cash or on hire purchase over 20 or 38 weeks. The tables show the weekly payments for hire purchase.

A choice of carpeting READY-CUT into ROOM-SIZE PIECES

✱ So easy to trim to your own requirements ✱ Cushion foam backing

- Fibre: nylon
- Available in blue or tan

Number	Size	Price	20 weeks	38 weeks
SG 023	3 m × 4 m	£42·50	£2·33	–
SG 026	4 m × 4 m	£55·00	£2·95	£1·65
SG 029	4 m × 5 m	£67·50	£3·58	£1·98
SG 032	4 m × 6 m	£79·99	£4·20	£2·31
SG 035	4 m × 7 m	£89·99	£4·70	£2·57

- Fibre: polypropylene
- Available in rose or brown

Number	Size	Price	20 weeks	38 weeks
SG 126	3 m × 4 m	£40·50	£2·13	–
SG 129	4 m × 4 m	£50·00	£2·75	£1·50
SG 132	4 m × 5 m	£61·50	£3·38	£1·83
SG 135	4 m × 6 m	£72·60	£4·00	£2·16
SG 138	4 m × 7 m	£80·90	£4·50	£2·42

a What is the cash price of a blue nylon carpet 4 m by 5 m?

b What is the size and price of carpet with the code number SG 135? What is the carpet made from?

c Find the cash price of a brown polypropylene carpet 4 m by 4 m. If it is bought on hire purchase over 20 weeks, what is the price paid for it? How much is saved by paying cash?

d Is it cheaper to buy a nylon carpet 4 m by 6 m on hire purchase over 20 weeks or 38 weeks? By how much is it cheaper?

e How much *per square metre* are the nylon carpets SG 023 and SG 035? Which carpet is *relatively* cheaper?

f Mr & Mrs Dawson have a large sitting room $5\frac{1}{2}$ m by 6 m. They want to cover it wall-to-wall with nylon carpet. What different ways can they do this with the sizes available in this advertisement? Which is the cheapest way?

8

SUPERSAVERS

This list shows just a few of the many items on sale now at Supersavers. Your local store has many, many more sale offers.

STEREO RADIO/CASSETTES

Item	Was	Now
KIMARI XL300 Twin Cass. B/M	~~£69·99~~	£59.99
KIMARI XL350 Cube style B/M	~~£64·99~~	£54.99
KIMARI XL400 Twin Cass. B/M	~~£89·99~~	£79.99
PANASONIC RXFM25R 4 Band B/M	~~£69·99~~	£59.99
TOSHIBA RT7025R 4 Band B/M	~~£89·99~~	£69.99

Tables, charts and matrices

a What is the sale price of a Kimari XL400 stereo radio/cassette player? By how much has it been reduced?

b Which stereo radio/cassette player has been reduced by the largest amount?

c What is the sale price of a Sharp VC651 video recorder? By how much has it been reduced?

d List the names of all the video recorders which have been reduced by £20 in the sale?

e Which item on this advertisement has been reduced by
(i) the largest amount of money
(ii) the largest percentage of the original pre-sale price?

9 A company buys a number of new cars each year. This bar chart shows, over a three-year period, where the cars were made.

Copy and complete this table to show the same information numerically.

Year	1984	1985	1986
British European Non-European			
Total			

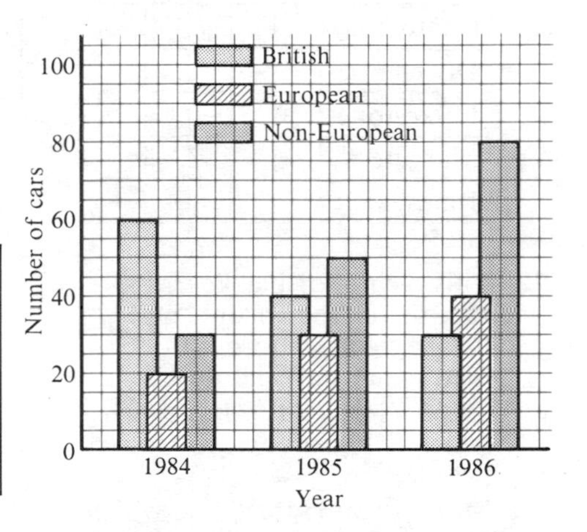

10

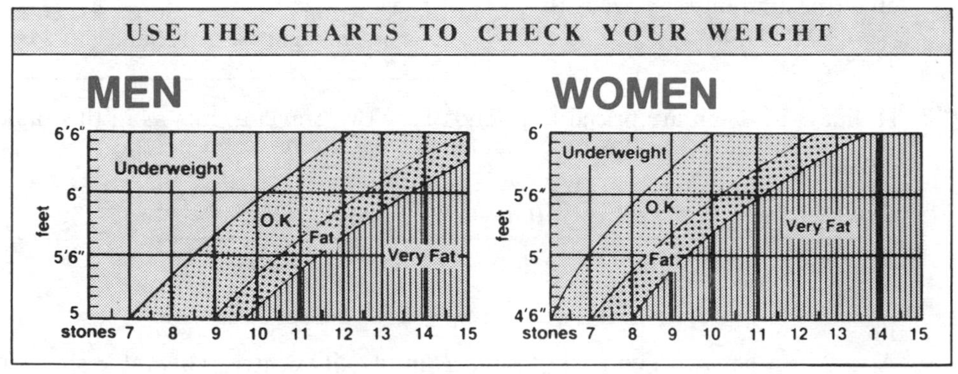

These two charts give some indication of how over- and under-weight an adult might be.

a Mr Robson is 5 ft 6 in tall and weighs 10 stones. What category does the chart place him in?

b Mrs Robson is 5 ft 2 in tall and weighs 9 stones. On what borderline does the chart place her?

c John Sadler is overweight. He weighs 14 stones and is 5 ft 7 in tall. How much has he to lose before he is just 'OK'?

d Linda Bentley is 5 ft 6 in tall and weighs $9\frac{1}{2}$ stones. How much weight can she gain on holiday before the chart classes her as 'fat'?

e Between what limits can the weight of a 6-foot tall man vary so that he is classed as 'OK'?

Tables, charts and matrices

11 This map shows mileage between five major towns. Study the map to find the *shortest* distances between each pair of towns, and enter your answers on a copy of this mileage chart.

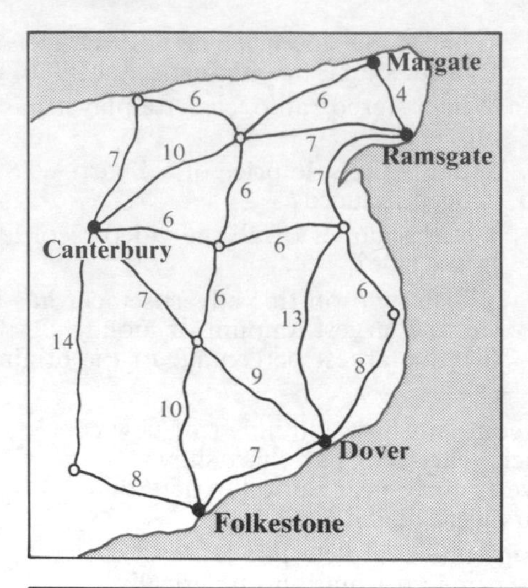

	Canterbury	Dover	Folkestone	Margate	Ramsgate
Canterbury	—				
Dover		—			
Folkestone			—		
Margate				—	
Ramsgate					—

12 This recipe for blackcurrant flan serves four people and lists ingredients in both metric and imperial units.
Sadie Hirons has a large dish which is big enough for six people.
Make a recipe for Sadie which gives the amounts she needs to serve six portions.

BLACKCURRANT FLAN – serves 4

Pastry	*Metric*	*Imperial*
Plain flour	180 g	6 oz
Butter	120 g	4 oz
Caster sugar	24 g	1 oz
Ground cinnamon	2 teaspn	
Water	2 teaspn	
Egg	1 yolk	
Filling		
Blackcurrants	480 g	16 oz
Demerara sugar	120 g	4 oz

13 Holidays in Spain are priced in pounds £ by two travel agents as in this table. If £1 = 200 pesetas, construct a table showing the prices in pesetas.

	1 week	10 days	2 weeks	1 extra week
Espantours	100	140	180	60
Iberiaplan	120	150	175	50

14 A taxi firm has two taxi ranks A and B in the city centre. This table shows the mileages from these ranks to various places.

Distance, miles	Railway station	Airport	Police station	Hospital
Rank A	1	9	$\frac{1}{2}$	3
Rank B	2	7	1	2

This flow diagram calculates the cost of a journey by taxi. Construct another table to show the costs of the journeys given in this table.

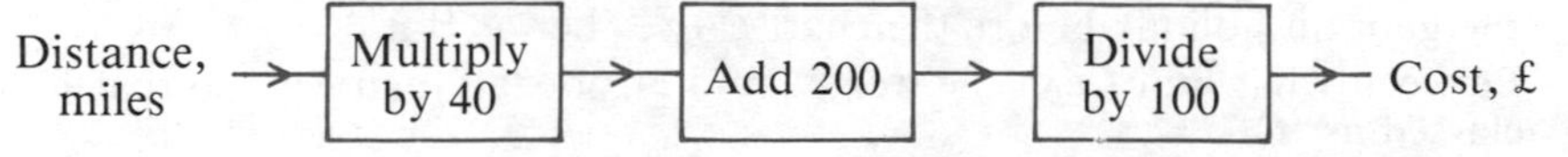

Tables, charts and matrices

15 A man owns two shops, one on Market Street and the other on Kirkgate. For the first four months of the year he monitors the expenditure on services with these results.

Market street		Jan	Feb	Mar	Apr
Electricity	£	750	820	680	650
Water	£	40	32	39	28
Telephone	£	16	20	13	18

Kirkgate		Jan	Feb	Mar	Apr
Electricity	£	620	670	630	520
Water	£	18	24	28	20
Telephone	£	11	9	13	15

Construct a table for his total expenditure on services for both shops together.

16 Matrix A gives the amount of various products in stock at the start of the month in a small shop. During the month there are two equal deliveries of stock, each given by matrix B. Matrix C shows the stock in the shop at the end of the month.

Numbers in stock (columns: Tins, Packets; rows: Beans, Peas, Carrots)

$$A = \begin{pmatrix} 12 & 8 \\ 15 & 24 \\ 21 & 10 \end{pmatrix} \quad B = \begin{pmatrix} 15 & 10 \\ 15 & 30 \\ 15 & 20 \end{pmatrix} \quad C = \begin{pmatrix} 18 & 5 \\ 21 & 32 \\ 14 & 15 \end{pmatrix}$$

Construct the matrix $A + 2B - C$ and explain in words what information it gives you.

Part 2 Rows and columns

Without writing any words, copy each pair of matrices in questions **1**, **2** and **3**, and then calculate the total required.

1 A football club plays several games with the results given in the first matrix. The second matrix gives the points awarded. Find the team's total number of points.

Columns: Away wins, Home wins, Draws, Losses. Points awarded: each away win, each home win, each draw, each loss. Total points.

$$\text{Number of games} \begin{pmatrix} 4 & 6 & 3 & 5 \end{pmatrix} \begin{pmatrix} 3 \\ 2 \\ 1 \\ 0 \end{pmatrix} = \begin{pmatrix} \cdots \end{pmatrix}$$

2 Two firms, Caedmon Flex and Hild Electrics, employ workers on three grades of pay in numbers as in the first matrix, earning basic weekly wages as in the second matrix. Calculate the total weekly wage bills for the two firms.

Columns: Grade A, Grade B, Grade C; rows: Caedmon Flex, Hild Electrics. Weekly wage, £: each Grade A, each Grade B, each Grade C. Total weekly wage bill, £: Caedmon Flex, Hild Electrics.

$$\begin{pmatrix} 8 & 24 & 16 \\ 6 & 28 & 12 \end{pmatrix} \begin{pmatrix} 120 \\ 100 \\ 95 \end{pmatrix} = \begin{pmatrix} \cdots \\ \cdots \end{pmatrix}$$

Tables, charts and matrices

3 Two neighbours both decide to make some fruit cakes using different recipes which list ingredients in 100-gram portions, as shown in the first matrix. The second matrix gives the cost of the ingredients at two local grocers' shops.

Calculate how much the ingredients will cost at both shops for both people.

Number of 100g portions (columns: flour, sugar, fat, fruit); *Cost (pence) of each 100g portion* (columns: First shop, Second shop; rows: flour, sugar, fat, fruit); *Total cost, pence* (columns: First shop, Second shop)

$$\begin{matrix}\text{Mrs Metcalfe}\\ \text{Mr Firth}\end{matrix}\begin{pmatrix} 4 & 3 & 1 & 10 \\ 3 & 2 & 2 & 6 \end{pmatrix}\begin{pmatrix} 3 & 4 \\ 5 & 4 \\ 12 & 15 \\ 20 & 18 \end{pmatrix}\begin{matrix}\text{flour}\\ \text{sugar}\\ \text{fat}\\ \text{fruit}\end{matrix} = \begin{pmatrix} \dots & \dots \\ \dots & \dots \end{pmatrix}\begin{matrix}\text{Mrs Metcalfe}\\ \text{Mr Firth}\end{matrix}$$

4 Multiply these matrices together by combining their rows and columns.

a $\begin{pmatrix} 1 & 3 & 2 \end{pmatrix}\begin{pmatrix} 4 \\ 2 \\ 2 \end{pmatrix} = \begin{pmatrix} \dots \end{pmatrix}$

b $\begin{pmatrix} 1 & 3 & 2 \\ 5 & 0 & 1 \end{pmatrix}\begin{pmatrix} 4 \\ 2 \\ 2 \end{pmatrix} = \begin{pmatrix} \dots \\ \dots \end{pmatrix}$

c $\begin{pmatrix} 1 & 3 & 2 \\ 5 & 0 & 1 \\ 4 & 1 & 0 \end{pmatrix}\begin{pmatrix} 4 \\ 2 \\ 2 \end{pmatrix} = \begin{pmatrix} \dots \\ \dots \\ \dots \end{pmatrix}$

d $\begin{pmatrix} 1 & 3 & 2 \\ 5 & 0 & 1 \\ 4 & 1 & 0 \end{pmatrix}\begin{pmatrix} 4 & 2 \\ 2 & 6 \\ 2 & 0 \end{pmatrix} = \begin{pmatrix} \dots & \dots \\ \dots & \dots \\ \dots & \dots \end{pmatrix}$

5 $P = \begin{pmatrix} 1 & 2 & 3 \\ 4 & 5 & 6 \end{pmatrix}$ $Q = \begin{pmatrix} 0 & 2 \\ 1 & 1 \\ 2 & 0 \end{pmatrix}$ $R = \begin{pmatrix} 2 & 4 \\ 3 & 1 \end{pmatrix}$ $S = \begin{pmatrix} 2 & 1 \\ 0 & 5 \end{pmatrix}$

Multiply these matrices together. If any multiplication is impossible, then say so.

a PQ **b** RS **c** SR **d** QP

e PR **f** QS **g** RP **h** RQ

Tables, charts and matrices

Part 3 Transformations

1 For each matrix, draw and label both axes from -8 to 8,
draw each object shape using the given points,
transform each object and draw its image on the same diagram,
and describe, as fully as possible, the transformation which has taken place.

	Matrix	Object shape
a	$\begin{pmatrix} 2 & 0 \\ 0 & 2 \end{pmatrix}$	(0, 2), (4, 3), (3, 1), (2, 2)
b	$\begin{pmatrix} -1 & 0 \\ 0 & -1 \end{pmatrix}$	(2, 2), (6, 2), (7, 4), (5, 4), (7, 8), (5, 8)
c	$\begin{pmatrix} 0 & 1 \\ 1 & 0 \end{pmatrix}$	(2, 1), (2, 7), (5, 7), (3, 5), (3, 1)
d	$\begin{pmatrix} 0 & -1 \\ -1 & 0 \end{pmatrix}$	(2, 2), (6, 2), (7, 4), (5, 4), (7, 8), (5, 8)
e	$\begin{pmatrix} 0.7 & 0.7 \\ -0.7 & 0.7 \end{pmatrix}$	(0, 0), (0, 6), (4, 8), (4, 6), (2, 6)

2 For each matrix, draw and label both axes from -6 to 6.

a The matrix $\begin{pmatrix} 2 & 0 \\ 0 & -2 \end{pmatrix}$ transforms the object shape (2, 1), (3, 1), (2, 2), (3, 3), (0, 3) onto its image. Draw both object and image on the same diagram and describe the *two* simple transformations which this matrix has performed simultaneously.

b Draw on one diagram the object shape (0, 1), (1, 2), (2, 2), (2, -2) and its image after being transformed by the matrix $\begin{pmatrix} 0 & 3 \\ -3 & 0 \end{pmatrix}$. Describe the *two* simple transformations which have occurred simultaneously.

3 Draw and label both axes from -6 to 6.

a A five-sided shape has corners $A(1, -2)$, $B(2, -2)$, $C(3, -1)$, $D(3, 2)$, $E(2, -1)$. It is transformed by the matrix $\begin{pmatrix} 2 & 0 \\ 0 & 2 \end{pmatrix}$ onto $A'B'C'D'E'$. Draw both the object and the image on the same diagram and describe fully the transformation which has taken place.

b The matrix $\begin{pmatrix} 0 & -\frac{1}{2} \\ \frac{1}{2} & 0 \end{pmatrix}$ transforms the shape $A'B'C'D'E'$ onto a new image $A''B''C''D''E''$. Calculate the position of this new image and draw it on the same diagram.

c Calculate the matrix which will transform $ABCDE$ directly onto $A''B''C''D''E''$ in one step. Describe in words the transformation involved.

Tables, charts and matrices

4 Draw and label both axes from -6 to 6.

a L is the shape with corners (2, 2), (3, 3), (1, 5), (-1, 3), (1, 3) joined in this order. It is mapped onto its image L' by the transformation given by the matrix $S = \begin{pmatrix} 2 & -1 \\ 1 & 0 \end{pmatrix}$. Calculate the coordinates of L.
On one diagram, draw the shape L and its image L'.

b L' is now mapped onto L'' under the transformation given by the matrix $T = \begin{pmatrix} 0 & -1 \\ 1 & -2 \end{pmatrix}$. Calculate the position of L'' and draw it on the same diagram.

c Calculate the single matrix TS which maps L directly onto L'' and describe this single transformation in words.

5 Draw and label both axes from -6 to 6.

a The matrix $\begin{pmatrix} 1 & 0 \\ 0 & -1 \end{pmatrix}$ defines the transformation **M**, and the matrix $\begin{pmatrix} 0 & -1 \\ -1 & 0 \end{pmatrix}$ defines the transformation **N**. An object shape S with vertices (2, 2), (2, 3), (-1, 3), (-3, 5), (-2, 2) maps onto its image S' such that $S' = \mathbf{M}(S)$. Calculate the positions of the vertices of S' and draw both S and S' on one diagram. Describe the transformation **M** in words.

b S' is now transformed onto S'' such that $S'' = \mathbf{N}(S')$. Calculate the positions of the vertices of S'' and draw S'' on the same diagram. Describe the transformation **N** in words.

c Find the single matrix which maps S directly onto S'' in one step, and describe the transformation involved.

6 Find the determinants of these matrices.

a $\begin{pmatrix} 4 & 5 \\ 2 & 3 \end{pmatrix}$ **b** $\begin{pmatrix} 2 & 6 \\ 0 & 1 \end{pmatrix}$ **c** $\begin{pmatrix} 4 & 2 \\ 1 & 3 \end{pmatrix}$ **d** $\begin{pmatrix} 2 & 3 \\ -1 & 2 \end{pmatrix}$

7 An object shape of area 5 cm^2 is transformed by the matrix $\begin{pmatrix} 3 & 6 \\ 2 & 5 \end{pmatrix}$.
Find the area of its image.

8 The matrix $\begin{pmatrix} 2 & 3 \\ -1 & 1 \end{pmatrix}$ transforms an object of area 6 cm^2. What is the area of its image?

Part 4 Inverse matrices

1 a The square (2, 1), (4, 3), (6, 1), (4, -1) is transformed using the matrix $\begin{pmatrix} \frac{1}{2} & 1 \\ 0 & 2 \end{pmatrix}$ onto its image.
Calculate the co-ordinates of the corners of this image and draw both the square and its image on axes labelled from -2 to 6.

b The parallelogram (2, 2), (5, 6), (4, 2), (1, -2) is transformed by the matrix $\begin{pmatrix} 2 & -1 \\ 0 & \frac{1}{2} \end{pmatrix}$ onto its image.
Calculate the position of its image and draw both the parallelogram and its image on axes labelled from -2 to 6.

c What is the relation between the matrices $\begin{pmatrix} \frac{1}{2} & 1 \\ 0 & 2 \end{pmatrix}$ and $\begin{pmatrix} 2 & -1 \\ 0 & \frac{1}{2} \end{pmatrix}$?
Check your answer by multiplying them together.

Tables, charts and matrices

2 The transformation **M** is given by the matrix $\begin{pmatrix} -2 & 4 \\ -2 & 2 \end{pmatrix}$.

The transformation **N** is given by the matrix $\begin{pmatrix} \frac{1}{2} & -1 \\ \frac{1}{2} & -\frac{1}{2} \end{pmatrix}$.

a On axes labelled from -2 to 6, draw the square $S(1, 1), (2, 1), (2, 2), (1, 2)$ and its image **M**(S).

b On similar axes, draw the parallelogram $P(2, 0), (6, 2), (4, 0), (0, -2)$ and its image **N**(S).

c What is the relation between these two transformations? Check your answer by multiplying the two matrices together.

3 Find (i) the determinant
(ii) the inverse, if it exists, of each of these matrices.

Check your answers by multiplying each matrix with its inverse.

a $\begin{pmatrix} 3 & 7 \\ 2 & 5 \end{pmatrix}$ **b** $\begin{pmatrix} 4 & 2 \\ 5 & 3 \end{pmatrix}$ **c** $\begin{pmatrix} 5 & 10 \\ -2 & -3 \end{pmatrix}$ **d** $\begin{pmatrix} 6 & 3 \\ 4 & 2 \end{pmatrix}$

4 The simultaneous equations $\begin{aligned} 4x + y &= 7 \\ 7x + 2y &= 13 \end{aligned}$ can be written as the matrix equation $\begin{pmatrix} 4 & 1 \\ 7 & 2 \end{pmatrix}\begin{pmatrix} x \\ y \end{pmatrix} = \begin{pmatrix} 7 \\ 13 \end{pmatrix}$.

Find the inverse of $\begin{pmatrix} 4 & 1 \\ 7 & 2 \end{pmatrix}$ and hence solve the equations for x and y.

5 Use the same method to solve these simultaneous equations.

a $\begin{aligned} 7x + 5y &= 12 \\ 4x + 3y &= 7 \end{aligned}$ **b** $\begin{aligned} 5x + 3y &= 10 \\ 3x + 2y &= 6 \end{aligned}$ **c** $\begin{aligned} 7x + 6y &= 10 \\ 2x + 2y &= 3 \end{aligned}$

d $\begin{aligned} x - 3y &= 1 \\ 3x - 4y &= 8 \end{aligned}$ **e** $\begin{aligned} 2x - 3y &= 4 \\ x + y &= 7 \end{aligned}$ **f** $\begin{aligned} 3x - 4y &= 1 \\ 2x - 2y &= 1 \end{aligned}$

6 The point $P(x, y)$ maps onto its image $P'(3, 2)$ under a transformation given by the matrix

$M = \begin{pmatrix} 3 & -5 \\ -1 & 2 \end{pmatrix}$.

In short, $\begin{pmatrix} 3 & -5 \\ -1 & 2 \end{pmatrix}\begin{pmatrix} x \\ y \end{pmatrix} = \begin{pmatrix} 3 \\ 2 \end{pmatrix}$.

Use the method of **4** above, to find the values of x and y.

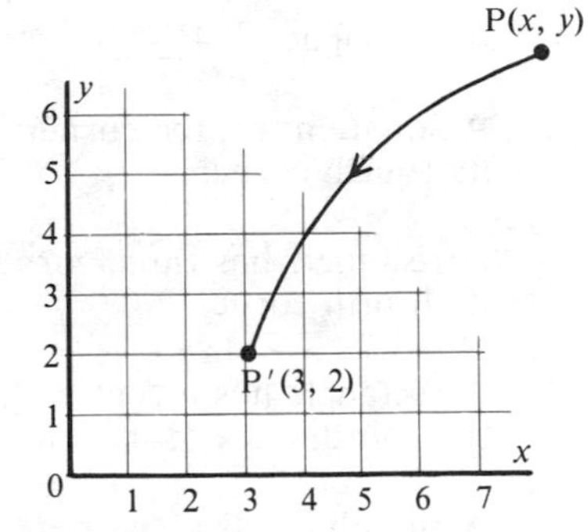

7 The point $Q(x, y) \rightarrow Q'(2, 2)$ under the matrix $\begin{pmatrix} 3 & -5 \\ -2 & 4 \end{pmatrix}$.

Find the values of x and y.

8 The point $R(x, y) \rightarrow R'(3, 6)$ under the matrix $\begin{pmatrix} -1 & 4 \\ -2 & 5 \end{pmatrix}$.

Find the values of x and y.

Coordinates and graphs

Part 1 Points and axes

1 This map shows a group of four islands connected by three ferries F.

Write the coordinates of

a Greypike **b** Carnbay

c Thorcrick **d** Crowhead.

Which places on the map have these coordinates?

e $(2, 1)$ **f** $(-2, -3)$

g $(-3, -1)$ **h** $(-3\frac{1}{2}, 2\frac{1}{2})$

Give the coordinates of the two terminals of the ferry between

i Crawlin and Tusker

j Muckle and Ramsay

k Ramsay and Crawlin.

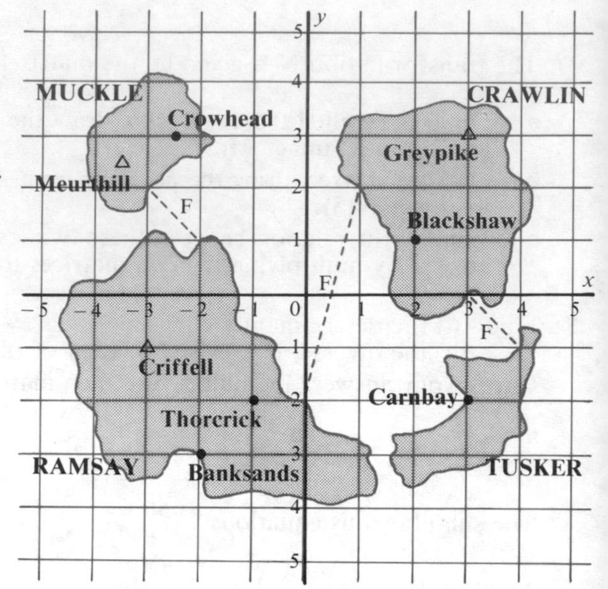

2 Draw and label both axes from -5 to 5.

Draw part of the Scottish coast by joining these points in order.

$(-5, 4)$, $(-4, 4)$, $(2, 4\frac{1}{2})$, $(2, 3)$, $(-2, -1)$, $(-3, -1)$, $(-1, -1\frac{1}{2})$, $(-2\frac{1}{2}, -3\frac{1}{2})$, $(\frac{1}{2}, -2\frac{1}{2})$, $(5, -3)$

Label the positions of these places on your map.

a Wick $(2, 3)$ **b** Elgin $(1, -3)$

c Nairn $(-1, -3)$ **d** Inverness $(-3, -4)$

e Bonar Bridge $(-4, -1)$ **f** Lairg $(-4\frac{1}{2}, 0)$

g Tongue $(-4\frac{1}{2}, 3\frac{1}{2})$

3 A square has three corners $(1, 2)$, $(-2, 1)$, $(-1, -2)$. Find the coordinates of its fourth corner.

4 A rectangle has three corners $(1, 3)$, $(-1, 2)$, $(1, -2)$. Find the coordinates of its fourth corner.

5 A rectangle has a centre $(\frac{1}{2}, 1)$ and two corners $(-1, -1)$ and $(-2, 1)$. Find the coordinates of its other two corners.

6 A rhombus has a centre $(1, 1)$ and two corners $(-1, 2)$ and $(-1, -3)$. Find the coordinates of its two other corners.

7 A square has a centre $(2, 2)$ and one corner $(1, 4)$. Find its other three corners.

8 A parallelogram has three corners $(-1, 0)$, $(1, 2)$, $(2, -1)$. Find *three* different possible positions of its fourth corner.

9 Find any three points which are equidistant from the two points $P(-1, 2)$ and $Q(3, 0)$. Describe the locus of a point which moves so that it is always equidistant from points P and Q.

Coordinates and graphs

10 This map shows a stretch of coastline. Contours on the map are at 25-metre intervals.

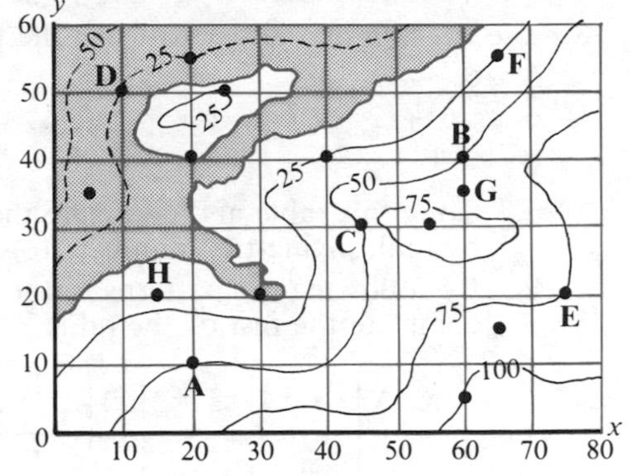

a Give the coordinates of the lettered points A to H as ordered triples.

b Copy and complete these ordered triples for the unlettered points.

(60, 5, . . .)
(40, . . ., 25)
(20, 40, . . .)
(. . ., 50, 25)
(55, . . ., 80)
(20, 55, . . .)
(5, 35, . . .)
(. . ., 15, 85)

Part 2 Directed numbers

1

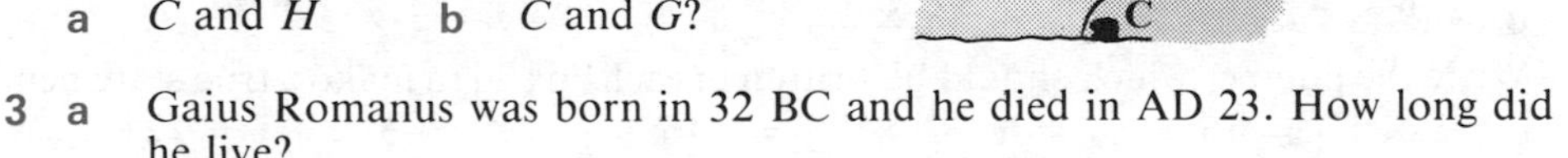

a This thermometer shows the temperature at 8 p.m. one evening in winter. By midnight, it has fallen 5 degrees. What is the temperature at midnight?

b Newcastle is 6 degrees colder than Bristol. If it is −4°C in Newcastle, what is the temperature in Bristol?

c On Christmas Day it is −3°C in Manchester and −10°C in Perth. Which city is colder and by how many degrees is it colder?

2 A hill top H stands 250 metres above sea-level and a coastguard station G is 50 metres below the top of the hill.

An underwater cave C lies 60 metres below sea-level.

What is the difference in heights between

a C and H **b** C and G?

3 a Gaius Romanus was born in 32 BC and he died in AD 23. How long did he live?

b His daughter, Livia, was born in AD 5 and she lived for 64 years. In what year did she die?

4 The balance in Ruth Auty's bank account is shown in this table. A negative sign indicates that she is 'overdrawn'.

Date	Balance, £
May 4	46·30
May 7	5·00
May 12	−7·00
May 15	−5·60
May 21	12·60

a How much was spent from this account between
(i) May 4 and May 7
(ii) May 7 and May 12?

b How much was put into this account between
(i) May 12 and May 15
(ii) May 15 and May 21?

Coordinates and graphs

5 Write the values on this number ladder to which the lettered arrows *a* to *j* are pointing.

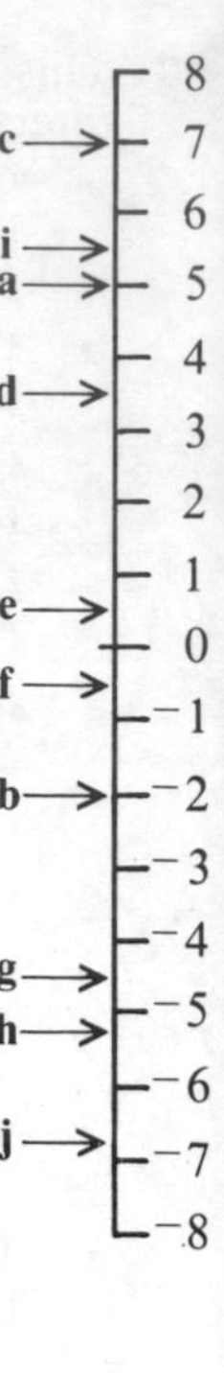

6 Use the number ladder to work the following.

a $6 - 4$ **b** $3 - 5$ **c** $^{-}2 + 7$
d $^{-}6 + 2$ **e** $^{-}1 - 4$ **f** $3 + 2 - 4$
g $2 + 1 - 5$ **h** $3 - 7 + 4$ **i** $6 - 7 - 2$

7 **a** Copy this table and complete the top left-hand quarter by multiplying two numbers together.
b By following the patterns in the rows and columns, complete the rest of the table.

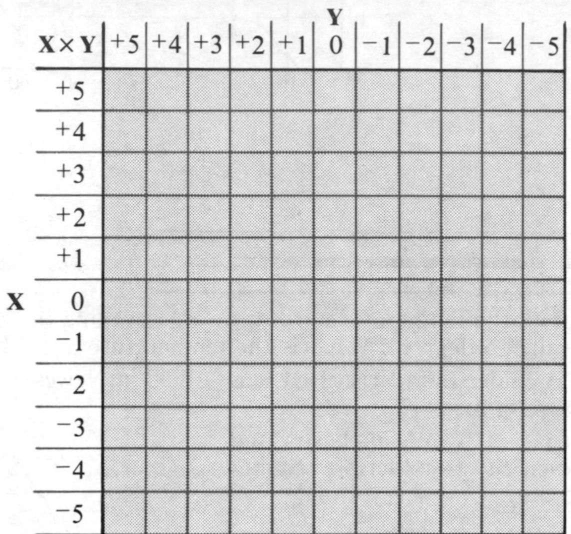

X×Y	+5	+4	+3	+2	+1	0	−1	−2	−3	−4	−5
+5											
+4											
+3											
+2											
+1											
X 0											
−1											
−2											
−3											
−4											
−5											

Use your results to write the answers to these multiplications.

c $4 \times {}^{-}3$ **d** $^{-}4 \times {}^{-}3$ **e** $2 \times {}^{-}3$
f 3×5 **g** $^{-}5 \times 2$ **h** $^{-}3 \times {}^{-}1$

8 Write the answers to these multiplications.

a $3 \times {}^{-}7$ **b** $^{-}6 \times {}^{-}4$ **c** $^{-}5 \times 6$
d $^{-}9 \times {}^{-}2$ **e** 8×7 **f** $3 \times {}^{-}2$

Write the number which should be written in each box □ to make a true statement.

g $2 \times \square = 16$ **h** $2 \times \square = {}^{-}16$ **i** $^{-}3 \times \square = 12$
j $\square \times 4 = {}^{-}20$ **k** $\square \times {}^{-}7 = 21$ **l** $\square \times 5 = {}^{-}15$

9 Use either a number ladder or the multiplication table to write the answers to these.

a $3 + 7$ **b** $3 - 7$ **c** $^{-}2 \times {}^{-}4$ **d** $3 \times {}^{-}6$
e $^{-}5 + 3$ **f** $^{-}8 \times 3$ **g** $7 - 9$ **h** $7 \times {}^{-}9$
i $^{-}8 + 13$ **j** $^{-}2 \times {}^{-}8$ **k** $^{-}2 - 8$ **l** 5×7

10 Find which number should be placed in each of these boxes □ to make true statements.

a $3 \times \square = 12$ **b** $5 \times \square = {}^{-}20$ **c** $\square + 3 = 7$
d $\square - 2 = 4$ **e** $\square + 5 = 1$ **f** $^{-}4 \times \square = {}^{-}20$
g $7 \times \square = 35$ **h** $\square - 3 = {}^{-}1$ **i** $\square + 2 = {}^{-}4$
j $\frac{\square}{3} = {}^{-}2$ **k** $\frac{\square}{^{-}2} = 8$ **l** $\frac{\square}{^{-}5} = {}^{-}10$

Coordinates and graphs

Part 3 Relations

1 This diagram shows the relation "*. . . is the mother of . . .*".

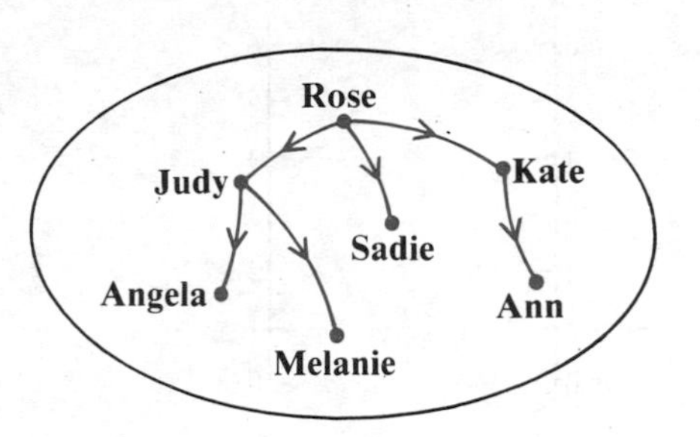

- **a** Who is Judy's mother?
- **b** Who are Rose's other daughters?
- **c** How many granddaughters has Rose?
- **d** Who are Sadie's sisters?
- **e** What relation is Ann to Judy?
- **f** What relation is Ann to Melanie?

2 Copy these arrow diagrams and draw arrows for the given relations.

a

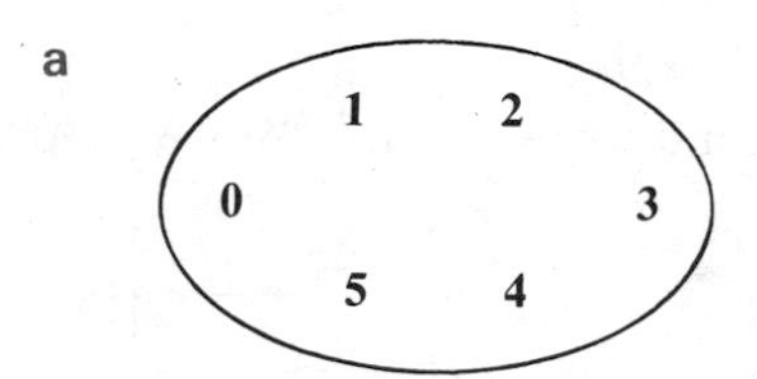

"*. . . is more than the double of . . .*"

b

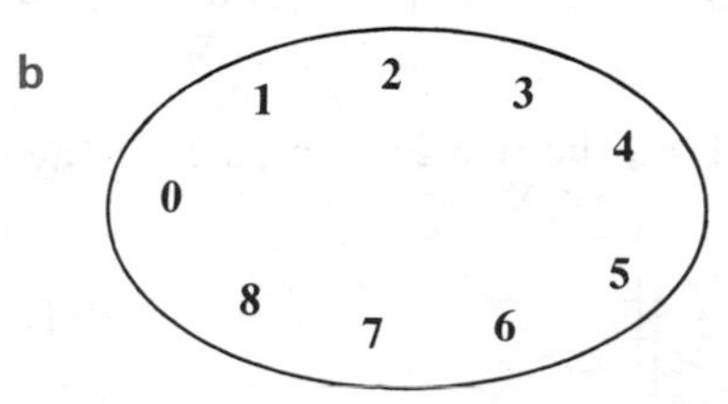

"*. . . gives a total of 8 with . . .*"

3 Copy each of these diagrams and map set X onto set Y using the relation given.

a

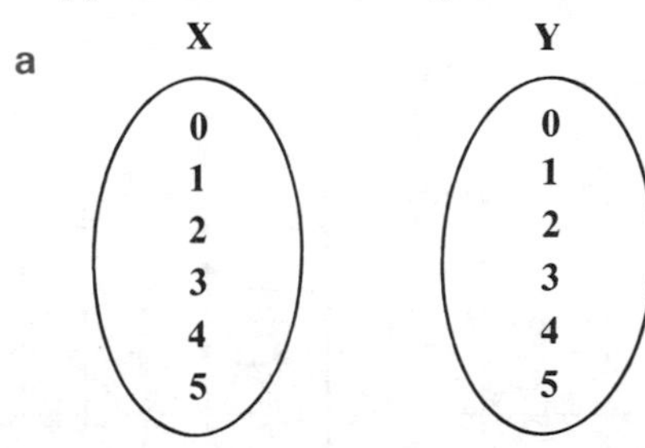

"*. . . is 2 more than . . .*"

b

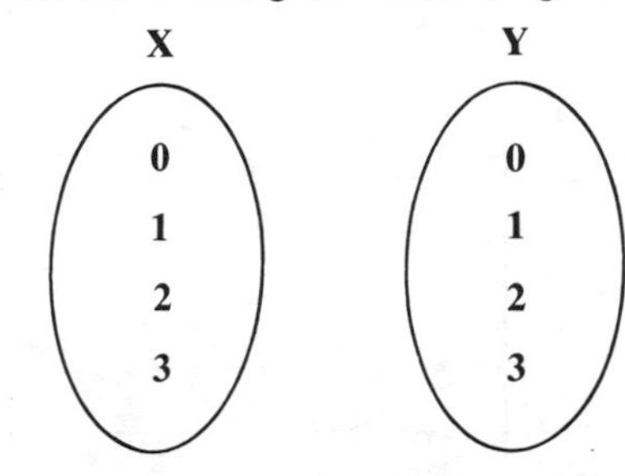

"*. . . is less than half of . . .*"

4 Two department stores are having a sale. These two diagrams map the sale prices of goods onto their original prices.
Find the value of the letters a to h in the diagrams.

a

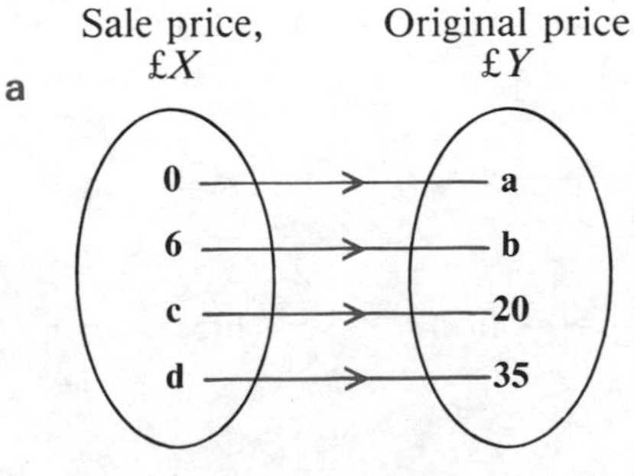

"*. . . is 3 less than . . .*"

b

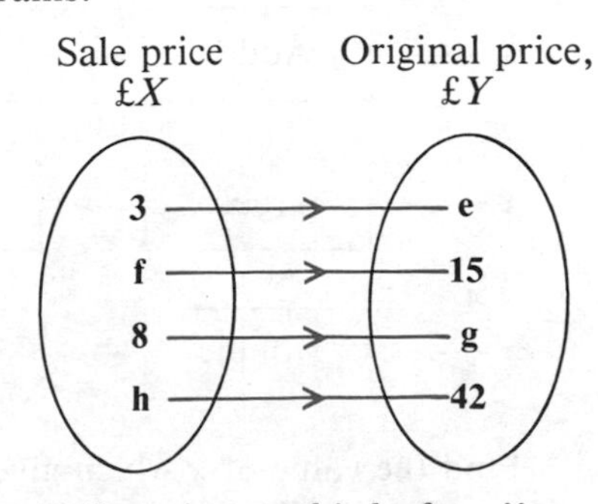

"*. . . is one third of . . .*"

Coordinates and graphs

5 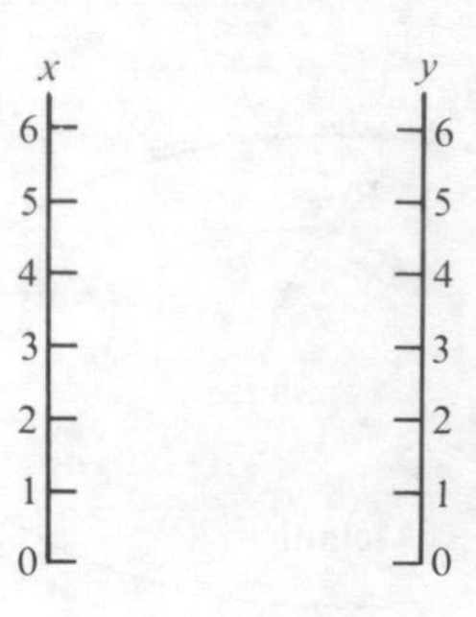

Copy and complete these two diagrams, one with arrows and the other with points, to illustrate each of these mappings.

a $x \to x + 1$ **b** $x \to x - 1$ **c** $x \to 2x$ **d** $x \to 6 - x$

6 Find the relation for each of these mappings, and write it in the form $x \to$ ▒

a 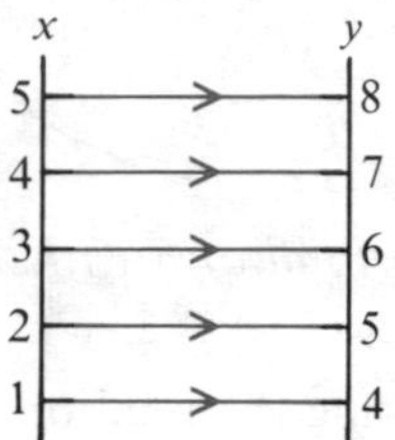

b 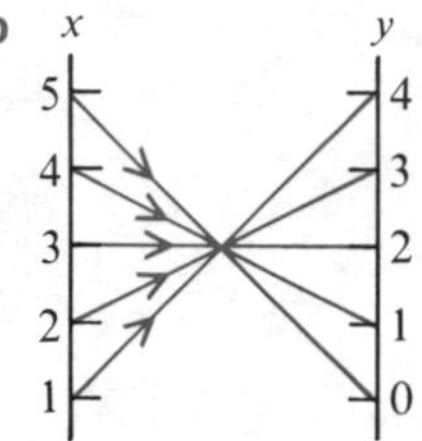

c 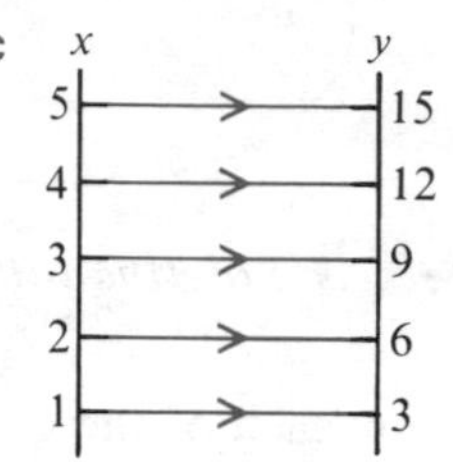

d 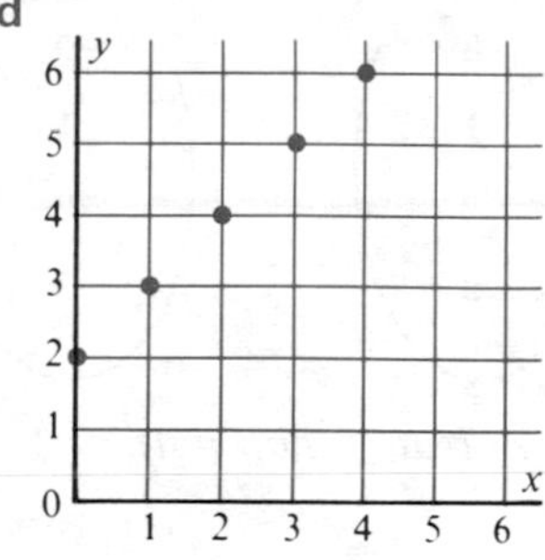

e 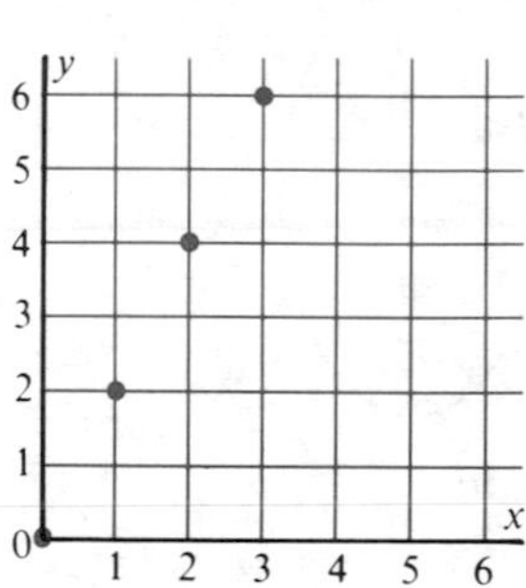

f 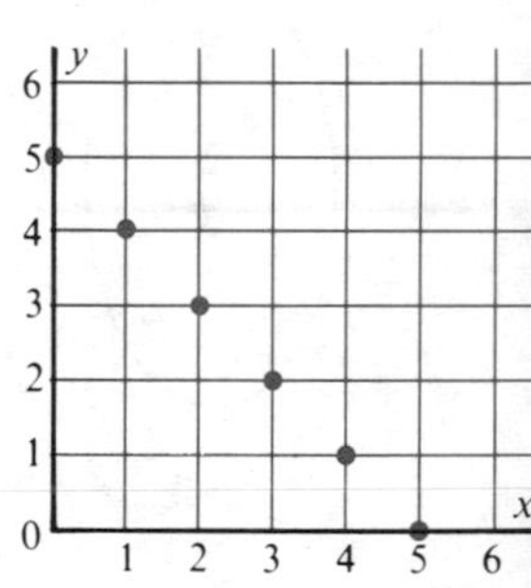

7 Use the flow diagrams to calculate the required values of these functions.

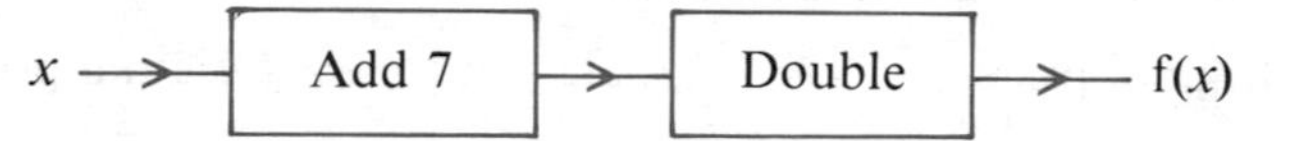

a f(3) **b** f(5)

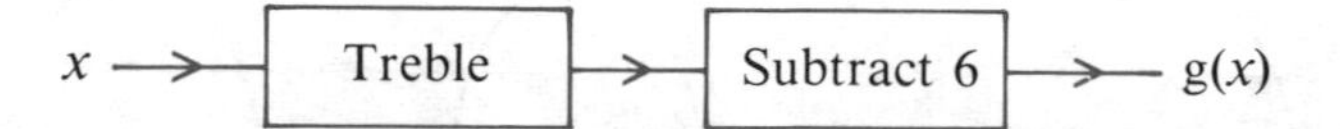

c g(7) **d** g(2)

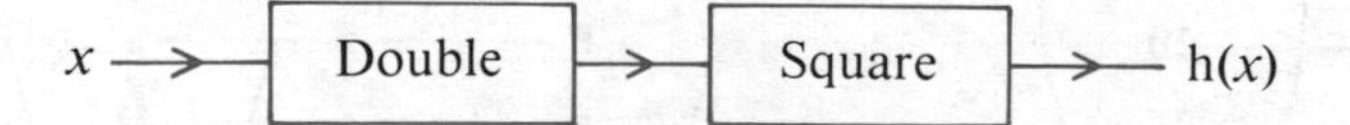

e h(5) **f** h(1)

Find the value of x which gives

g $f(x) = 30$ **h** $g(x) = 21$ **i** $h(x) = 9$.

Coordinates and graphs

8 Mr and Mrs Peace take their family on a holiday to Germany by car. Distances on their map are given in kilometres, but the car's odometer measures distance in miles. This flow diagram converts kilometres to miles.

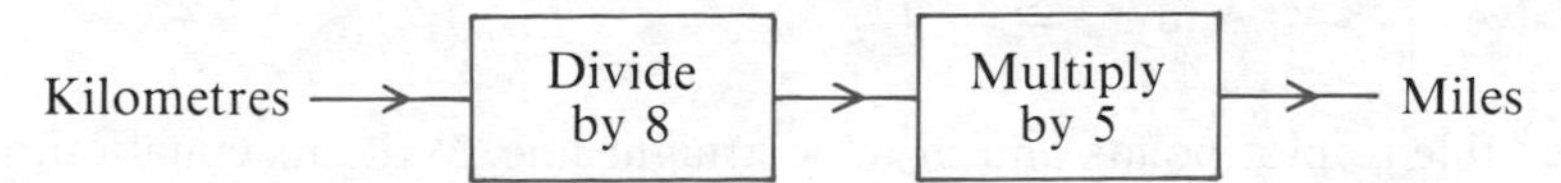

a It is 176 km from Rabensdorf to Hagenbach. How far is this in miles?

b The Peace family make a long journey of 243 km on the autobahn. How far is this to the nearest mile?

c During all their time in Germany, they travel 1240 miles. How far is this in kilometres?

9 A car has to stop as fast as possible on a dry road. If it has good brakes, then the distance required to come to a stop can be found as follows:

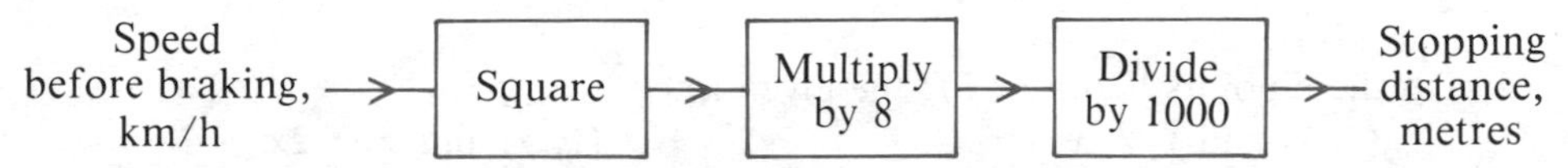

a What is the stopping distance for a car travelling at 20 km/h?

b A car is travelling on a straight road at 95 km/h, use your calculator to find its stopping distance (to the nearest 10 metres).

c To avoid an accident, a car stops as quickly as possible in a distance of 115 metres. At what speed was it initially travelling?

10 Miss Elsworth works for a firm which pays a mileage allowance when she uses her own car on the firm's business. This flow diagram shows how to calculate her allowance, £A.

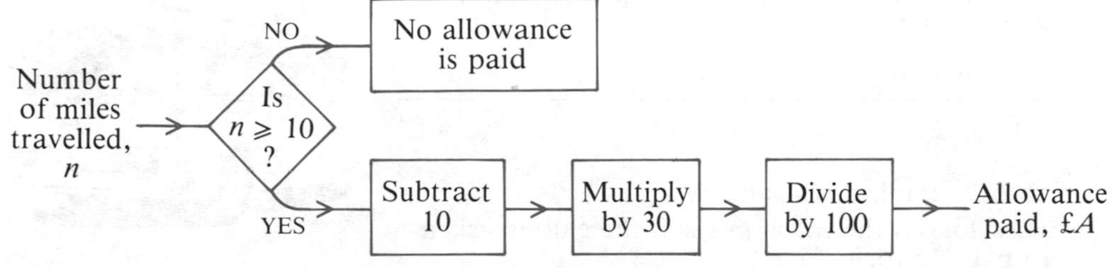

a What allowance is she paid when she travels (i) 8 miles (ii) 18 miles?

b How far does she have to travel to earn an allowance of £6?

c How much is she paid for each mile that she can claim for?

Part 4 Straight lines

1 a Use this flow diagram to complete these ordered pairs by finding their y values.

x → Double → Subtract 1 → y

(0, . . .), (1, . . .), (2, . . .), (3, . . .), (4, . . .), (5, . . .)

b Draw axes as shown, plot the ordered pairs and draw a straight line through them. Use the flow diagram to write its equation.

y: 10, 0, −2; x: 5

Coordinates and graphs

2 a Copy and complete this table by using this flow diagram.

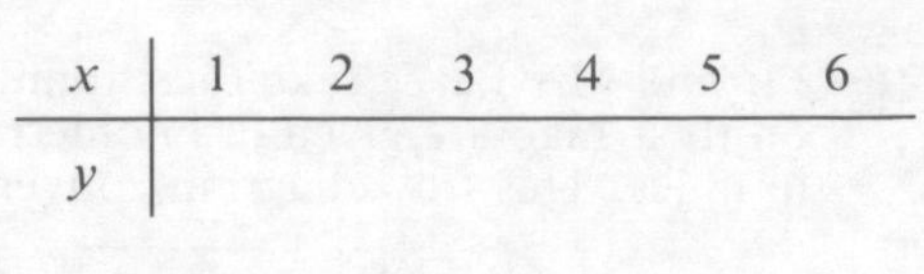

x	1	2	3	4	5	6
y						

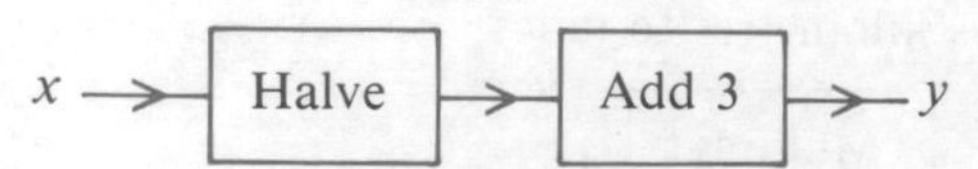

b Use the table to plot points and draw a straight line. Write its equation.

3 a Copy and complete this table for the equation $y = 2x - 1$.

x	-1	0	1	2	3	4
y						

b Draw the graph of $y = 2x - 1$ labelling the x-axis from -1 to 4 and the y-axis from -4 to 8.

c Use the same table to draw the graph of $y = x + 2$ on the same diagram. Write the point of intersection of the two lines.

4 Do these points lie *on* or *off* the given line?

a (4, 1) and $y = x - 3$ **b** (1, 5) and $y = 2x + 3$

c (2, -4) and $y = 6 - x$ **d** (10, -1) and $x + y = 9$

5 On which of these lines does the point (3, -1) lie?

a $y = x - 4$ **b** $x + y = 4$ **c** $y = 2x - 5$

d $x + 2y = 5$ **e** $y = 2 - x$ **f** $2x - y = 7$

6 A youth club hires a coach for which the cost of the hire £y depends on the number of hours x the driver is away from the depot.
Copy and complete this table using the equation $y = 2x + 5$.

Number of hours, x	0	1	2	3	4	5
Cost, £y						

Draw a graph of y against x.
How long is the coach in use if the club is asked to pay £12 for the hire?

7

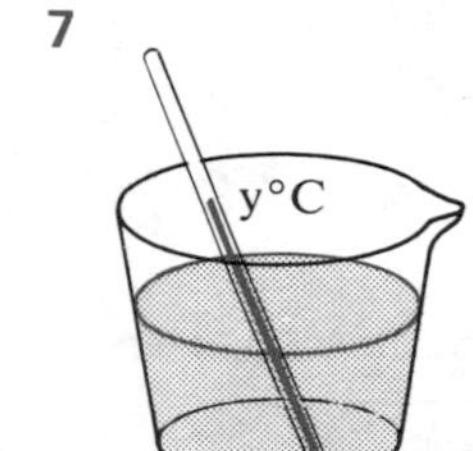

In a science experiment, the temperature y°C of a beaker of water is gradually reduced from room temperature over a period of t minutes, as given by the equation $y = 20 - \frac{1}{2}t$.
Copy and complete this table and draw the graph of y against t.

Time, t min	0	10	20	30	40	50
Temperature, y°C						

a How many minutes does it take for the temperature of the water to fall to 5°C?

b After how long does ice start to form?

c What is room temperature?

Coordinates and graphs

8 A beaker, partly filled with a liquid, is placed under a dripping pipette so that after a time of t minutes the volume $V\ \text{cm}^3$ in the beaker is given by $V = 10 + 2t$.

Copy and complete this table and draw the graph of V against t.

t	0	2	4	6	8	10
V						

a What is the volume in the beaker after 5 minutes?
b How long did it take for the volume to reach $24\ \text{cm}^3$?
c How much liquid was in the beaker at the start?

9 A length L cm of a spring is increased by adding masses of M kg to one end so that $L = 20 + \frac{1}{5}M$.

Copy and complete this table and draw the graph of L against M.

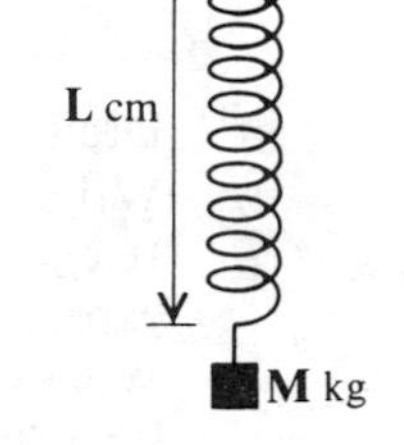

Mass, M kg	0	10	20	30	40	50
Length, L cm						

a What length is the spring when 35 kg are hanging on the end?
b What mass is needed for the length to be 23 cm?
c What is the *unstretched* length of the spring?

10 For each of the straight lines A to F, find
a the gradient
b the y-intercept
c the equation.

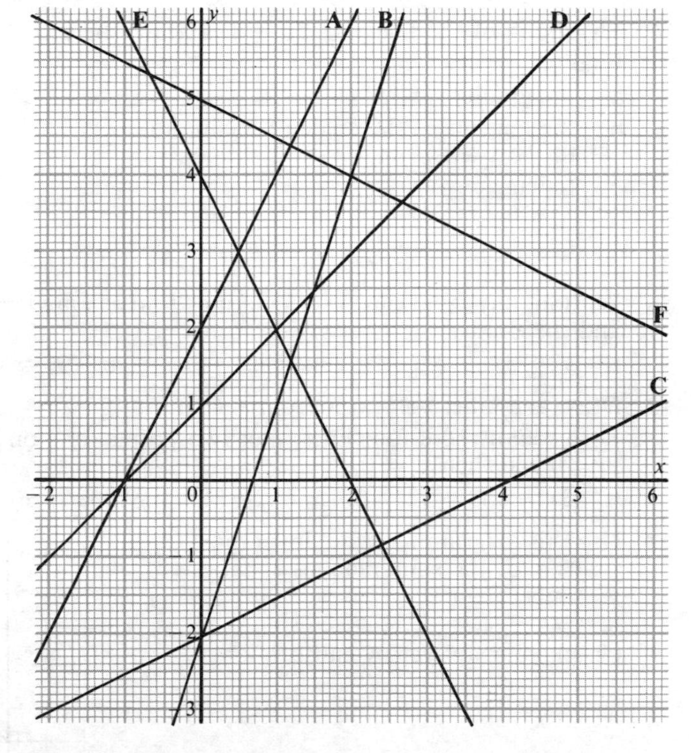

11 Find the gradient, y-intercept and hence the equation of the straight lines through these pairs of points. Label both your axes from -2 to 6.

a $(-1, 0)$ and $(2, 6)$
b $(-2, 1)$ and $(2, 5)$
c $(-2, 5)$ and $(4, 2)$
d $(-1, 1)$ and $(2, -2)$

Coordinates and graphs

12 The electric current, y amps, which flows through a circuit increases gradually over a period of time as shown in this table.

Time, t min	0	1	2	3	4	5
Current, y amps	5	8	11	14	17	20

a Draw the graph of y against t and find its gradient and y-intercept.
b Write down the equation relating y and t.
c What would you expect the current to be when $t = 10$?

13 During a school trip to Germany, pupils have to change their money into German marks on several occasions.
A record is kept of the money exchanged and it is shown in this table.

English pounds, £x	5	8	10	12	15
German marks, y DM	15	27	35	43	55

a Draw a graph of y against x and find its gradient and y-intercept.
b Write down the equation relating y and x.
c If you changed £28 into German marks, how many marks would you expect to receive?
d What does the bank do to explain why your graph does not pass through the origin (0, 0)?

14 The graph of the line $y = 2x - 1$ is drawn here.
Find

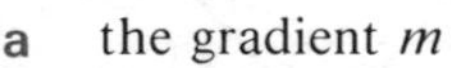

a the gradient m
b the tangent of the angle of inclination α to the x-axis.
c the angle α.

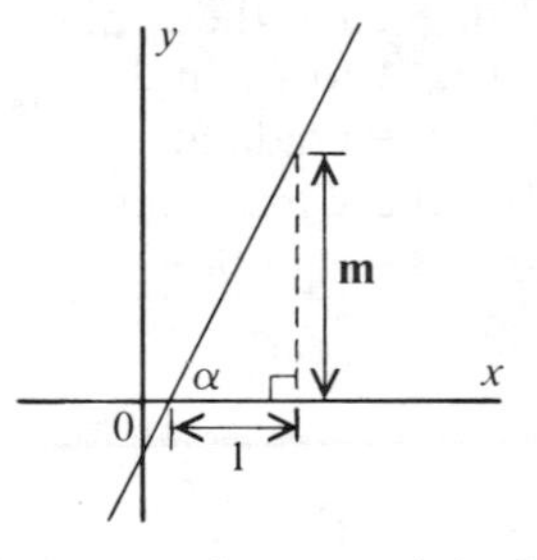

15 Without any diagrams or calculation, find the angle between these straight lines and the x-axis.

a $y = 3x - 1$ **b** $y = 4x + 3$ **c** $y = \frac{1}{2}x - 2$ **d** $y = x + 4$

16 Draw and label both axes from -6 to 6.
For each of these four lines, plot two points, one on each axis, by taking $x = 0$ and then $y = 0$.

a $3x + 2y = 12$ **b** $x + 2y = 4$ **c** $5x + 6y = 30$ **d** $3x - 5y = 15$

17 Use this diagram to write down the solution of the simultaneous equations
$y = x + 1$
$y = 2x - 1$.
Check your answer by substitution.

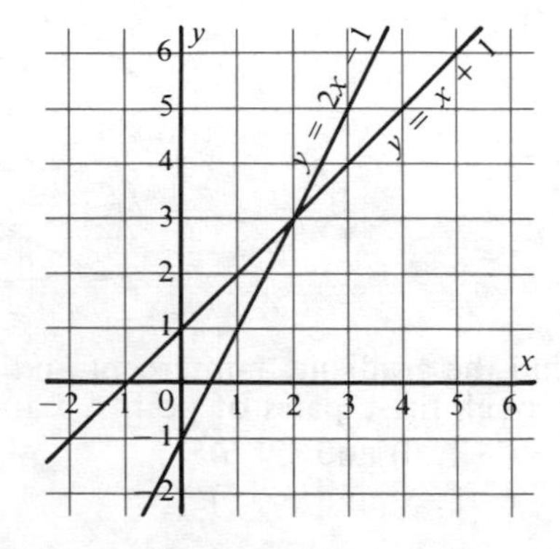

Coordinates and graphs

18 a Copy and complete this table and draw the graphs of the straight lines given by the equations $y = x + 2$ and $y = 2x - 1$ on one diagram.

x	-1	0	1	2	3
$y = x + 2$					
$y = 2x - 1$					

b Solve these simultaneous equations $\begin{matrix} y = x + 2 \\ y = 2x - 1 \end{matrix}$ by using
(i) your diagram (ii) an algebraic method.

19 On axes labelled from -1 to 8, find how many solutions there are for each of these pairs of simultaneous equations.

a $y = 2x - 1$, $y = 2x + 1$ **b** $x + 2y = 6$, $2x + 4y = 12$

20 Two hire companies, Hyravan and Flexicar, charge different rates depending on the number of days n for which you need a van. The costs £C are worked out using these equations.

Hyravan $C = 10n + 20$
Flexicar $C = 15n + 5$.

a Draw two graphs on the same axes to show the costs for up to one week of hire.

b Which firm would you choose if you had to hire a van
(i) for 6 days (ii) for 2 days? Give reasons for your answers.

Part 5 Parabolas

1 Copy each table and use the flow diagram to complete it.

Use your tables of values with axes as shown to draw three parabolas on one diagram.

20 y; x; -5; 0; 10; -5

a $y = x^2$

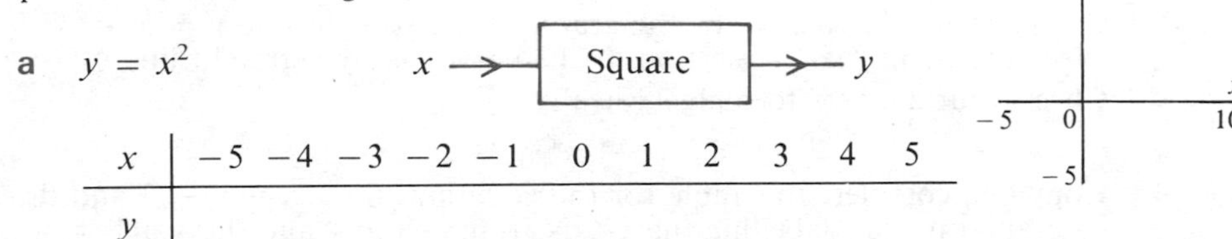

x	-5	-4	-3	-2	-1	0	1	2	3	4	5
y											

b $y = x^2 - 4$

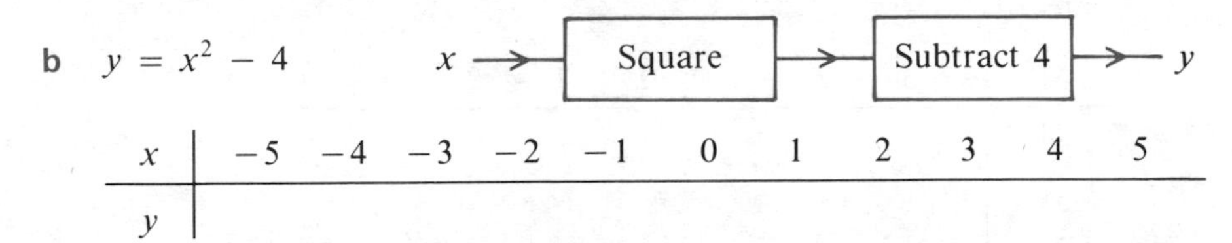

x	-5	-4	-3	-2	-1	0	1	2	3	4	5
y											

c $y = (x - 4)^2$

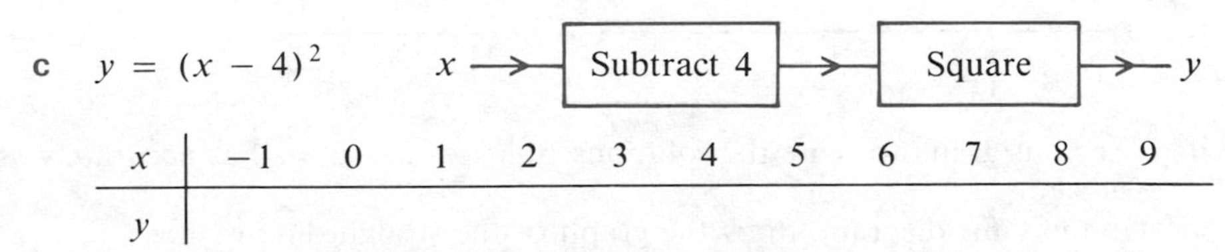

x	-1	0	1	2	3	4	5	6	7	8	9
y											

d Describe the transformation which maps
(i) $y = x^2 - 4$ onto $y = x^2$ (ii) $y = x^2$ onto $y = (x - 4)^2$.

Coordinates and graphs

2 a Copy and complete this table for $y = x^2 - 4x + 3$ and draw its graph on axes with the x-axis labelled from -1 to 5 and the y-axis from -2 to 10.

x	-1	0	1	2	3	4	5
x^2							
$-4x$							
$+3$							
y							

b Use the graph to write down the solutions of these equations.
(i) $x^2 - 4x + 3 = 0$ (ii) $x^2 - 4x + 3 = 3$ (iii) $x^2 - 4x + 3 = -1$

c On the same diagram, draw the graph of the straight line $y = x - 1$. Write down the points of intersection of the curve and the straight line.

d Check your answers to **c** algebraically.

3 a Copy and complete this table for the equation $y = x^2 - 2x - 3$ and draw its graph on axes, labelling the x-axis from -2 to 4 and the y-axis from -5 to 6.

x	-2	-1	0	1	2	3	4
x^2							
$-2x$							
-3							
y							

b Write the solutions of the equation $x^2 - 2x - 3 = 0$ from your graph.

c On the same diagram, draw the graph of the straight line $y = x - 3$. Write the points of intersection of the curve and the straight line.

d Check your answers to **c** algebraically.

4 a Copy and complete this table for the equation $y = x^2 - x - 3$ and draw its graph on axes, labelling the x-axis from -3 to 4 and the y-axis from -4 to 10.

x	-3	-2	-1	0	1	2	3	4
x^2								
$-x$								
-3								
y								

b Use your graph to write the solutions of $x^2 - x - 3 = 0$ as accurately as you can.

c On the same diagram, draw the graph of the straight line $y = x$. Write the solutions of the equation $x^2 - x - 3 = x$ from your graph.

d Check your answers to **c** algebraically.

Coordinates and graphs

5 **a** Draw and label the x-axis from -2 to 4 and the y-axis from -5 to 5. Draw the graph of $y = 3 + 2x - x^2$ on these axes by copying and completing this table.

x	-2	-1	0	1	2	3	4
3							
$+2x$							
$-x^2$							
y							

b On the same diagram, draw the graph of the line $y = x + 2$. For what range of x-values is $3 + 2x - x^2 > x + 2$?

6 A rocket is fired from a cliff top to take a life-line to a boat in distress out at sea. The height, y metres, of the rocket above the sea is given by $y = 48 - 2x - x^2$, where x is the horizontal distance (in tens of metres) of the rocket from the foot of the cliff.

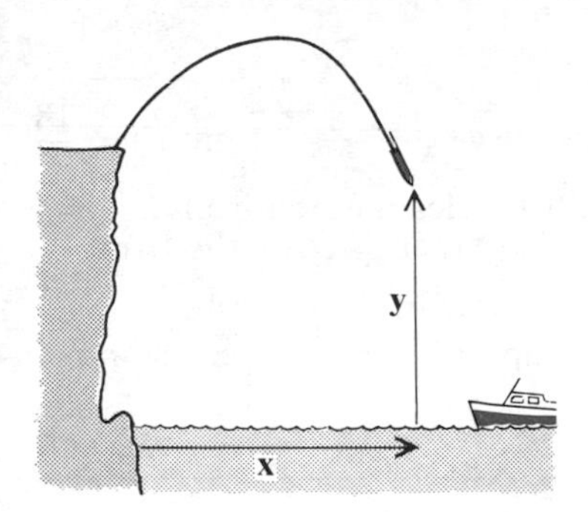

a Copy and complete this table and draw the path of the rocket.

x	0	1	2	3	4	5	6
48							
$-2x$							
$-x^2$							
y							

b How high is the cliff top above the sea?

c How far out from the foot of the cliff is the boat?

7 A gun is fired from the top T of a hill 80 metres high so that the shot strikes the hill further down the slope. The height y of the shot above the base of the hill is given by $y = 8 + 4x - x^2$, where x is the horizontal distance travelled and x and y are both measured in tens of metres.

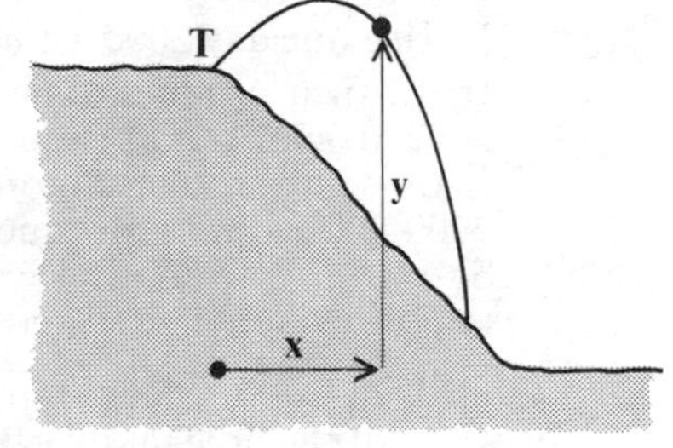

a Copy and complete these tables, and use them to draw the path of the shot and the slope of the hill.

For the shot, $y = 8 + 4x - x^2$

x	0	1	2	3	4	5
8						
$+4x$						
$-x^2$						
y						

For the slope, $x + y = 8$

x	0	1	2	3	4	5	6	7	8
y									

b What is the greatest height above T which is reached by the shot?

c What is the vertical distance below T at which the shot strikes the slope?

Coordinates and graphs

Part 6 Other curves

1 For each equation (i) copy and complete the table
(ii) draw axes with scales of your choice
(iii) plot points and draw a curve.

a $y = \frac{18}{x}$

x	-9	-6	-3	-2	-1	0	1	2	3	6	9
y											

b $y = \frac{18}{x - 4}$

x	-5	-2	1	2	3	4	5	6	7	10	13
$x - 4$											
y											

c $y = \frac{18}{x} - 4$

x	-9	-6	-3	-2	-1	0	1	2	3	6	9
$\frac{18}{x}$											
y											

d Describe the transformation which maps
(i) the curve $y = \frac{18}{x}$ onto $y = \frac{18}{x - 4}$ (ii) the curve $y = \frac{18}{x} - 4$ onto $y = \frac{18}{x}$.

2 A firm has a machine which manufactures certain articles automatically. If the machine produces x thousand articles each day, then it costs the firm y thousands of pounds, where $y = \frac{12}{x}$ provided $1 \leqslant x \leqslant 6$.

a Copy and complete this table and draw the graph of $y = \frac{12}{x}$ over the range $1 \leqslant x \leqslant 6$.

No. of articles produced (thousands) x	1	2	3	4	5	6
Production costs (£ thousands) y						

b How much does it cost to produce 1500 articles per day?

c During one day 2000 articles are made. What is the average cost of each of them?

d If the firm produces fewer than 1000 articles, then the cost is the same as if 1000 articles were produced.
If the firm produces more than 6000 articles, then the cost is the same as if 6000 articles were produced.
Show this information on your diagram by extending the graph into the regions where $x < 1$ and $x > 6$.

3 Water flows out of a large tap, and when fully open the width of the hole in the valve is 6 cm. The width, x cm, of this hole determines the water pressure, y units, such that $y = \frac{24}{x + 2}$.

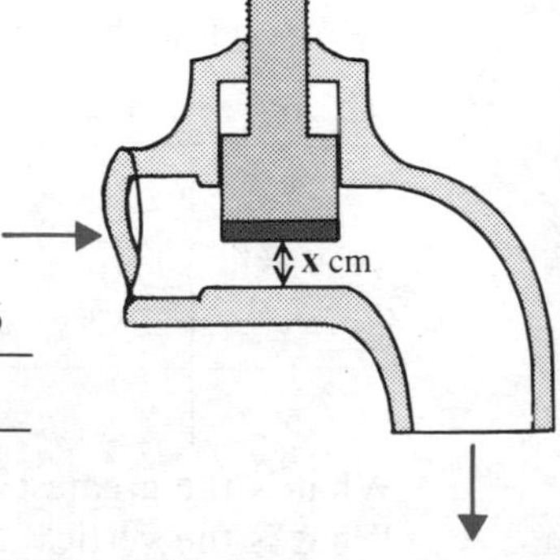

a Copy and complete this table and draw the graph of $y = \frac{24}{x + 2}$.

Width, x cm	0	1	2	3	4	5	6
$x + 2$							
Pressure, y units							

What is the water pressure when the valve is

b fully open **c** fully closed **d** half open?

Coordinates and graphs

4 For each equation (i) copy and complete the table
(ii) draw axes, labelled as shown
(iii) plot points and draw the curve.

a $y = x^3 - 4x$

x	-3	-2	-1	0	1	2	3
x^3							
$-4x$							
y							

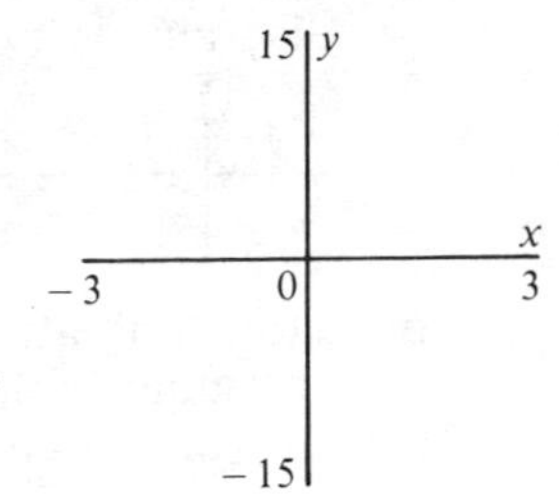

b $y = x^3 - 5x + 2$

x	-3	-2	-1	0	1	2	3
x^3							
$-5x$							
$+2$							
y							

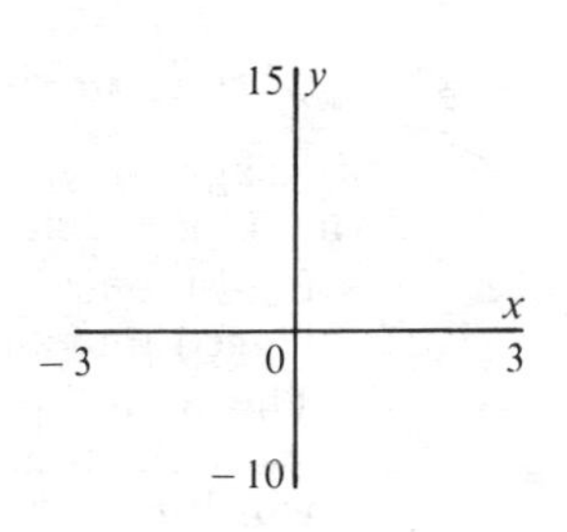

c $y = x^3 - 3x^2 + 4$

x	-1	0	1	2	3	4
x^3						
$-3x^2$						
$+4$						
y						

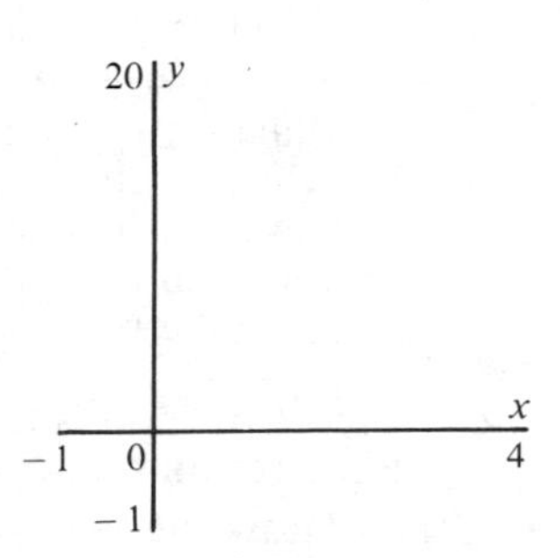

5 In a science experiment, liquid in a flask has a temperature of 10°C. Its temperature, θ°C, is then controlled so that it changes according to the equation $\theta = t^3 - 8t^2 + 15t + 10$ where t is the time in minutes from the start of the experiment.

a Copy and complete this table and draw the graph of $\theta = t^3 - 8t^2 + 15t + 10$ on axes where the t-axis is labelled from 0 to 5 and the θ-axis from 0 to 20.

t	0	1	2	3	4	5
t^3						
$-8t^2$						
$+15t$						
$+10$						
θ						

b What is (i) the highest, (ii) the lowest, temperature of the liquid during the first 5 minutes of the experiment?

c After how long was the temperature of the liquid the same as its initial temperature?

Coordinates and graphs

d The temperature, ϕ°C, of the air surrounding the flask rises during the experiment such that $\phi = 2t + 8$.

Copy and complete this table, and draw the graph of ϕ on the same diagram.

t	0	1	2	3	4	5
ϕ						

e How long after the start of the experiment is the temperature of the liquid the same as that of the surrounding air?

f When is the liquid 8 degrees warmer than the air?

Part 7 Areas under lines and curves

1 This diagram shows the shaded area between a jagged line and the x-axis, divided into rectangles and triangles.

a Find the area between the jagged line and the x-axis.

b This area represents the space in front of a house which is to be covered with concrete for parking cars.

If each square on the diagram represents 1 m^2 which costs £1·20 to cover with concrete, find the total cost of the concrete required.

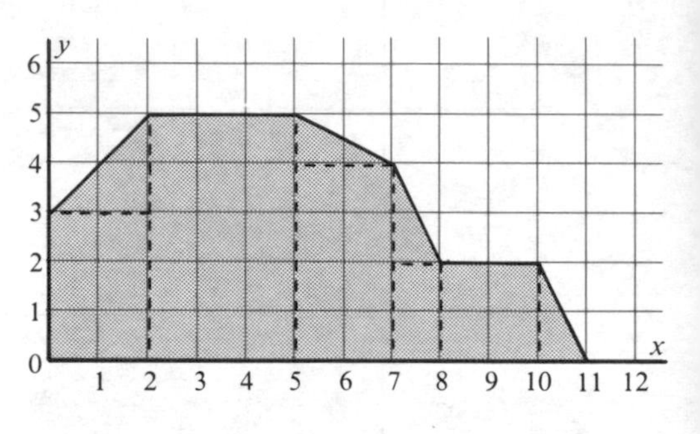

2 A piece of land is to be planted with trees. A map shows two edges of the land 16 metres and 8 metres long running along the x- and y-axes.

a Draw the x-axis from 0 to 16 and the y-axis from 0 to 10. Plot these points and join them in this order to draw a map of this piece of land.

(0, 8), (2, 6), (4, 6), (7, 8), (8, 10), (10, 10), (12, 4), (14, 3), (16, 0)

Calculate the area of this land in square metres.

b If each square metre is planted with a tree costing £4·25, find the total cost of the trees required.

c A fence is to be erected around all the plot. Use your ruler to find the perimeter of the plot, and calculate the cost of the fence if each metre length costs £2·40.

3 The area between this curve and the x-axis has been divided into strips of *equal* widths.

Approximate each strip by a rectangle and hence estimate the shaded area.

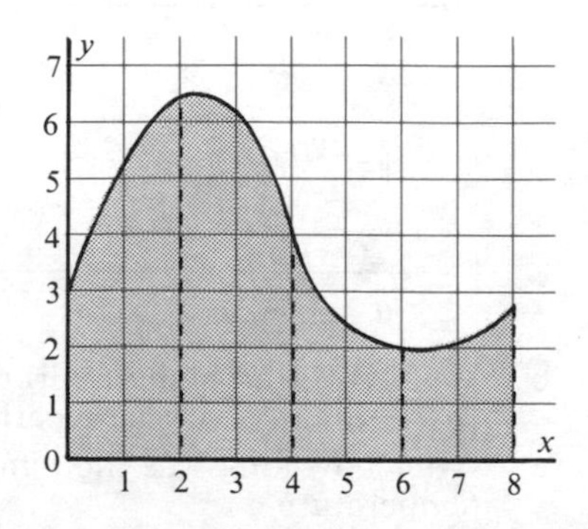

Coordinates and graphs

4 Draw axes, labelling the x-axis from 0 to 12 and the y-axis from 0 to 100. Draw a smooth curve through each set of points given in these tables. Estimate the area under each curve by approximating it with rectangles or trapezia.

a

x	0	2	4	6	8	10	12
y	40	60	70	70	60	30	10

b

x	0	2	4	6	8	10	12
y	100	40	30	50	65	50	60

c

x	0	3	6	9	12
y	20	80	95	90	70

d

x	0	3	6	9	12
y	40	65	60	70	90

Estimate the average height of each curve and illustrate this average by a straight line on the diagram.

5 The temperature on the shop-floor of a factory during the morning is measured every half-hour with these results.

Time	08.00	08.30	09.00	09.30	10.00	10.30	11.00	11.30	12.00
Temperature, °C	4	10	15	17	18	18	17	16	16

a Draw a graph of temperature against time and estimate the area between the curve and the time axis.

b Estimate the average temperature during this period and illustrate it on your diagram.

6 Find the area beneath each graph and give your answer the correct units.

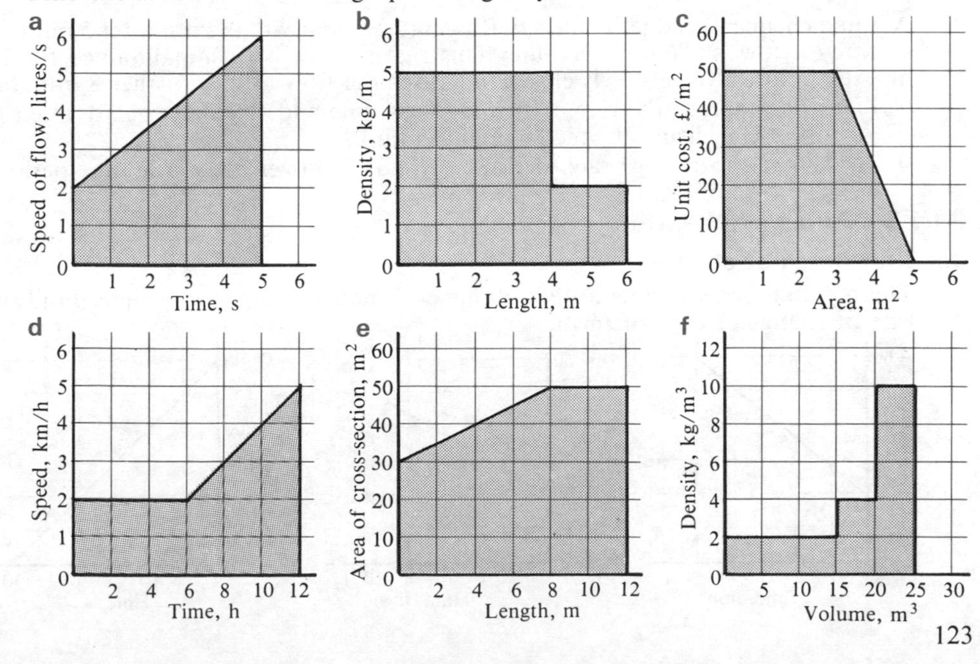

Coordinates and graphs

7 A train leaves one station and three minutes later stops at the next. Its speed varies with time as shown in this diagram.
Calculate
a the distance between the two stations
b the average speed of the train during the three minutes.

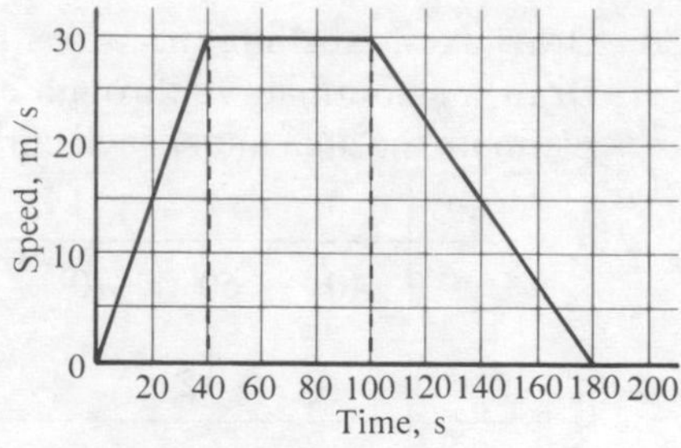

8

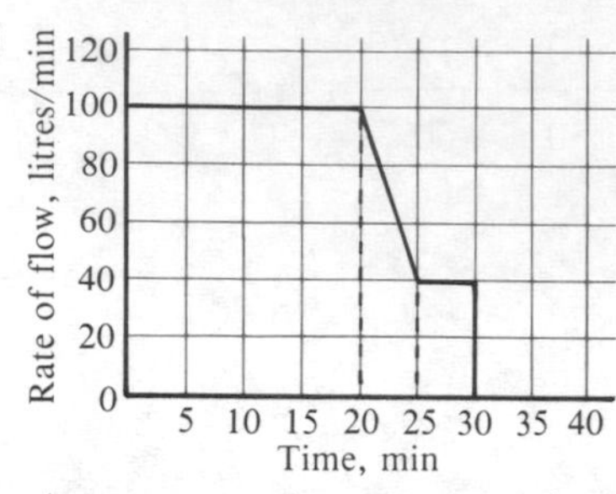

Oil is being pumped into a storage tank over a period of 30 minutes. A meter measures the rate of flow as illustrated in this diagram.
Calculate
a the volume of oil entering the tank
b the average rate of flow into the tank.

9 A passenger in a car noted the speed of the car every 5 seconds for one minute.

Time, seconds	0	5	10	15	20	25	30	35	40	45	50	55	60
Speed, m/s	0	4	8	12	18	16	17	18	18	17	16	15	14

a Draw a graph of speed against time and estimate the distance travelled in this minute.
b Estimate the average speed of the car during the whole minute and draw a straight line on your diagram to illustrate it.

10 A lift starts from rest and gradually reaches a speed of 2 m/s after moving for 4 seconds. It keeps this speed steady for another 6 seconds, before gradually coming to rest over a final 5 seconds.
a Draw axes, labelling the time axis in seconds from 0 to 15, and draw a graph of the speed against time.
b Calculate the distance travelled by the lift, and hence its average speed.

11 A liquid chemical gradually starts to flow into an empty tank until after 5 minutes its rate of flow is 600 litres/minute. This rate of flow is then maintained for 12 minutes before a valve slowly closes and stops the flow after a further 8 minutes.
a Draw a graph of the rate of flow against time and calculate the volume (in litres) of the chemical which enters the tank.
b Calculate the average rate of flow of the liquid over the whole time period.

Part 8 Rates of change

Constant rates of change

1 These graphs show temperature, volume and mass changing with time. Find the rate of change of each of them.

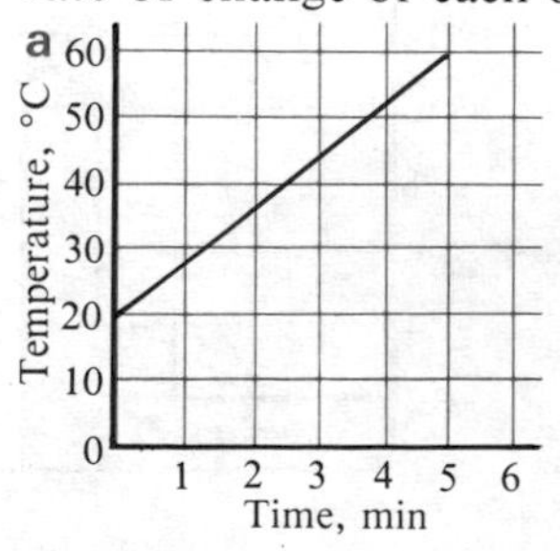

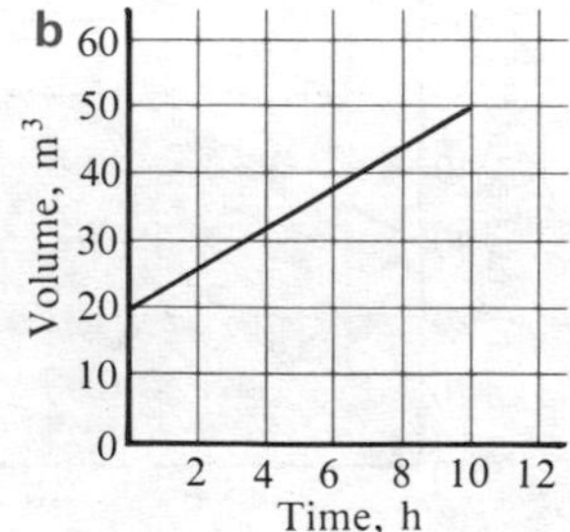

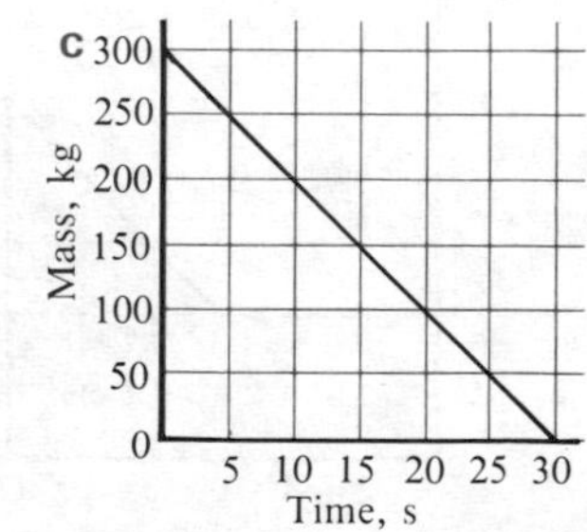

Coordinates and graphs

2 Heating oil is used at a steady rate of 20 litres per day. How much is used in two weeks?

3 A machine produces lengths of plastic sheeting at a rate of 12 m/s. What length is produced in one minute?

4 A beaker of water has a temperature of 16°C. It is then heated so that its temperature rises at a steady rate of 2 degrees per minute. Find its temperature after 5 minutes.

5 A boat starts a journey lasting 4 hours with 100 litres of diesel in the tank. It uses the fuel at a rate of 15 litres per hour. How much diesel is left in the tank at the end of the trip?

Average rates of change

6 A liquid is heated for a few minutes before being left to cool. Its temperature is given in the table and shown on the graph.

Time, t min	0	1	2	3	4	5	6	7	8	9	10
Temperature, °C	20	21	25	30	40	65	75	75	70	61	50

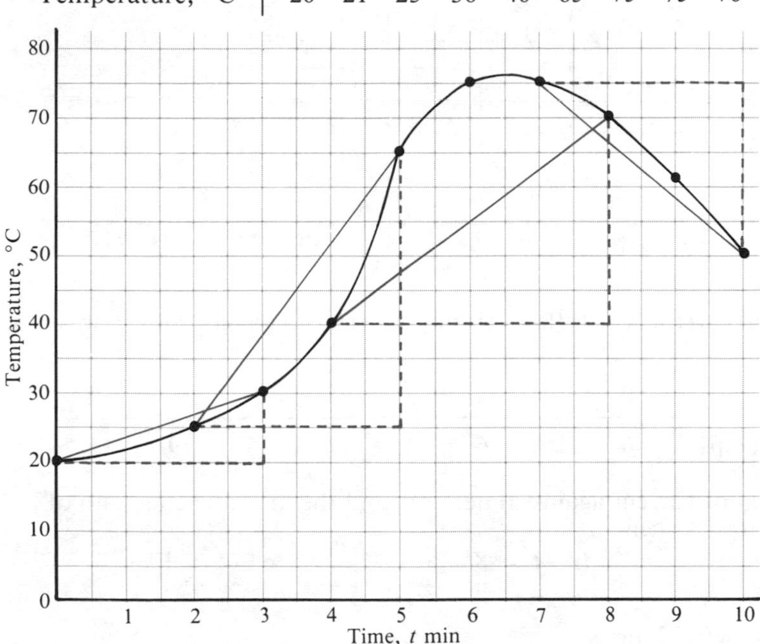

Use the table to find the average rate of change of temperature

a from $t = 0$ to $t = 3$

b from $t = 2$ to $t = 5$

c from $t = 4$ to $t = 8$

d from $t = 7$ to $t = 10$.

7 The winner of the 200-metres race covered the distance in the times shown in this table.

Time, t s	0	5	10	15	20	25	30	35
Distance, metres	0	50	75	90	100	120	140	200

Draw the graph of distance against time.
Find the average speed of the athlete

a during the first 10 seconds

b between $t = 5$ and $t = 25$

c during the last 10 seconds

d over the whole race.

Coordinates and graphs

Instantaneous rates of change

8 This graph shows the temperature of water in a beaker which is heated and then left to cool.

Find the rate of change of temperature at the instant when

a $t = 3$ **b** $t = 7$ **c** $t = 12$

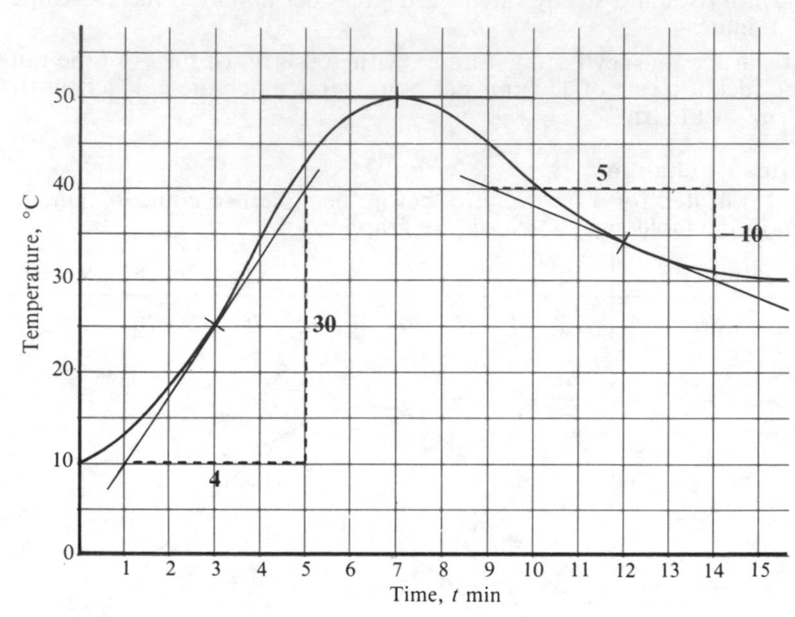

9 The electrical current which flows through a circuit changes with time as shown in this table.

Time, t s	0	2	4	6	8	10	12	14
Current, amps	40	15	6	8	17	25	19	15

Draw a graph of current against time, and find the instantaneous rate of change of current when

a $t = 2$ **b** $t = 8$ **c** $t = 10$.

10 A test bed 8 metres long is used in experiments with a trolley driven by an electric motor. The distance y metres of the trolley from one end of the bed is controlled so that t seconds after the start of the experiment

$$y = t^2 - 5t + 8.$$

Draw a graph of y against t and find the instantaneous rate of change of y when

a $t = 1$ **b** $t = 2\frac{1}{2}$

c $t = 3$.

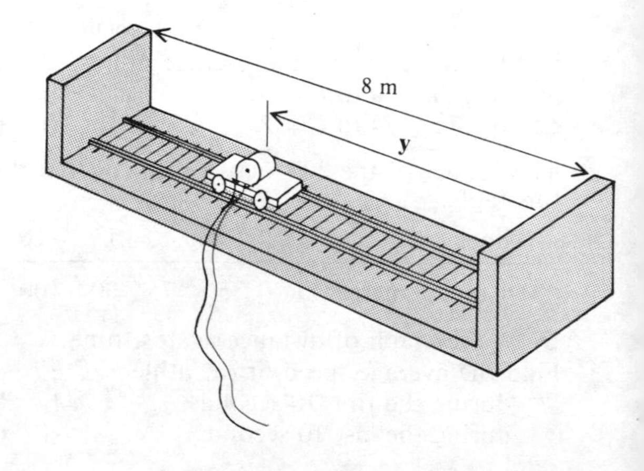

Coordinates and graphs

Part 9 Distance, speed and acceleration

Distance – time graphs

1

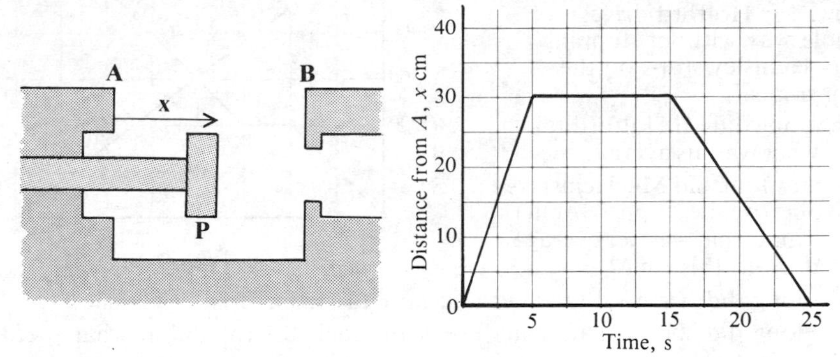

A piston P moves backwards and forwards between two faces A and B in a metal block once every 25 seconds, so that it covers a hole in face B for a certain time.

The graph illustrates the distance x of P from A during the 25 seconds.

- **a** How far apart are A and B?
- **b** How long does it take for P to move from A to B?
- **c** What is the speed of P when moving from A to B?
- **d** For how long does P cover the hole in B?
- **e** How long does P take to return from B to A?
- **f** What is the speed of P during the return to A?

2

Mrs Naylor leaves home and walks to the local shop where she spends some time making purchases. She then continues to the post-box to post a letter and then immediately returns home by the same route.

The graph illustrates her journey.

- **a** How far is the shop from her home and how long does she take to get there?
- **b** What is her speed in going from her home to the shop?
- **c** How long does she spend at the shop?
- **d** How far is the post-box from (i) her home (ii) the shop?
- **e** What is her speed in going from the shop to the post-box?
- **f** How long does she take to return home from posting the letter and at what speed does she return?

Coordinates and graphs

3 Bristol to Glasgow is 600 km by motorway. Mr Holford leaves Bristol as Mrs Hemsley leaves Glasgow. Mr Holford drives the whole way without stopping, but Mrs Hemsley stops on the way for a meal.

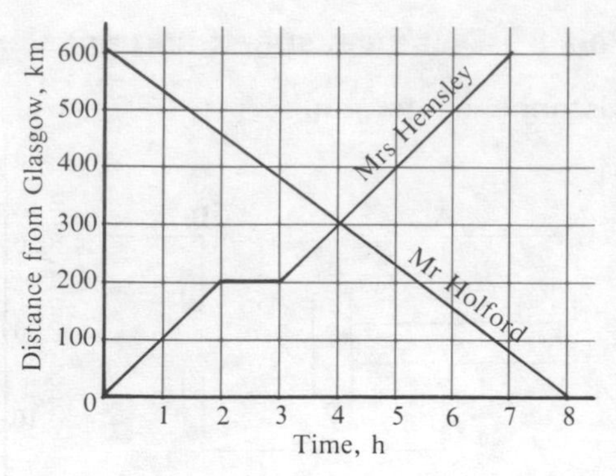

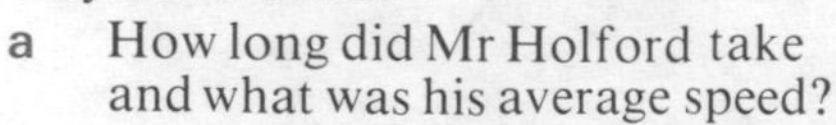

a How long did Mr Holford take and what was his average speed?

b For how long did Mrs Hemsley drive before she stopped for her meal, and what was her average speed up to this time?

c How long did her meal take her and how far from Bristol was she?

d How long did she take after her break to reach Bristol and at what speed did she travel?

e How long after the start of their journeys did Mr Holford and Mrs Hemsley pass each other and how far from Glasgow did it happen?

4 Draw axes, labelling the time axis in minutes from 0 to 12 and the distance axis in kilometres from 0 to 6.

A car is driven 6 km from its garage in 4 minutes at a constant speed and then it stands at rest for 2 minutes. It then takes 6 minutes to return steadily to its garage by the same route.

Draw a graph of its distance from the garage against time, and find its speed during

a the first 4 minutes **b** the last 6 minutes.

5 A piston travels down a channel at a steady speed of 6 cm/s for 4 seconds, before returning immediately to its starting point in a further 3 seconds.

Draw a time axis labelled in seconds from 0 to 7 and a distance axis labelled in centimetres from 0 to 25.

Draw a graph of the distance from the start against time and find its speed on the return journey.

6 An object moves so that its distance, d metres, from a fixed point P depends on the time, t seconds, such that

$$d = 3 + 2t.$$

Copy and complete this table, and hence draw a graph of d against t.

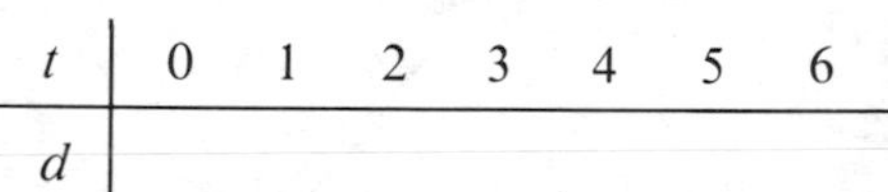

t	0	1	2	3	4	5	6
d							

a Find the speed of the object and explain why it is a *steady* speed.

b How far was the object from P at the start of its journey?

7

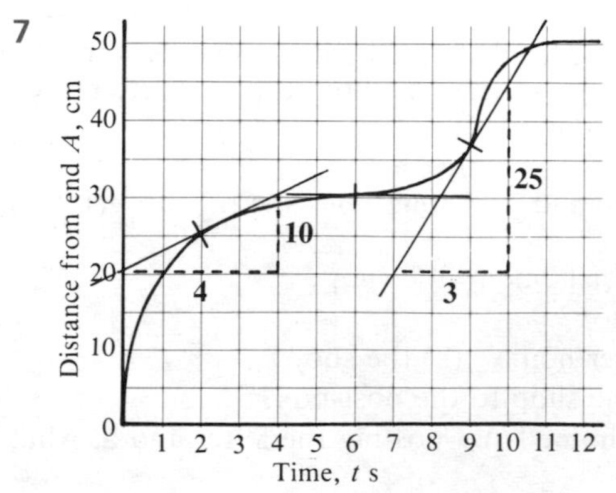

A shuttle travels in a groove 50 cm long from end A to end B. The graph shows how its distance from end A changes with time, t seconds.

Find the speed of the shuttle, when

a $t = 2$ **b** $t = 6$ **c** $t = 9$.

d How long does it take to reach the other end B?

e With what speed does it reach the other end B?

f How far is it from end A after travelling for half of its total time?

Coordinates and graphs

8 A hoist is used to lift materials vertically up the outside of a building which is under construction. It takes about 10 seconds for the hoist to travel from ground level to the top of the building, and the height risen is given by this table.

Time, s	0	1	2	3	4	5	6	7	8	9	10
Height, m	0	25	41	47	49	53	63	73	75	75	75

Draw a graph of height against time, and draw tangents to the curve to find the speed of the hoist at the instant

a 2 seconds after the start **b** 5 seconds after the start

c 9 seconds after the start.

d What is the *average* speed of the hoist during the whole journey?

9 A boy catapults a stone from ground level so that its height, h metres, varies with the time, t seconds, such that

$$h = 12t - 2t^2.$$

a Copy and complete this table and draw the graph of h against t.

t	0	1	2	3	4	5	6
$12t$							
$-2t^2$							
h							

b Estimate the speed of the stone when (i) $t = 1$ (ii) $t = 3$ (iii) $t = 4$.

c What is the greatest height of the stone and how long did it take to reach this height?

d Estimate the speed of impact when the stone hits the ground.

Speed-time graphs

10 A cyclist starts from rest and steadily increases his speed for 4 seconds. He then continues at a constant speed. Find

a the constant speed he reaches

b his acceleration during the first 4 seconds

c the distance travelled during the first 4 seconds

d his acceleration during the next 4 seconds

e the distance travelled over the first 6 seconds.

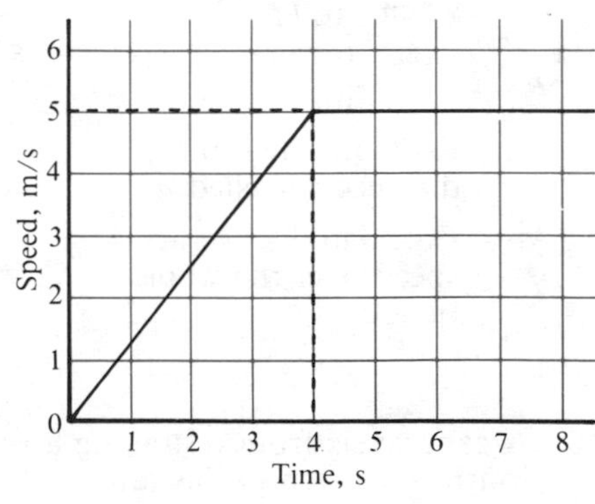

11

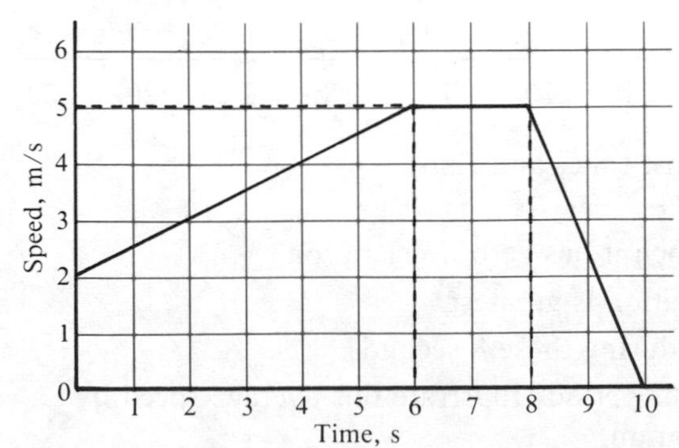

An object moves so that its speed varies with time as shown on this diagram. Find

a the distance travelled during the first 6 seconds

b the total distance travelled before coming to rest

c the average speed during the whole 10 seconds

d the acceleration during
(i) the first 6 seconds
(ii) the next 2 seconds
(iii) the final 2 seconds.

Coordinates and graphs

12 A train starts from rest at a station and gradually increases its speed up to 2 km/min over 2 minutes. It then maintains this speed for a further 6 minutes before gradually slowing down in the final 4 minutes to rest at the next station.

Draw a graph of speed against time, labelling both axes from 0 to 12. Find

a the total distance travelled between the two stations

b the average speed of the train during its journey

c its acceleration during the first 2 minutes

d its deceleration (or retardation) during the last 4 minutes.

13 An object moves at a steady speed of 4 m/s for 2 seconds. It then accelerates for 4 seconds to reach a speed of 8 m/s. It then immediately starts to slow down and gradually comes to rest after a further 4 seconds.

Draw a graph of its speed against time on axes labelled from 0 to 10. Find

a its acceleration (i) during the first 2 seconds
(ii) during the next 4 seconds
(iii) while it is slowing down

b the total distance it travels before coming to rest

c its average speed during the whole 10 second period. Illustrate this average speed on your graph.

14 The speed of an object varies with time as shown in this diagram.

a Find its acceleration when (i) $t = 3$
(ii) $t = 7$
(iii) $t = 11$

b Estimate the total distance travelled.

c Calculate its average speed over the whole period.

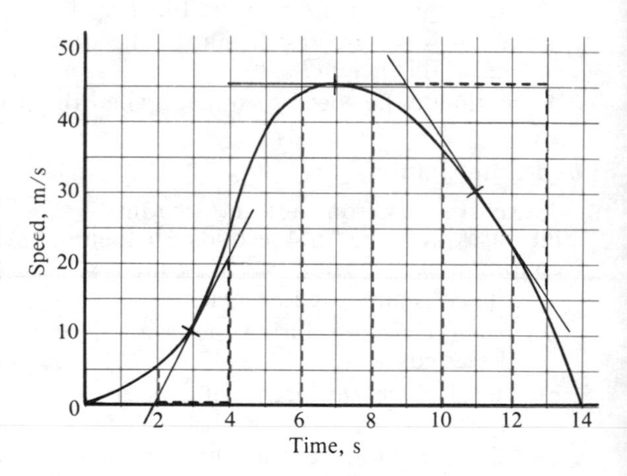

15 A car travels from rest along a road and its speed for the first 8 seconds of its journey is given in this table.

Time, t s	0	1	2	3	4	5	6	7	8
Speed, v m/s	0	1	4	10	15	18	20	20	20

Draw a graph of its speed against time, and find

a its acceleration when $t = 2$

b the time interval during which it has zero acceleration

c the instant when its acceleration is greatest

d the total distance travelled during these 8 seconds

e the average speed during this period. Illustrate this average speed by drawing a line on your diagram.

Coordinates and graphs

16 The end of a linkage in a machine travels along a straight rod so that its speed, v cm/s, after a time, t seconds, is given by this table.

Time, t s	0	2	4	6	8	10	12
Speed, v cm/s	0	1.2	2.4	3.4	4.0	4.4	4.6

Draw a graph of v against t and hence estimate

a its acceleration when $t = 8$

b the period when the acceleration is least

c the distance travelled during the whole 12 seconds

d its average speed during the whole period. Illustrate the average speed on your diagram.

17 An object moves in a straight line such that its speed, v m/s, after a time, t seconds, is given by $v = 7 + 6t - t^2$.

Copy and complete this table, and draw the graph of v against t. Find

t	0	1	2	3	4	5	6	7
7 $+6t$ $-t^2$								
v								

a its initial speed

b how long it takes to come to rest

c the value of t when its acceleration is zero

d its acceleration at the instant when $t = 4$

e the distance travelled during the seven seconds

f its average speed over this period.

18 The flow of water in a uniform pipe is controlled by a valve. As the valve closes, the speed of the water, v m/s, varies with time, t seconds, such that $v = 64 - t^2$.

Copy and complete this table and draw the graph of v against t.

t	0	1	2	3	4	5	6	7	8
64 $-t^2$									
v									

a What is the speed of the water as the valve begins to close?

b How long does it take for the valve to close?

c What is the acceleration of the water when (i) $t = 0$ (ii) $t = 4$?

d What length of pipe has the water filled in the time taken for the valve to close?

e If the pipe has an area of cross-section of 0.1 m^2, what volume of water passes through the valve as it is closing?

Algebra

Part 1 Simplification and substitution

1 Simplify these expressions.

a $5x + 7y - 3x - 2y + 4x$ b $7m + 6n + 2m - n$
c $8(x + 2) - 5(x + 3)$ d $4(3y - 2z) + 2(y + 6z)$
e $3(a - 3b) - 2 + 4a - 3b + 7$ f $5(y - z) + 6z - 8 + 2y - 4$

2 Find the values of these expressions when $p = 4$, $q = 2$ and $r = 5$.

a $3p - 4q + r$ b $5p + q - 8$ c $p - 5q + 7$
d $2p - q - r$ e $3(p + q) - 2r$ f $p^q + r$

3 Simplify and then find the values of these expressions, given that $x = 6$, $y = 2$ and $z = 3$.

a $4x - 3y + 7x + 4y$ b $9z + y - 3y - 4z$
c $4(2x - 3) + 5(2 - x)$ d $3(x + y - z) + 2(x - y - z)$
e $5(2x - y) + 3(y - 2z) - 4(x + z)$
f $2(x - y + z) + 3(x + 2) - 5(x + 1)$

4 A book has p pages of print and g pages of glossy photographs. The total mass m grams of the book is given by the formula $m = 5p + 10g$.
Find m when a $p = 20, g = 5$ b $p = 100, g = 15$ c $p = 42, g = 8$.

5 The cost £C of a taxi ride depends on the number m of miles of the journey and the number b of pieces of baggage carried.
If $C = 2m + \frac{1}{2}b$, find C when
a $m = 2, b = 4$ b $m = 4, b = 8$ c $m = 9, b = 1$.

6 The time, T minutes, to make a cup of tea depends on the volume, V litres, of water in the kettle and the time, S minutes, which the tea is allowed to brew, where $T = 4V + S$.
Find T when a $V = 1, S = 2$ b $V = 2, S = 5$ c $V = 1\frac{1}{2}, S = 4$.

7 A box holds n nuts and b bolts. The cost, C pence, of the box is given by $C = 4n + 8b + 5$.
Find C when a $n = 10, b = 10$ b $n = 8, b = 4$ c $n = 12, b = 8$.

8 The total mass m grams of a parcel depends on the mass C of the contents, the area A of the brown paper and the length L of the string tying it.
If $m = C + \frac{1}{5}A + \frac{1}{10}L$, find m when
a $C = 200, A = 5, L = 10$ b $C = 500, A = 10, L = 20$
c $C = 120, A = 5, L = 5$.

9 Simplify these expressions.

a $5 \times p \times 3 \times q \times p$ b $3x \times 5y \times 2x$
c $4m \times 3n \times 2$ d $7a \times b \times 3c$
e $2p \times 3q \times 4q$ f $4s \times 3t \times s \times t$
g $3a(b + c) + 2ab - 2ac$ h $2m(p + 2q) - 2mp - mq$
i $x(3y - 2z) + xy + 3xz$ j $t(u + 3v) - t(u + v)$
k $c(2a - 4b) + c(a + 5b)$ l $xy - 3yz + 2y(x + 3z)$

10 Find the values of these expressions when $x = 4$, $y = 2$ and $z = 5$.

a $3xy + xz$ b $3yz - 5x$ c $\frac{1}{2}xyz$
d $y(x + z)$ e $3x(z - y)$ f $xy(z - 5) + x^y$

Algebra

11 A package holiday for A adults and C children costs £P where $P = 80(2A + C)$.
Find P when **a** $A = 2, C = 2$ **b** $A = 1, C = 3$ **c** $A = 3, C = 5$.

12 The focal length f of a lens is related to the object distance u and image distance v by the formula $f = \frac{uv}{u + v}$.
Find f when **a** $u = 4, v = 4$ **b** $u = 2, v = 6$ **c** $u = 6, v = 4$.

13 I travel t miles to work by train and b miles by bus. If both run on time, the total time T minutes it takes me to get to work is given by $T = 3(t + 2b) + 8$.
Find T when **a** $t = 6, b = 2$ **b** $t = 11, b = \frac{1}{2}$ **c** $t = 20, b = 1$.

14 The income £I of a ferry operator depends on the number C of cars on the ferry, the number P of passengers and the average amount £A which each passenger spends on board.
If $I = P(7 + A) + 50C$, find I when

a $P = 50, A = 3, C = 10$ **b** $P = 20, A = 2, C = 6$
c $P = 160, A = 13, C = 40$.

15 **a** This pentagon has two sides of length x cm. The other sides all have length y cm.
Write down an expression for the perimeter of the pentagon in terms of x and y.
b If $x = 5$ and $y = 2$, what is the perimeter of the pentagon?
c If the perimeter is 25 cm, what is the value of x when $y = 3$?

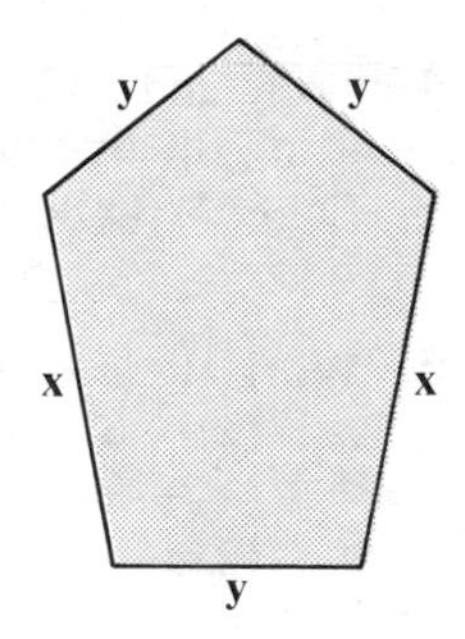

16

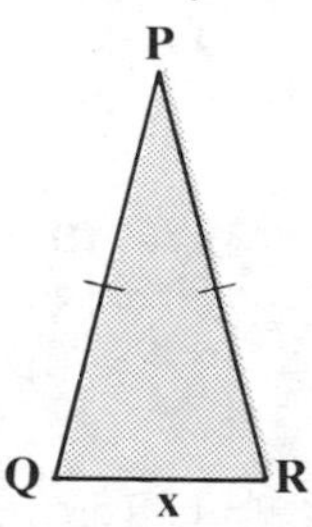

In this isosceles triangle, side PQ is twice the length of side QR.
If QR is x cm long, write down an expression for the perimeter of the triangle in terms of x.
What is the length of the perimeter when $x = 4$?

17 Mr and Mrs Watson move into a new house and they decide to pave part of the back garden.
Each row of paving slabs is made from hexagons (of area h m^2 each) and triangles (of area t m^2 each) as shown in the diagram.
a Write down the area of one row in terms of h and t.
b If 5 rows are laid, what is the total paved area in terms of h and t?
c The area of one triangular slab is $\frac{1}{2}$ m^2 and of one hexagonal slab is 3 m^2. What is the total paved area of 5 rows in m^2?

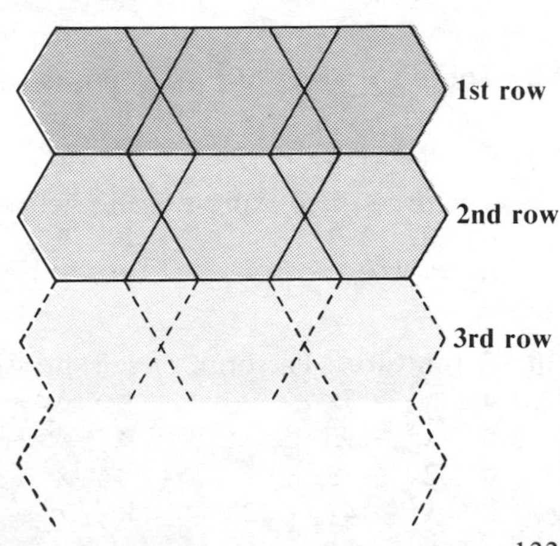

Algebra

Part 2 Expansions and factors

1 Copy and complete each statement, using the diagrams to help you.

a $x(x + 6) = \dots\dots$

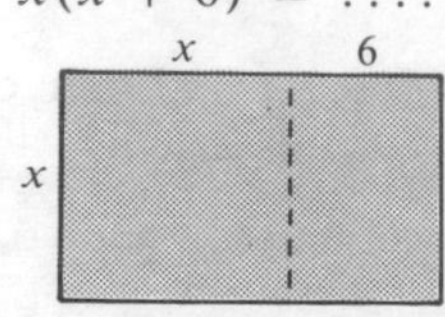

b $x(x + 3) = \dots\dots$

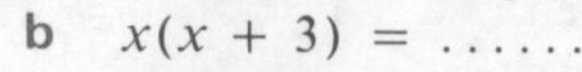

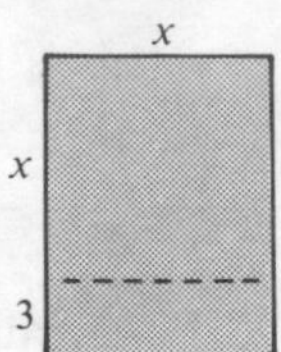

2 Work these multiplications.

a $x(x - 5)$ **b** $3(x + 2)$ **c** $4(2x - 3)$
d $3x(2x + 1)$ **e** $x(x^2 + 3x - 2)$ **f** $2x(x^2 - 4x + 1)$
g $\frac{1}{2}x(4x^2 - 2x + 6)$ **h** $5x(3y - 4)$ **i** $3x(x + 2y + 3z)$

3 Common factors Factorise these expressions.

a $2x + 6$ **b** $2x - 10$ **c** $8x + 2$
d $x^2 - 3x$ **e** $y^2 + 5y$ **f** $2y^2 - 8y$
g $6z^2 + 9z$ **h** $2xy - 6xz$ **i** $x^2y + 2xy$

4 Copy and complete this statement using the diagram to help you.
$(x + 2)(x + 5) = \dots\dots$

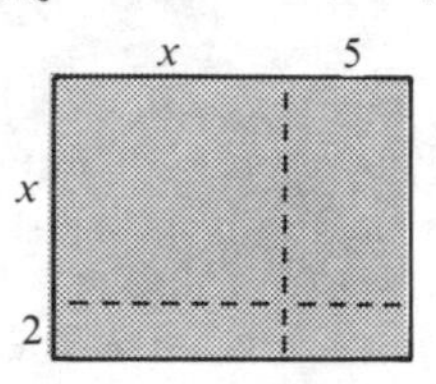

5 Work these multiplications.

a $(x + 3)(x + 7)$ **b** $(x - 4)(x + 6)$ **c** $(x - 5)(x - 4)$
d $(x - 3)^2$ **e** $(x + 4)(2x + 3)$ **f** $(3x + 1)(2x + 5)$
g $(2x - 3y)(4x + 3y)$ **h** $(3x - 4y)(2x + 3y)$ **i** $(2x - y)^2$

6 Quadratic factors Factorise these expressions.

a $x^2 + 6x + 5$ **b** $x^2 + 5x + 4$ **c** $x^2 + 3x - 10$
d $x^2 - 3x - 10$ **e** $x^2 + 2x - 15$ **f** $x^2 - 7x - 18$
g $x^2 - 6x + 8$ **h** $x^2 - 11x + 30$ **i** $x^2 - 30x + 144$

7 Work these multiplications.

a $(x + 5)(x - 5)$ **b** $(x - 3)(x + 3)$ **c** $(3x + 4)(3x - 4)$

8 The difference of two squares Factorise these expressions.

a $x^2 - 36$ **b** $x^2 - 16$ **c** $4x^2 - 49$
d $9x^2 - 25$ **e** $x^2 - 1$ **f** $x^2 - \frac{1}{4}$

9 Factorise these expressions to help calculate their values.

a $97^2 - 3^2$ **b** $873^2 - 127^2$ **c** $53^2 - 51^2$
d $7.85^2 - 2.15^2$ **e** $(2\frac{1}{4})^2 - (1\frac{3}{4})^2$ **f** $(3\frac{2}{5})^2 - (1\frac{3}{5})^2$

10 A mixture Factorise these expressions.

a $x^2 - 3x$ **b** $6x + 4$ **c** $x^2 + 6x + 8$
d $x^2 - 9$ **e** $x^2 - 9x$ **f** $x^2 - 9x + 14$
g $2x^2 - 8x$ **h** $x^2 - 4$ **i** $x^2 - 4x$
j $x^2 - 4x - 12$ **k** $x^2 - x$ **l** $x^2 - x - 72$

Algebra

Part 3 Equations

1 Write the solutions only to these equations.

a	$x + 2 = 8$	**b**	$x - 2 = 8$	**c**	$2x = 8$	**d**	$\frac{1}{2}x = 8$
e	$x + 3 = 12$	**f**	$x - 3 = 12$	**g**	$3x = 12$	**h**	$\frac{1}{3}x = 12$
i	$2x - 1 = 7$	**j**	$4x + 5 = 17$	**k**	$\frac{1}{2}x + 1 = 5$	**l**	$\frac{1}{4}x - 2 = 1$

2 Solve these equations.

a	$3x + 5 = 2x + 9$	**b**	$6x - 1 = 5x + 3$
c	$4x + 7 = 2x + 13$	**d**	$2(4x - 1) = 6x + 1$
e	$5(3x + 2) = 2(6x + 7)$	**f**	$8x - 3 = 5(2x - 1)$
g	$2(3x + 1) + 5(x - 2) = 14$	**h**	$6(x - 2) + 4(2x + 1) = 13$
i	$4(x - 7) + 3x = 0$	**j**	$5(3x + 1) - 3(x - 3) = 20$
k	$2x + 3(x - 1) = 4x + 5$	**l**	$4(x - 7) + 9 = 2x + 1$

3 In triangle PQR, angle P is $x°$, angle Q is 20° bigger than angle P and angle R is 40° bigger than angle P. Find the value of x.

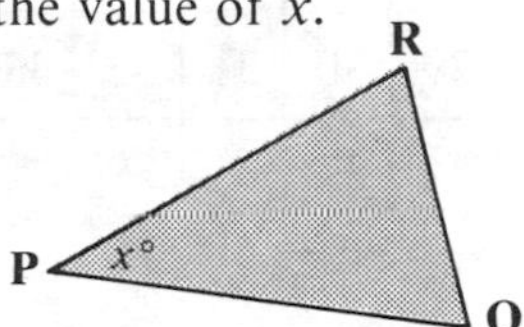

4 Triangle LMN is isosceles with angle L of $x°$. If angle M is three times the size of angle L, find the value of x.

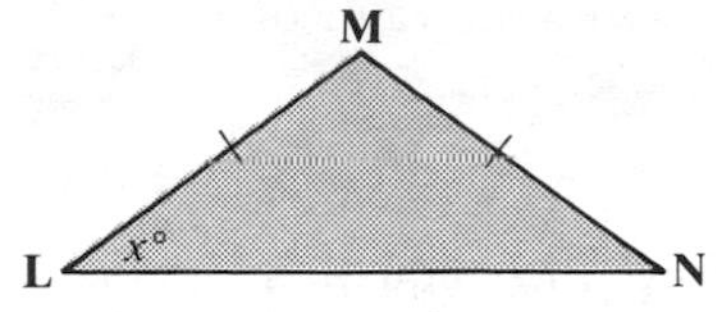

5 John spent £x on holiday. His brother, Alan, spent twice as much as John. His sister, Lucy, spent £5 more than John. If they spent £29 altogether, find the value of x.

6 A firm has x lorries and three times as many cars. The number of vans is three more than the number of lorries.

a If there are 33 vehicles altogether, find the value of x.

b How many cars and how many vans has the firm?

7 A gardener plants m blue flowers in each square and n white flowers in each octagon of the tesselation shown here.

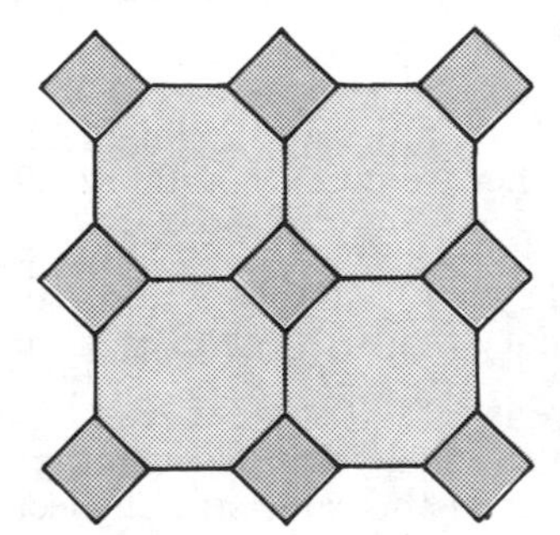

a Write down an expression for the total number of flowers planted.

b If $m = 5$ and $n = 10$, find the total number planted.

c If the total number is 68, find the value of m when $n = 8$.

8 A wheel is bolted to its axle by 3 small nuts of mass m grams and one large central nut of mass M grams.

a The 4 nuts are sold together in one plastic pack. What is the total mass of the nuts in one pack? Give your answer in terms of m and M.

b If a garage stocks 20 of these packs, what is their total mass, in terms of m and M?

c If a customer buys 5 packs and their total mass is 900 grams, calculate the value of M when $m = 40$ grams.

Algebra

9 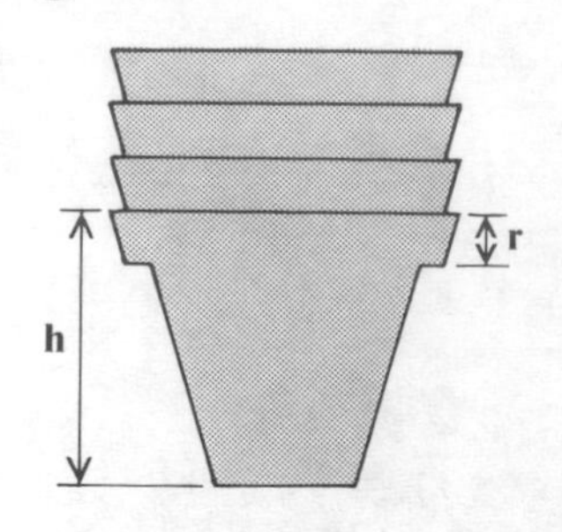

Plastic cups of height h cm have a rim of depth r cm. The cups are stacked inside each other as shown here, so that the rims rest on each other.

a If 6 cups are stacked in one pile, write down an expression for the height of the pile in terms of h and r.

b What is the height of a pile of 25 cups?

c If the total height of a pile of 25 cups is 60 cm and if $r = 2$ cm, what is the value of h?

10 An L-shape containing 3 numbers is placed on this grid. In the diagram, the L-shape covers 11, 18 and 19.

a If the corner of the L-shape contains the number x as shown here, write the two other numbers.

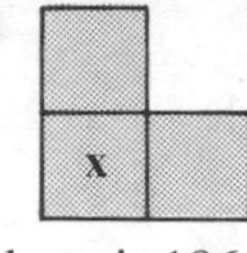

b If the total of the three numbers is 186, find the value of x.

1	2	3	4	5	6	7
8	9	10	11	12	13	14
15	16	17	18	19	20	21
22	23	24	25	26	27	28
29	30	31	32	33	34	35
36	37	38	39	40	41	42

11 Solve these equations.

a $\frac{1}{2}x + 4 = 5$ b $\frac{x + 4}{2} = 5$ c $\frac{1}{3}x - 1 = 4$ d $\frac{x - 1}{3} = 5$

e $\frac{2x}{3} + 1 = 4$ f $\frac{2x + 1}{3} = 4$ g $\frac{3x - 2}{5} = 2$ h $\frac{3x}{5} - 2 = 2$

i $\frac{x}{3} - \frac{1}{2} = 4$ j $\frac{2x}{3} - \frac{1}{5} = 2$ k $\frac{x}{4} = \frac{2x}{3} - 5$ l $\frac{3x}{4} = \frac{1}{3} - \frac{x}{2}$

12 I cycled x miles on my bike last week. You cycled only half this distance. Together we cycled a total of 60 miles. Find the value of x.

13 The shortest route from Shrewston to Yarbury is x miles long. One third of the journey is on motorways, one quarter is on dual-carriageways and the rest on ordinary roads. If 30 miles is on ordinary roads, find the value of x.

14 There is a fault somewhere in a cable x metres long. Two thirds of the cable were checked yesterday; another 24 metres were checked today. If there is a quarter of the cable still unchecked, find the value of x.

Simultaneous equations

15 The equations $y = 2x + 1$ and $x + y = 4$ have been plotted as straight lines.

Solve these equations simultaneously by looking at the diagram.

Check your answers by substituting them into the equations.

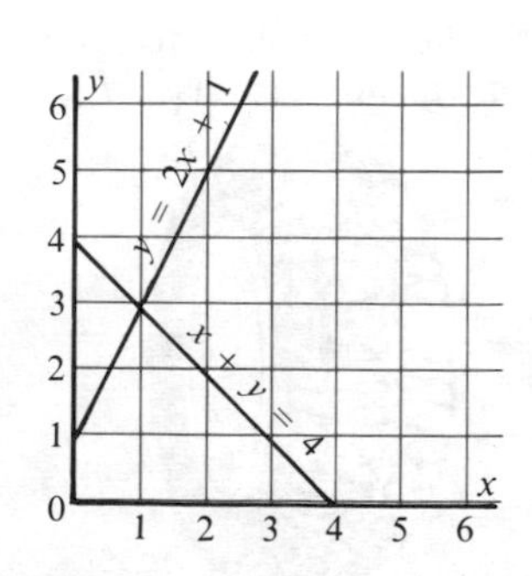

16 For each of these pairs of simultaneous equations, copy and complete the two tables and so draw two straight lines on axes labelled from -2 to 8.

Write down the solutions of the simultaneous equations.

a $y = x + 1$
$y = 3x - 1$

x	0	1	2	3
$y = x + 1$				
$y = 3x - 1$				

b $y = \frac{1}{2}x + 2$
$y = 2x - 1$

x	0	1	2	3	4
$y = \frac{1}{2}x + 2$					
$y = 2x - 1$					

c $y = x + 3$
$y = 6 - x$

x	0	1	2	3	4	5
$y = x + 3$						
$y = 6 - x$						

17 Solve these simultaneous equations.

a $3x + 8y = 14$
$7x - 8y = 6$

b $5x + 2y = 19$
$3x + 2y = 13$

c $x + 7y = 19$
$2x + 5y = 11$

d $4x - y = 9$
$5x + 3y = 7$

e $2x + 7y = 9$
$3x + 4y = 7$

f $5x - 2y = 8$
$7x + 3y = 17$

18 Solve these simultaneous equations by substituting one equation into the other.

a $y = x + 6$
$x + y = 14$

b $y = 3x - 1$
$2x + y = 4$

c $x = 2y - 3$
$3x + y = 5$

d $x = 2 - 5y$
$y - 3x = 10$

e $y = 2x + 5$
$6x - 2y = 1$

f $y = 7 - x$
$3x - y + 2 = 0$

19 Two numbers x and y add up to 61. Their difference is 13.
Find the two numbers.

20 A bus journey costs each adult a pence and each child c pence. On one trip, the conductor collected £2·70 from 6 adults and 2 children. On the return trip, £2·60 was collected from 5 adults and 4 children.

Find the values of a and c.

21 Plastic components for a machine cost p pence each, whereas metal components cost m pence each. A box holding 8 plastic and 5 metal parts costs a total of 75 pence. A larger box with 12 plastic and 10 metal parts costs a total of £1·30.
Find the cost of one of each type of component.

Algebra

Quadratic equations

22 **a** This diagram gives the graph of $y = x^2 - 4x + 3$.
Write the solutions of the quadratic equation $x^2 - 4x + 3 = 0$ by looking at this graph.
Check your answers by substituting them into the equation.

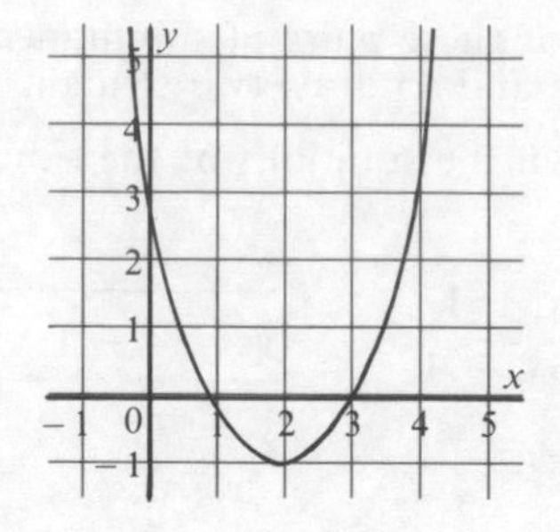

b Copy and complete this table for the equation $y = x^2 - 5x + 4$.
Draw its graph and write down the solutions of the quadratic equation $x^2 - 5x + 4 = 0$.

x	0	1	2	3	4	5
x^2						
$-5x$						
$+4$						
y						

c Copy and complete this table for the equation $y = x^2 - 7x + 9$.
Draw its graph and write down the solutions of the equation $x^2 - 7x + 9 = 0$ as decimals.

x	0	1	2	3	4	5	6
x^2							
$-7x$							
$+9$							
y							

23 Write the solutions of these equations.
a $(x - 2)(x - 7) = 0$ **b** $(x + 4)(x - 5) = 0$ **c** $(x + 6)(x + 5) = 0$
d $(x + 1)(x - 1) = 0$ **e** $x(x - 7) = 0$ **f** $x(x + 3) = 0$

24 Solve these quadratic equations by using factors.
a $x^2 + 6x + 5 = 0$ **b** $x^2 + 8x + 7 = 0$ **c** $x^2 - 3x + 2 = 0$
d $x^2 - 6x + 8 = 0$ **e** $x^2 - 7x + 10 = 0$ **f** $x^2 - 11x + 18 = 0$
g $x^2 + 6x - 7 = 0$ **h** $x^2 - 2x - 15 = 0$ **i** $x^2 - 7x - 8 = 0$
j $x^2 + 4x - 21 = 0$ **k** $x^2 - 9x + 20 = 0$ **l** $x^2 + 8x - 20 = 0$

25 Solve these quadratic equations by looking for either common factors or 'the difference of two squares'.
a $x^2 - 16 = 0$ **b** $x^2 - 5x = 0$ **c** $x^2 - 7x = 0$
d $x^2 - 4x = 0$ **e** $x^2 - 4 = 0$ **f** $x^2 - 1 = 0$
g $x^2 - x = 0$ **h** $x^2 + x = 0$ **i** $x^2 + 2x = 0$
j $x^2 - 9 = 0$ **k** $x^2 - 9x = 0$ **l** $x^2 + 9x = 0$

26 When the square of a number is added to the double of the number, the total is 24. Find the number.

27 A rectangle of width x cm is 5 cm longer than it is wide. If its area is 24 cm^2, find the value of x.

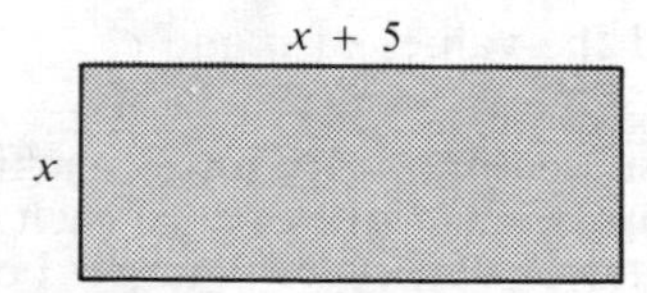

28 A room is x metres long. Its width is 4 metres less than its length. If the area of carpet needed to cover all the floor is 32 m^2, find the value of x.

29 A rectangular window in a large office has a frame 6 metres long and x metres high. A square section of side x metres can slide open. The rest of the window which cannot be opened, has an area of 8 m^2.
Write an expression for the area which cannot be opened and so form a quadratic equation in x.
Solve this equation to find the height of the window.

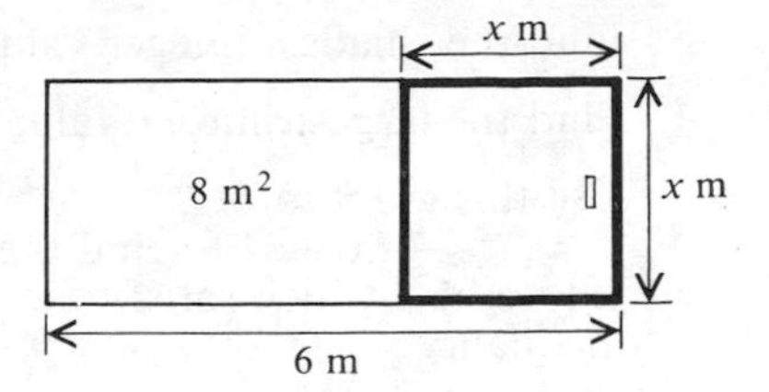

30 A water tank is 5 metres high and x metres wide. Its length is 5 metres longer than its width. If the capacity of the tank is 120 cubic metres, find the value of x.

31 Solve these harder equations by using factors.

a $2x^2 + 3x + 1 = 0$ **b** $5x^2 + 6x + 1 = 0$ **c** $6x^2 - 7x + 1 = 0$
d $6x^2 - 5x + 1 = 0$ **e** $5x^2 + 8x + 3 = 0$ **f** $3x^2 - 16x + 5 = 0$
g $2x^2 - x - 1 = 0$ **h** $3x^2 + x - 4 = 0$ **i** $8x^2 - 14x - 9 = 0$

32 **a** Copy and complete these tables and draw the graphs of both $y = x^2 - 4x + 2$ and $y = x - 2$ on the same axes.

For $y = x^2 - 4x + 2$

x	0	1	2	3	4	5
x^2						
$-4x$						
$+2$						
y						

For $y = x - 2$

x	0	1	2	3	4	5
y						

b Use the diagram to solve the equation $x^2 - 4x + 2 = x - 2$.
c Check your answers by solving the same equation algebraically.

Part 4 Inequalities

1 Write an inequality for each coloured section of the x-axis.
A coloured dot includes the end value, but an empty dot does not.

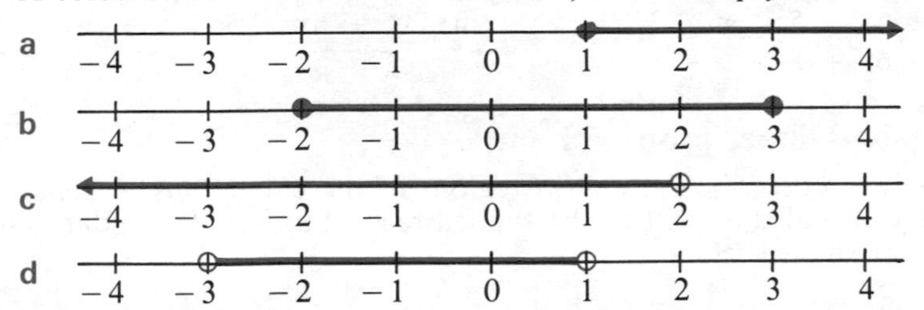

2 Find the values of x which satisfy these inequalities.

a $5x - 7 \geqslant 23$ **b** $\frac{1}{2}x + 1 \leqslant 4$
c $2(2x - 5) < 3$ **d** $4(2x - 1) + 5(2 - x) > 18$
e $6x - 4 \leqslant 4x + 9$ **f** $2x + 1 < 6 - 3x$

Remember to take care when multiplying or dividing by a negative number.

g $-2x + 5 \geqslant 9$ **h** $8 - 3x < 14$
i $8 - 6x < 5$ **j** $9 - \frac{1}{2}x \leqslant 6$
k $6 - \frac{1}{3}x > 4$ **l** $6x - 3 \geqslant 8x + 2$

Algebra

3 Find the smallest integer value of x which satisfies $4x - 6 > 7$.

4 Find the largest integer value of x which satisfies $8(x - 2) \leqslant 3x + 7$.

5 Use this graph of $y = x^2 - 5x + 4$ to find the values of x which satisfy the inequality $x^2 - 5x + 4 \leqslant 0$.

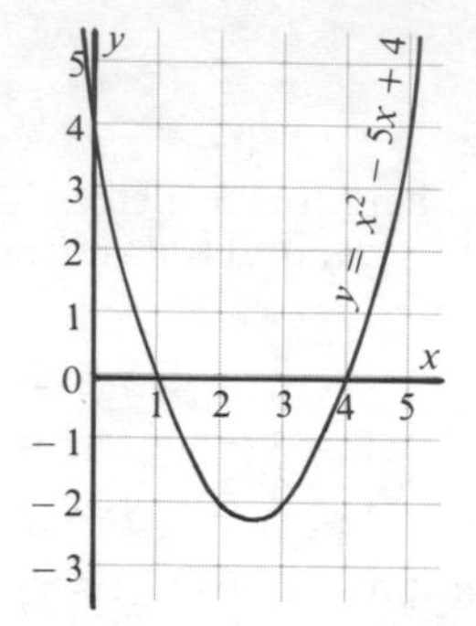

6 This diagram shows the graphs of $y = 4x - x^2$ and $y = 3$.
Find the values of x which satisfy these inequalities.

a $4x - x^2 \geqslant 0$ **b** $4x - x^2 \geqslant 3$
c $4x - x^2 \leqslant 0$ **d** $4x - x^2 \leqslant 3$

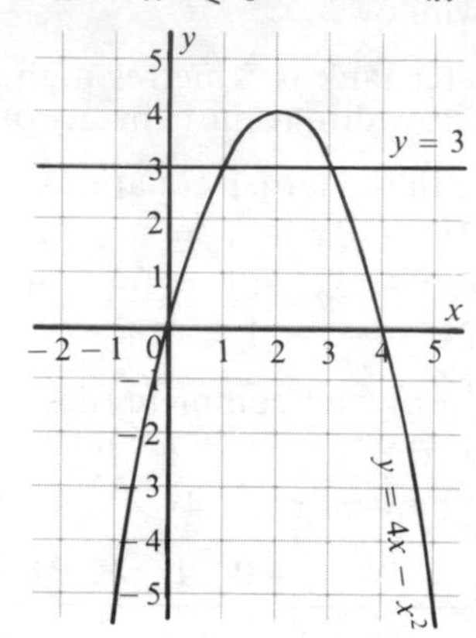

7 a Copy and complete both these tables.

$y = x^2 - 6x + 5$

x	0	1	2	3	4	5	6
x^2							
$-6x$							
$+5$							
y							

$y = x - 3$

x	0	1	2	3	4	5	6
y							

b Draw and label the x-axis from 0 to 6 and the y-axis from −4 to 6. Draw the graphs of $y = x^2 - 6x + 5$ and $y = x - 3$.

c From the diagram, write, as accurately as possible, the values of x for which
(i) $x^2 - 6x + 5 < x - 3$ (ii) $x^2 - 6x + 5 = x - 3$
(iii) $x^2 - 6x + 5 > x - 3$.

Part 5 Regions and linear programming

1 Write inequalities to describe the *shaded* regions of these diagrams. A *solid* boundary line is included as part of the shaded region; a *dotted* boundary line is not part of the shaded region.

a

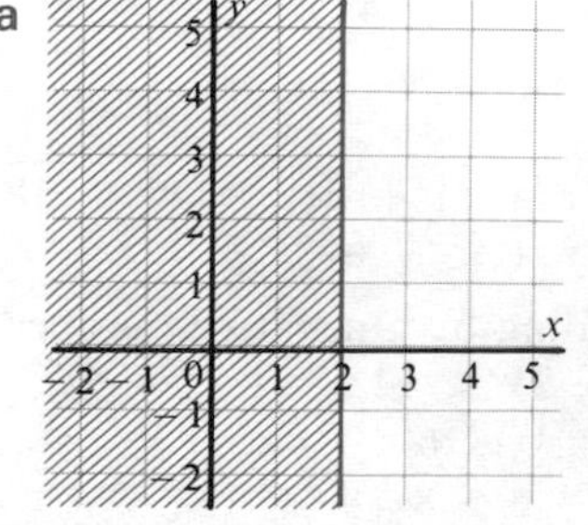

b

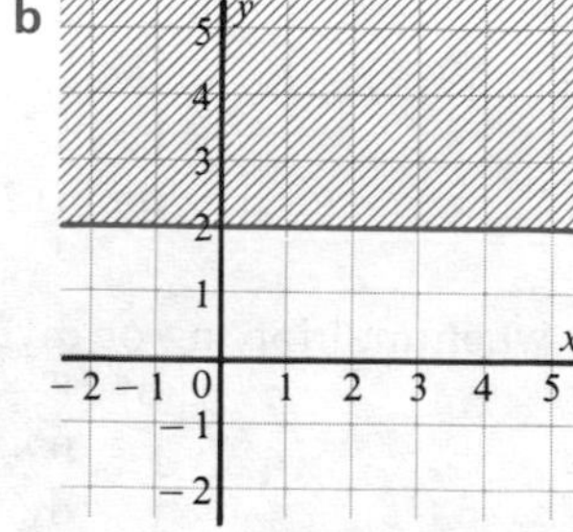

c

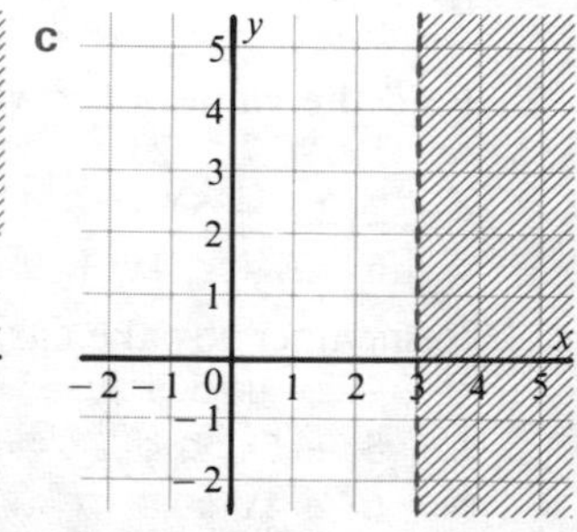

d

e

f

2 On axes labelled from -2 to 5, draw the boundaries of each of the regions described by these inequalities. Shade in the regions on your diagrams.

a $x \geqslant 3$ and $y \leqslant 2$

b $x \geqslant 0$ and $y \leqslant 4$

c $2 \leqslant x \leqslant 3$

d $1 \leqslant y \leqslant 4$ and $x \leqslant 2$

e $0 \leqslant x \leqslant 4$ and $3 \leqslant y \leqslant 4$

f $-1 \leqslant x \leqslant 1$ and $-1 \leqslant y \leqslant 2$

3 Describe these shaded regions by writing inequalities.

a

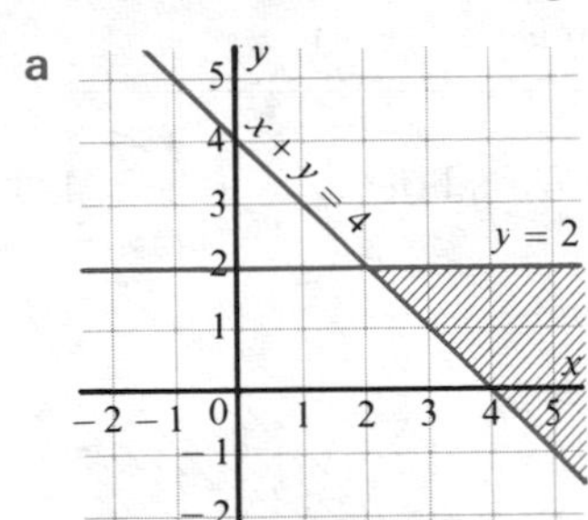

b

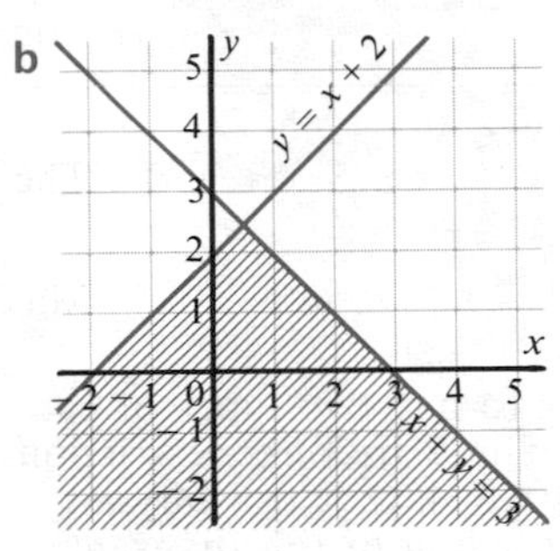

c

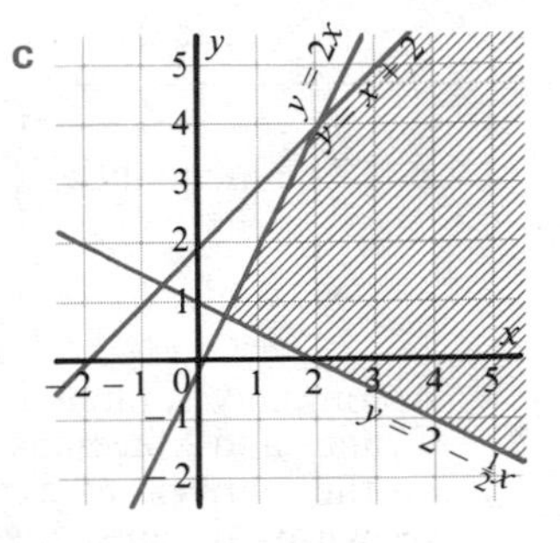

4 For each diagram, draw axes labelled from 0 to 12 and shade that part of the diagram described by these inequalities.

a $2 \leqslant x \leqslant 7$
$y \leqslant x + 4$

b $y \leqslant x + 1$
$x + y \leqslant 8$
$x \leqslant 6$

c $x + y \leqslant 9$
$x + y \geqslant 3$
$y \leqslant 2x + 1$

d $3x + 5y \geqslant 30$
$2x + y \geqslant 12$
$y \leqslant 9$

e $x + 6y \geqslant 12$
$x + y \geqslant 7$
$5x + 2y \geqslant 20$

f $2x + y \leqslant 10$
$3x + 8y \geqslant 24$
$x \geqslant 2$

5 Copy and complete the table for the extreme corners of the unshaded region of the diagram.

Corner	(,)	(,)	(,)
Value of $3x - y$			

At which corner has $3x - y$

a its maximum value

b its minimum value?

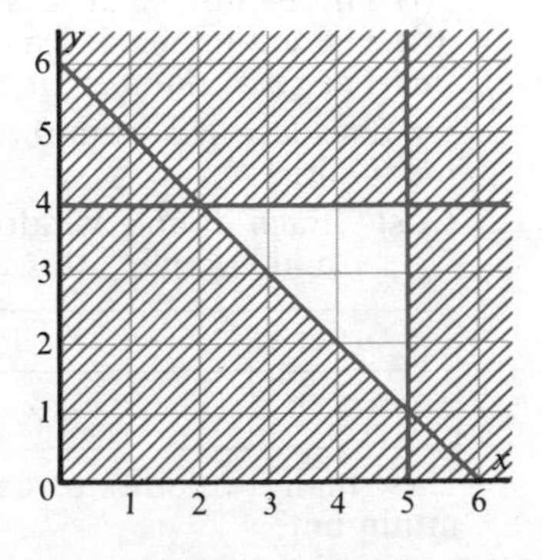

Algebra

6 Copy and complete this table to find which point of the unshaded region of this diagram has

a the maximum value

b the minimum value

of the expression $5x + 2y$.

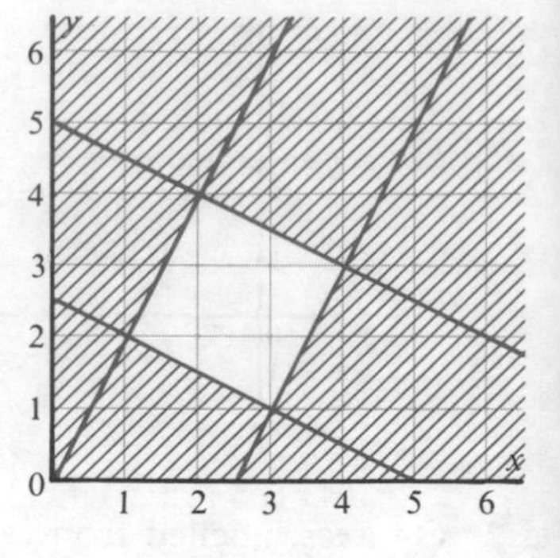

Corner	(,)	(,)	(,)	(,)
Value of $5x + 2y$				

7 For each diagram, draw axes labelled from 0 to 8. Shade the region *not* described by the inequalities. Then find which point in the permitted region gives the maximum or minimum value required.

a The inequalities $\begin{cases} x + y \leqslant 7 \\ x \geqslant 1 \\ y \geqslant 2 \end{cases}$

Minimise $2x + y$

b The inequalities $\begin{cases} y \leqslant x + 3 \\ y \geqslant 4 \\ x \geqslant 2 \end{cases}$

Minimise $3x + 4y$

c The inequalities $\begin{cases} y \leqslant x + 1 \\ x + y \geqslant 5 \\ x \leqslant 4 \end{cases}$

Maximise $5y - x$

d The inequalities $\begin{cases} x + y \leqslant 8 \\ x + y \geqslant 4 \\ 1 \leqslant x \leqslant 4 \end{cases}$

Maximise $4x - y$

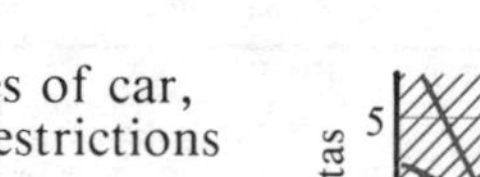

8 A company is intending to buy two types of car, Cyanes and Magentas. There are three restrictions on the numbers of cars which they can buy as shown by the unshaded region of this diagram.

a List the six possibilities shown in this diagram.

b If each Cyane costs £6000 and each Magenta costs £8000, find the total cost of each of the six possibilities.

c How many of each type of car gives the least total cost?

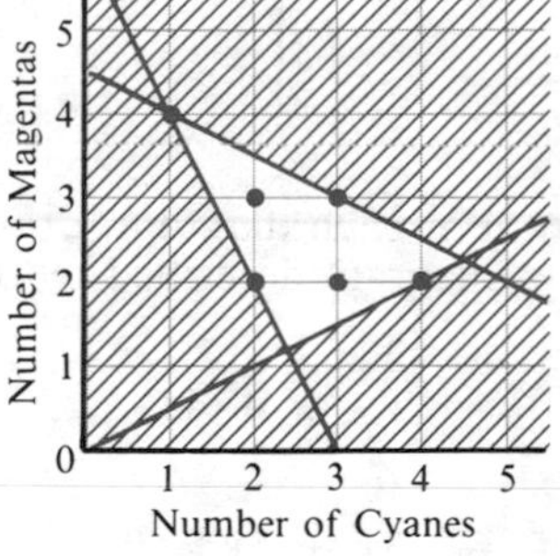

9 An architect designs a house to have x double windows and y single windows. Write inequalities for these statements.

(i) There must be at least two double windows.

(ii) There must be at least three single windows.

(iii) The total number of windows will be no more than eight.

On axes labelled from 0 to 10, indicate by shading the region given by your inequalities.

Cost Each double window costs £120 and each single window costs £80. Copy and complete this table for the extreme corners of the permitted region.

Corner	(,)	(,)	(,)
Cost, £			

How many windows of each type should be used to keep the cost to a minimum?

Algebra

10 A man is supplying electrical power to his garage using x metres of standard gauge wire and y metres of heavy gauge wire. He will need at least two metres of heavy gauge wire. The length of heavy gauge is no more than the length of standard gauge. The total length of wire is no more than eight metres.

a Write three inequalities from the above information. On axes labelled from 0 to 10, indicate the region given by the inequalities.

b One metre of standard gauge wire costs £1·50 and one metre of heavy gauge costs £2·00. Find the minimum cost of the wire which is needed and say how much of each type of wire should then be bought.

11 a A transport company has two types of lorry, Juggers and Nauts. Each Jugger can hold 8 tonnes per load and each Naut can hold 4 tonnes per load. The company has a delivery of 32 tonnes to make to a town so far away that each lorry can be used only once. If x Juggers and y Nauts are used, write an inequality involving x and y.

b If the company will use at least 6 drivers for its lorries, write another inequality involving x and y.

c If the company will use a minimum of 2 Nauts to make its delivery, write a third inequality for x and y.

On axes labelled from 0 to 10, show by shading the region which represents these inequalities. If each journey by a Jugger costs £25 and each journey by a Naut costs £10, find the minimum cost for the delivery and say how many of each type of lorry should then be used.

12 A gardener plants only potatoes and beans in his vegetable plot. He plants x m^2 of potatoes and y m^2 of beans. He must plant at least 10 m^2 of potatoes and at least 5 m^2 of beans. The area planted with beans is to be no more than twice the area planted with potatoes. He intends planting a total area of at least 20 m^2.

a Write four inequalities from the above information. On axes labelled from 0 to 25, indicate the region given by the inequalities.

b It costs £1 to buy seed potatoes to plant in each square metre and it costs 50 pence to buy bean seeds for each square metre. What is the minimum possible cost of seeds and how much would then be spent on each type of vegetable?

Part 6 Flow diagrams and computer programs

1 This flow diagram describes what you have to do before you arrive at school in a morning.
Copy it and enter these instructions in their correct order.

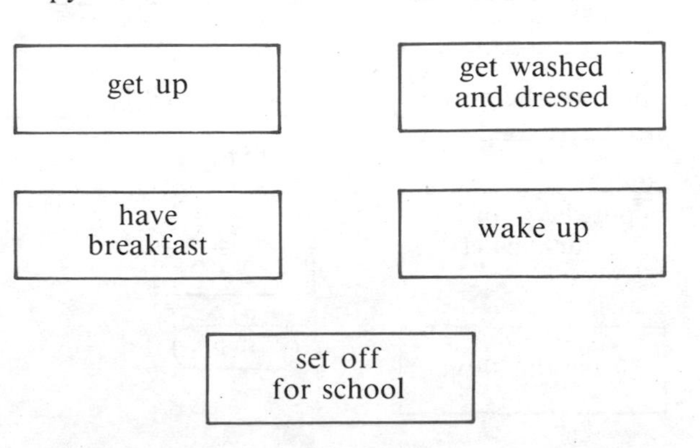

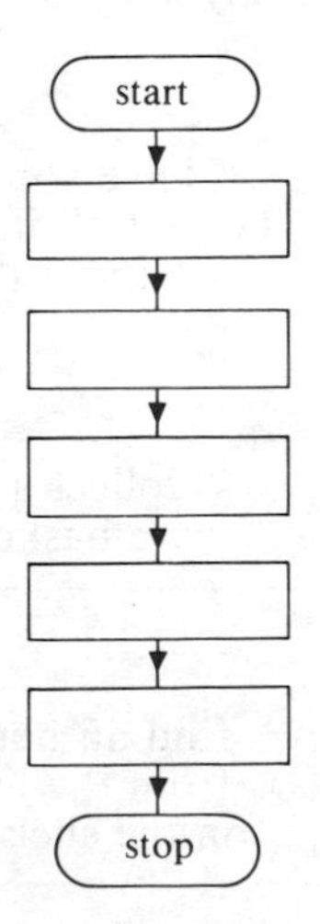

Algebra

2 These instructions are to be followed when planting a tree. Copy the flow diagram and insert the instructions in their correct order.

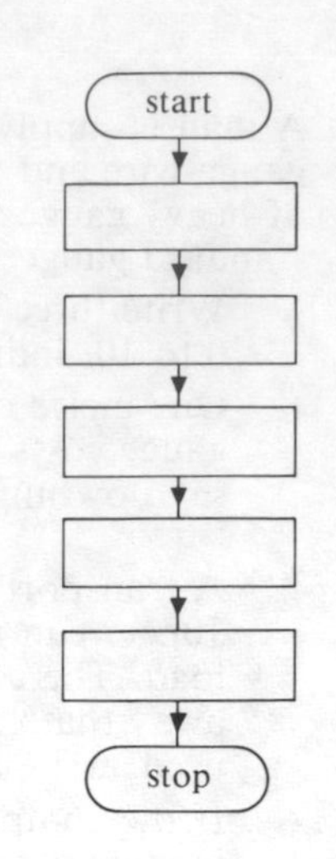

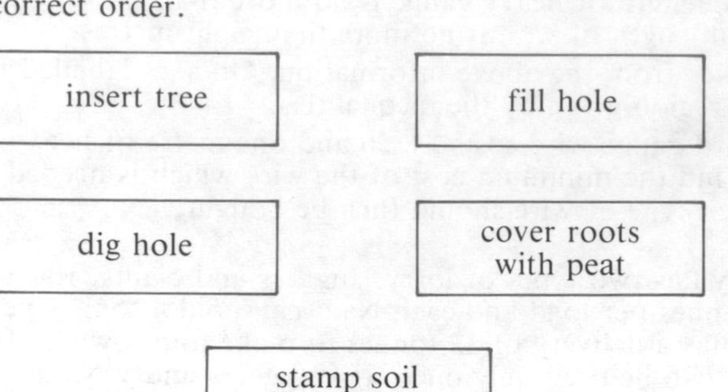

stamp soil down firmly

3

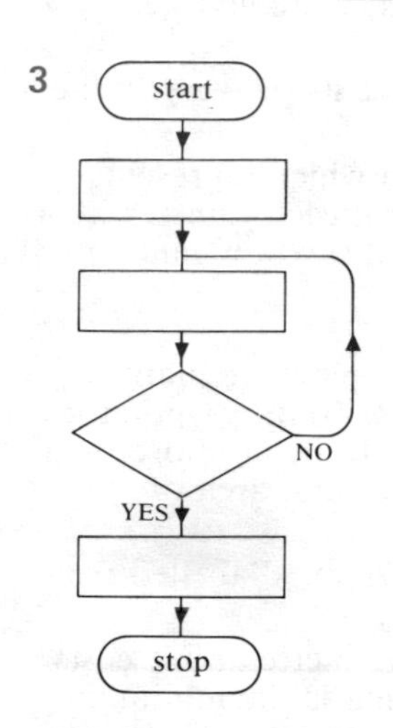

You have a bowl of cold water. You want to add enough hot water so that you can use it to wash up some plates.
Copy and complete this flow diagram using the given instructions.

run the hot water into the bowl

turn on the hot tap

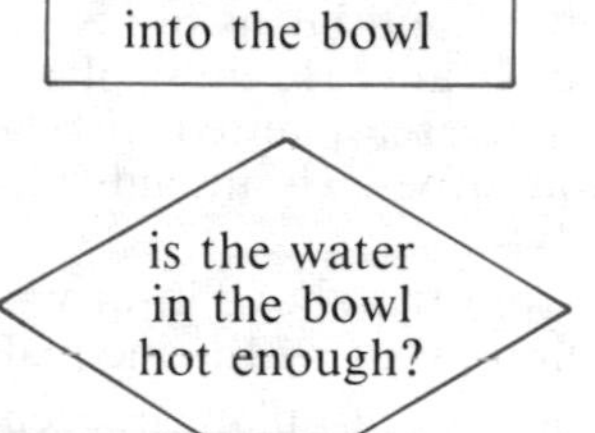

turn off the hot tap

4 You can follow these instructions when you use a shopping list in the supermarket. Copy and complete the flow diagram.

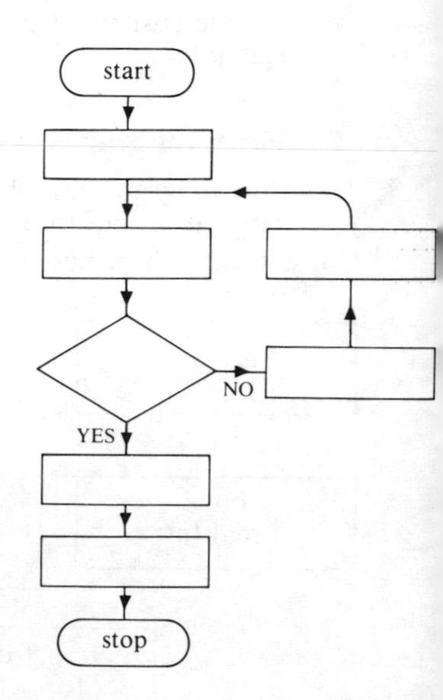

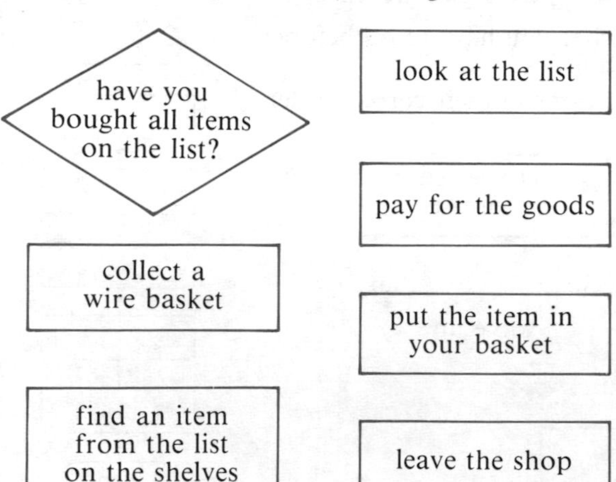

Algebra

5 This flow diagram describes the process of knocking a nail into a piece of wood.
Copy and complete this diagram using these instructions.

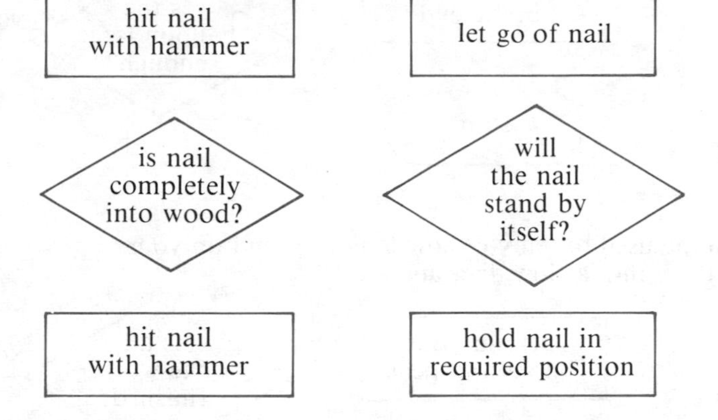

start
NO
YES
NO
YES
stop

6 These instructions can be used to describe the opening of a locked door. Arrange them to form a flow diagram.

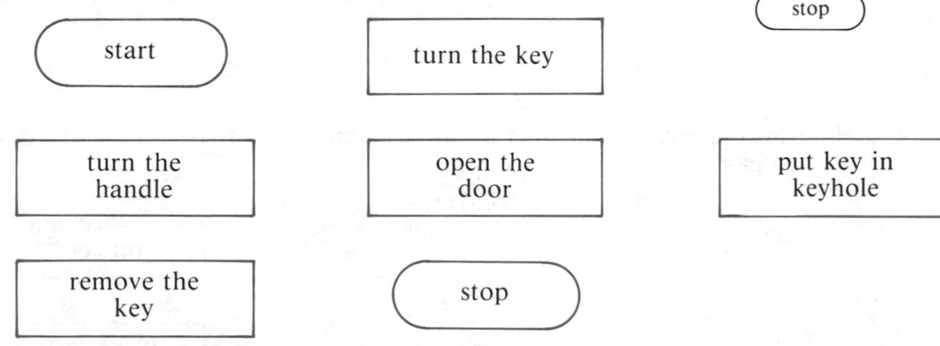

7 A car driver stops at a self-service garage to fill his tank with petrol. Arrange these instructions into a flow diagram to describe the operation.

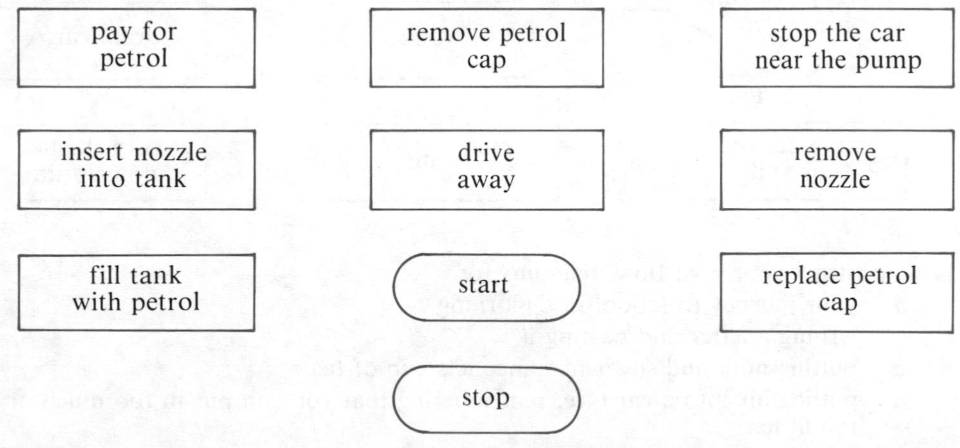

Algebra

8 Arrange these instructions into a flow diagram to describe the operation of blowing up a balloon.

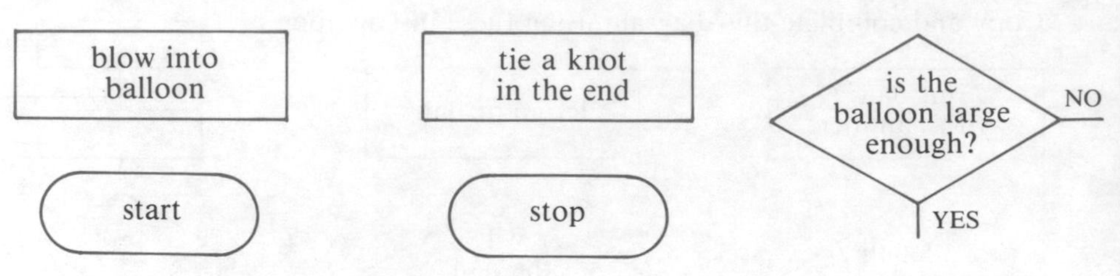

9 These instructions can be used to tell you how long to spend on your homework. Arrange them into a flow diagram.

10 Use these instructions to construct a flow diagram which tells you how to boil carrots in a pan.

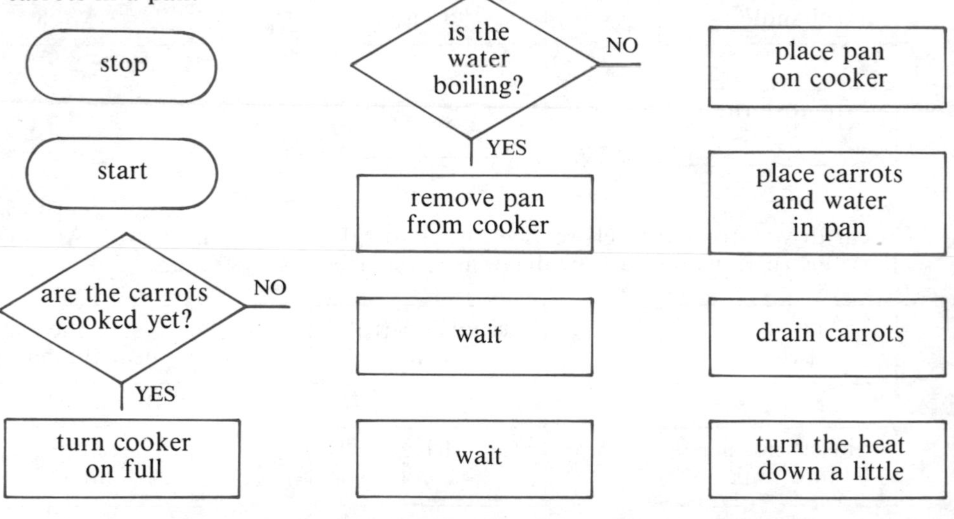

11 Construct your own flow diagrams for

- **a** your journey to school in a morning
- **b** writing a letter and posting it
- **c** putting milk and sugar in someone's cup of tea
- **d** putting air into a car tyre, remembering that you can put in too much or too little.

12 Julie (19), David (18), Stephen (14), Angela (9) and Pauline (5) are brothers and sisters.

- **a** Use their ages in this flow diagram and write down the final result.
- **b** What is it that the flow diagram calculates?
- **c** Can you describe in words the final value of T?

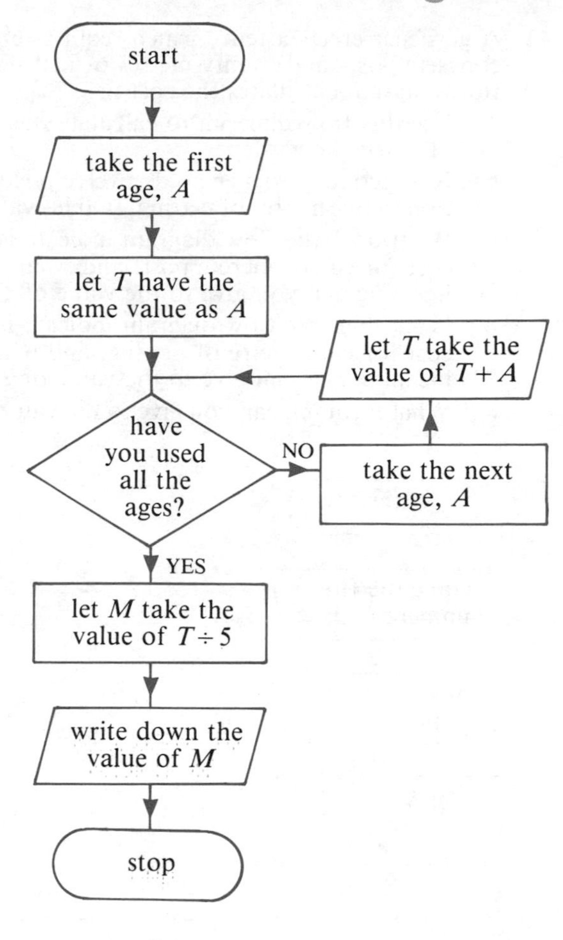

13

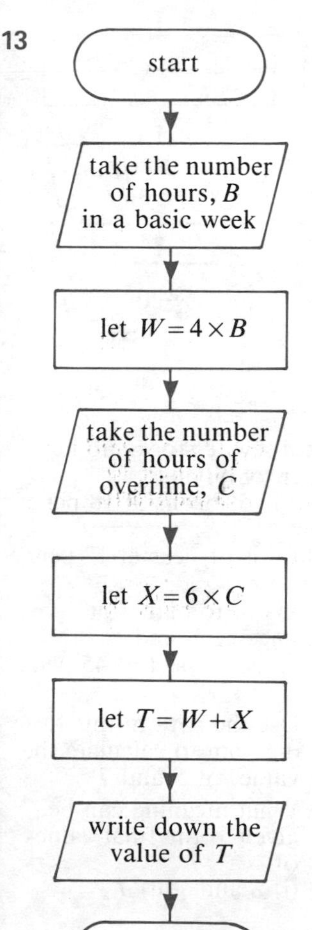

A man works a basic thirty-six hour week and also five hours overtime.

- **a** Use this flow diagram to calculate his wage £T for the week.
- **b** What meaning can be given to the symbols W and X?
- **c** How much does he earn per hour in the basic week?
- **d** How much does he earn per hour for overtime?

Algebra

14 A gardener erects a fence which requires eight concrete posts and twenty metres of netting. This flow diagram calculates the cost in £.

- **a** Use this flow diagram to calculate the value of T for the gardener.
- **b** If a second gardener needs twelve posts and twenty five metres of netting, find his value of T.
- **c** What does the flow diagram indicate is the cost for each concrete post, and what meaning can you give to the value of C?
- **d** What does the flow diagram indicate is the cost for each metre of netting, and what meaning can you give to the value of D?
- **e** What meaning can you give to the value of T?

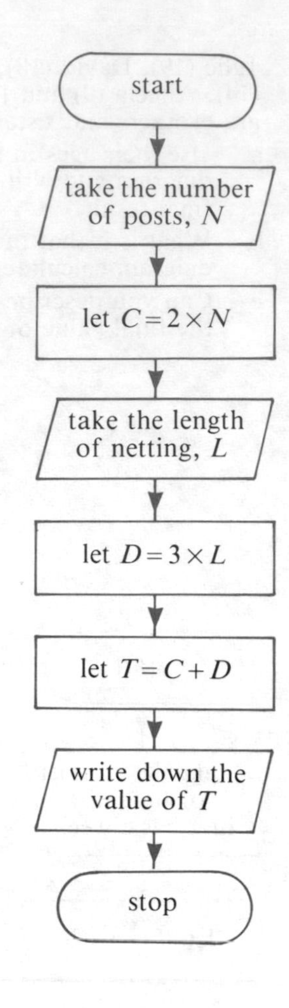

15

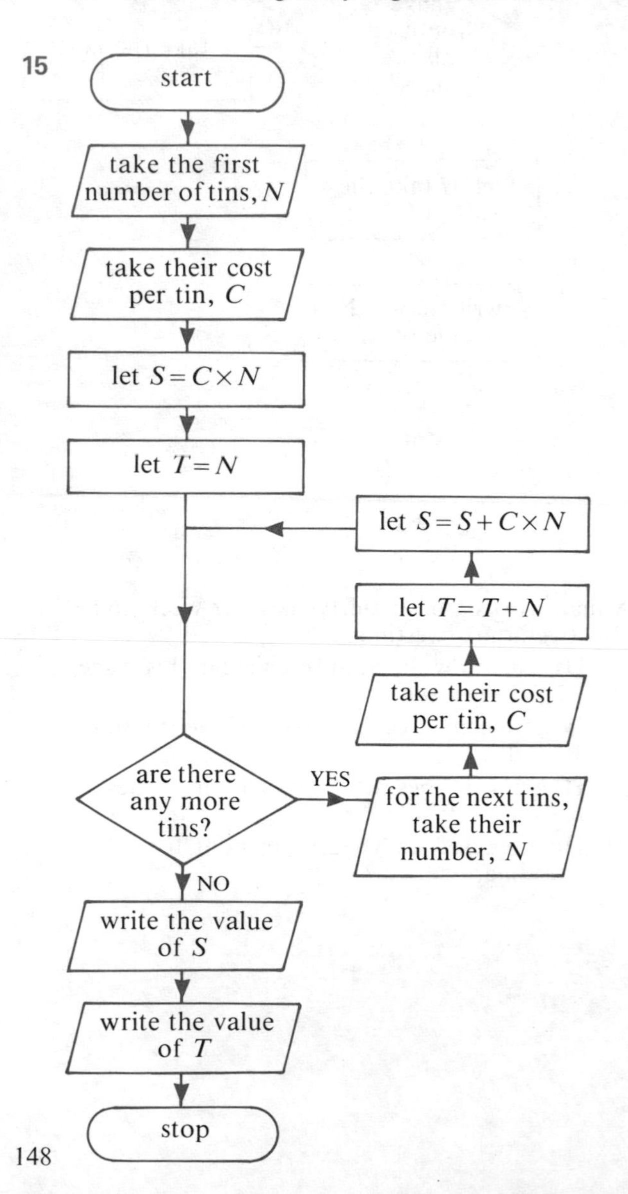

A housewife stocks up her larder by buying
ten tins of beans at 18 pence each,
eight tins of peas at 12 pence each,
six tins of tomatoes at 15 pence each and
four tins of meat at 45 pence each.

- **a** Use the data in this flow diagram to calculate the values of S and T.
- **b** What meaning can be given to the final values of
  (i) S and (ii) T?

Algebra

The next problems involve writing short computer programs. If you have a computer, you might wish to try the programs on it.

16 This flow diagram can be used to calculate the area A of a circle of radius r.

a Use the diagram to help you arrange these five instructions to give a program which calculates the value of A.

```
PRINT A
A=3.14×S
INPUT R
END
S=R*R
```

1Ø ....
2Ø ....
3Ø ....
4Ø ....
5Ø ....

b Find the value of A when $r = 5$ cm.

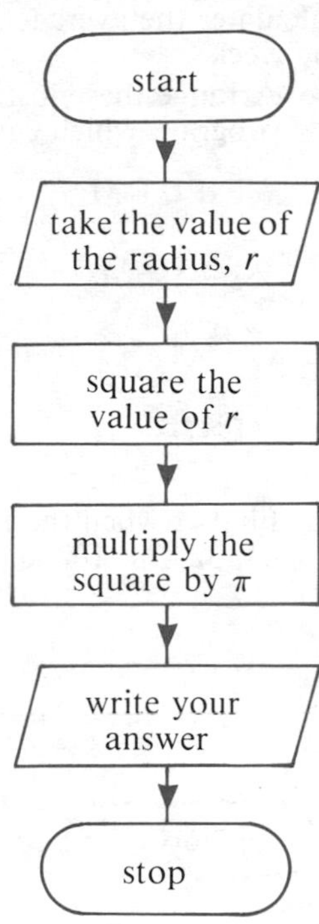

17

start

take the number of hours, H spent gardening

let $T = \frac{1}{2} \times H$

take the amount of pocket money, P

let $T = T + P$

write the value of T

stop

A boy's parents give him £P pocket-money each week, plus fifty pence for each of the H hours he spends gardening. This flow diagram can be used to calculate the total amount £T which he receives in a week.

a Arrange these six instructions into a program which calculates the value of T.

```
T=T+P
INPUT H
T=½*H
INPUT P
END
PRINT T
```

1Ø ....
2Ø ....
3Ø ....
4Ø ....
5Ø ....
6Ø ....

b Find T when $H = 4$ and $P = 2$.

Algebra

18 A salesman travels M miles each day of a five-day working week. This flow diagram calculates the average daily mileage A for the week.

a Arrange these instructions to form a program which calculates the value of A.

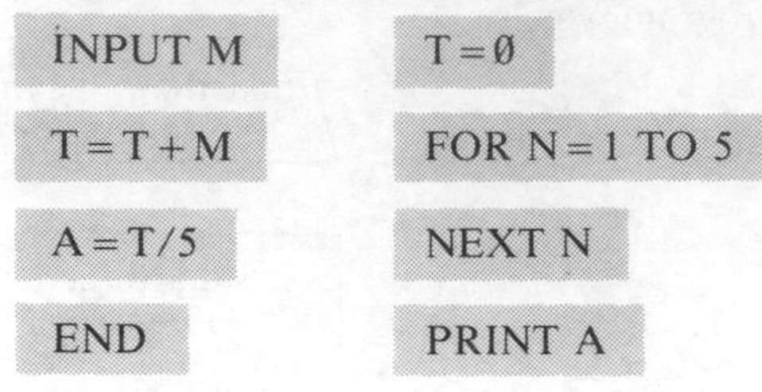

b Find A when the five values of M are 64, 32, 78, 102 and 24.

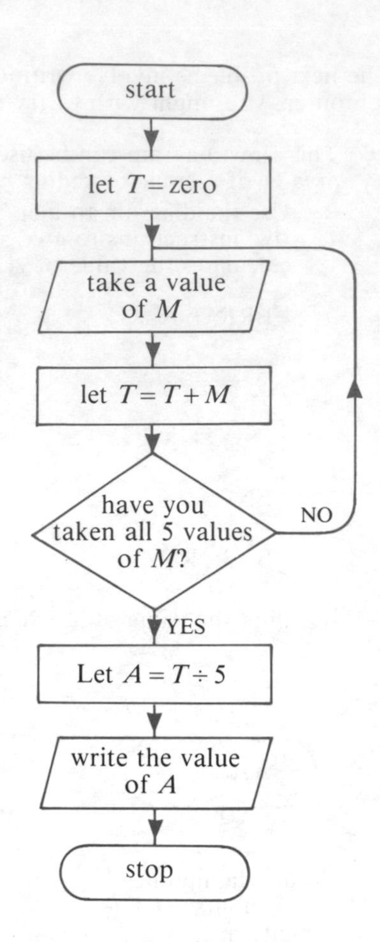

19

start
let T = zero
take one boy's contribution, C
let $T = T + C$
have all 4 boy's contributions been taken ?
NO
YES
let $X = 80 - T$
write the value of X
stop

Four boys do not have enough money to buy a box of fish bait costing 80 pence. This flow diagram calculates how much extra they need if they pool their money to buy one box.

a Arrange these instructions into a program which calculates the extra amount X which is needed.

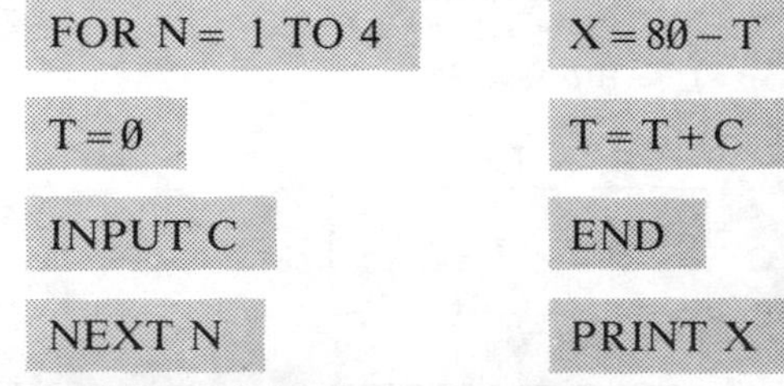

b Find the value of X when the four values of C are 12, 24, 18 and 14.

c What meaning can you give to the final value of T?

d Which line of the program would you need to alter if the cost of a box of bait increased to £1·20?

Algebra

20 A small primary school orders packets of six kinds of exercise book, each packet having the price given in this table.

Type of book	Number of packets, P	Cost per packet, C
wide lines	8	£1·50
narrow lines	2	£1·50
small squares	3	£1·40
large squares	4	£1·40
plain	12	£1·30
coloured	4	£1·60

a Arrange these instructions into a flow diagram which will calculate the total cost of the order.

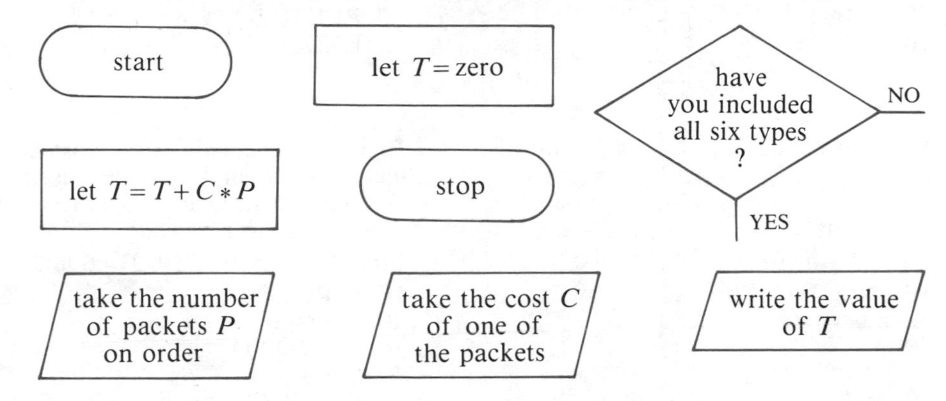

b Arrange these instructions into a program to find the total cost of the order, and calculate the total cost.

FOR $N=1$ TO 6 | $T=0$ | INPUT P | END

INPUT C | $T=T+C*P$ | NEXT N | PRINT T

Dry runs

All these programs will work on a Research Machine 380Z or a BBC model B computer. But you can do a *dry run* for each program without using a computer. Imagine the store for each lettered variable as a box into which you place its value.

For example, Program 1 can use boxes like this

M	N	P

Some of these values may change as you work through the program.

Program 1 calculates the bus fare P pence for a journey of M miles at a rate of 5p per mile with a basic charge of 25p.

Do a dry run with $M=6$ to find the value of P.

```
10 INPUT M
20 N=5*M
30 P=N+25
40 PRINT P
50 END
```

Program 2 finds the cost £C of covering a rectangular area L metres long and W metres wide with concrete costing £2 per m^2.

Do a dry run with $L=6$ and $W=4$ to find the value of C.

```
10 INPUT L
20 INPUT W
30 A=L*W
40 C=2*A
50 PRINT C
60 END
```

Algebra

Program 3 calculates the volume V cm^3 of a pyramid of height H cm with a square base of side L cm.

Do a dry run with $L=2$ and $H=6$ to find the value of V.

```
10 INPUT L
20 A=L*L
30 INPUT H
40 U=A*H
50 V=U/3
60 PRINT V
70 END
```

Program 4 finds the distance D metres travelled in time T seconds by a car starting from rest with an acceleration of A m/s^2.

Find D when $T=3$ and $A=8$.

```
10 INPUT T
20 Y=T*T
30 INPUT A
40 Z=A*Y
50 D=Z/2
60 PRINT D
70 END
```

Program 5 calculates the acceleration A m/s^2 of a car which increases its speed from U m/s to V m/s in a time T seconds.

Find A when $U=8$, $V=23$ and $T=2$.

```
10 INPUT U, V, T
20 Z=V-U
30 A=Z/T
40 PRINT A
50 END
```

Program 6 calculates the cost £C of electricity when N units are used costing P pence each, together with a standing charge of £S.

Find C when $N=250$, $P=6$ and $S=8$.

```
10 INPUT N, P, S
20 A=N*P
30 B=A/100
40 C=B+S
50 PRINT C
60 END
```

Program 7 works out the annual interest £I earned by an account holding £P at a rate of R% p.a. and also gives the total amount £Q in the account at the end of the year.

Find both I and Q when $P=200$ and $R=8$.

```
10 INPUT P, R
20 Z=P*R
30 I=Z/100
40 Q=P+I
50 PRINT I
60 PRINT Q
70 END
```

Program 8 calculates the wage £W of a man earning £X an hour for T hours and then 'time-and-a-half' for S hours.

Find W when $X=4$, $T=38$ and $S=6$.

```
10 INPUT X, T, S
20 A=X*T
30 B=1.5*X*S
40 W=A+B
50 PRINT W
60 END
```

Program 9 finds the cost £C of a gas bill when N therms are used costing P pence each together with a standing charge of £S.

Find C when $N=135$, $P=40$ and $S=12$.

```
10 INPUT N, P, S
20 A=N*P
30 B=A/100
40 C=B+S
50 PRINT"Standing charge is";S
60 PRINT"Cost of therms is ";B
70 PRINT"Total bill is      ";C
80 END
```

Algebra

Program 10 solves the simultaneous equations $Ax + By = C$
$Dx + Ey = F$.

Use the program to solve $3x + 4y = 17$
$2x + 5y = 16$.

```
10 INPUT A, B, C, D, E, F
20 G=A*E
30 H=D*B
40 I=G-H
50 G=E*C
60 H=B*F
70 J=G-H
80 G=A*F
90 H=D*C
100 K=G-H
110 G=J/I
120 PRINT"The value of x is";G
130 H=K/I
140 PRINT"The value of y is";H
150 END
```

Program 11 calculates a cricketer's batting average A after five innings when he scores R runs in each innings.

Find A for these values of R:
18, 42, 64, 102, 39.

```
10 T=0
20 FOR N=1 TO 5
30 INPUT R
40 T=T+R
50 NEXT N
60 A=T/5
70 PRINT"His average is";A
80 END
```

Program 12 calculates the 'goal difference D' for a football team which plays six matches.

Find D for these match results, where 'goals for' F are first and 'goals away' A are second:

3 – 1   2 – 0   0 – 1
1 – 1   2 – 3   0 – 1

```
10 T=0
20 U=0
30 FOR N=1 TO 6
40 INPUT F
50 T=T+F
60 INPUT A
70 U=U+A
80 NEXT N
90 D=T-U
100 PRINT"The goal difference is";D
110 END
```

Program 13 calculates the delivery charge £D made by a department store when transporting a customer's purchase for a distance of M miles at a rate of 20 pence per mile. But no charge is made if the cost of the goods £C is more than £50.

Use the program twice to find the value of D when

a $C = 35$, $M = 8$ **b** $C = 90$, $M = 8$.

```
10 INPUT C, M
20 A=20*M
30 D=A/100
40 IF C>50 THEN D=0
50 PRINT"Delivery charge is £";D
60 END
```

Program 14 tests to see if the point (A,B) lies on the line $y = Mx + C$.

Use the program twice to see if **a** the point (5,18) **b** the point (3,17) lies on the line $y = 4x + 5$.

```
10 PRINT"What are the values of M and C"
20 INPUT M,C
30 PRINT"What is your point (A,B)"
40 INPUT A,B
50 Y=(M*A)+C
60 IF Y=B THEN PRINT"The point is on the line"
70 IF Y<>B THEN PRINT"The point is NOT on the line"
80 END
```

Sets

1 This Venn diagram can be used to illustrate these sets:
$\mathscr{E}$ = {whole number from 1 to 15}
T = {multiples of 3}
E = {even numbers}
Copy the diagram and write in all the numbers.

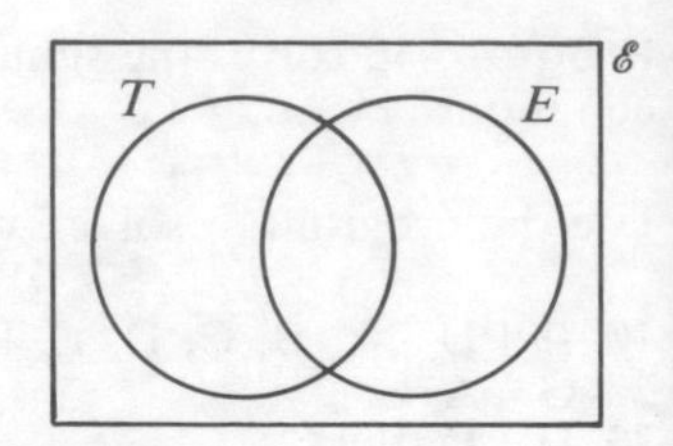

2 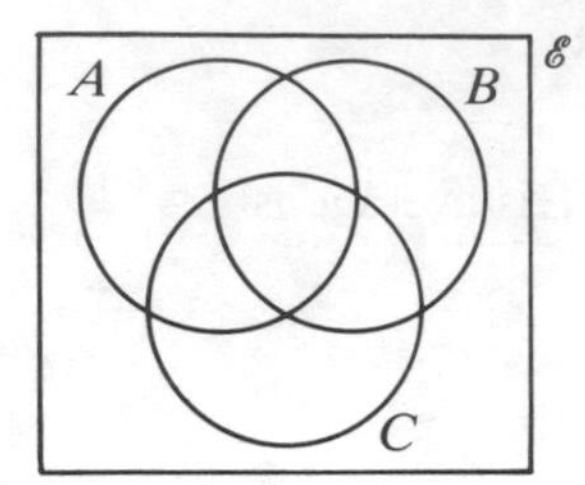

This diagram illustrates these sets:
$\mathscr{E}$ = {whole numbers from 1 to 20}
A = {odd numbers}
B = {multiples of 5}
C = {numbers greater than 10}
Copy the diagram and write in all the numbers.

3 A gardener classifies his plants into three sets:
S = {spring flowers}
B = {those grown from bulbs}
Y = {yellow flowers}
Copy this Venn diagram and write in these flowers:
snowdrop, primrose, buttercup, daffodil, rose.

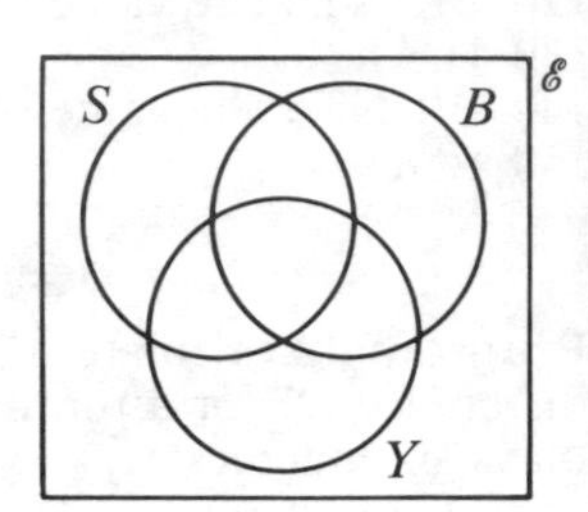

4 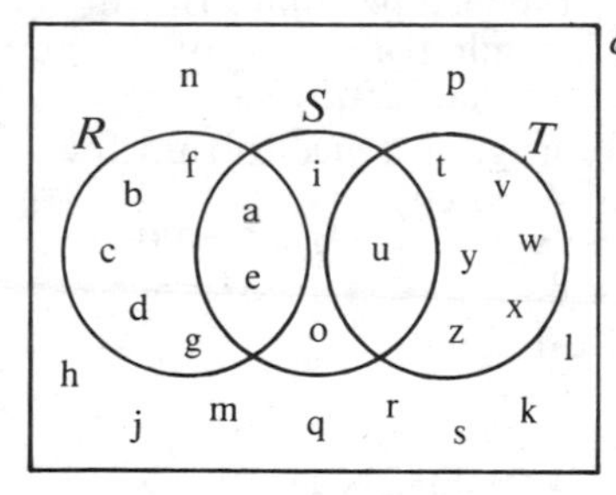

$\mathscr{E}$ = {the alphabet}
R = {the first seven letters}
S = {vowels}
T = {the last seven letters}

Use this diagram of sets R, S and T to say whether these statements are *true* or *false*.

a $c \in R$ **b** $i \in S$
c $u \in T$ **d** $o \notin S$
e $q \notin T$ **f** $n(R) = 5$
g $n(R \cup S) = 8$ **h** $n(S \cap T) = 1$
i $n(R') = 12$ **j** $n(R \cup S \cup T)' = 4$
k $R \cap S = \{u\}$ **l** $(S \cup T)' = \{m, n, p, q\}$

5 This diagram shows all the children who live on one street grouped to show who are members of three clubs: rugby R, swimming S and tennis T.
Copy and complete these statements.

a Simon $\in$ ... **b** Pam $\in$ ...
c Jim $\notin$ **d** Susan $\notin$
e $R \cap S = \{......\}$ **f** $S \cap T = \{......\}$
g $n(R \cup S) = ...$ **h** $n(S \cup T) = ...$
i $n(R') = ...$ **j** $n(S') = ...$
k $n(R \cup S)' = ...$ **l** $T \subset ...$

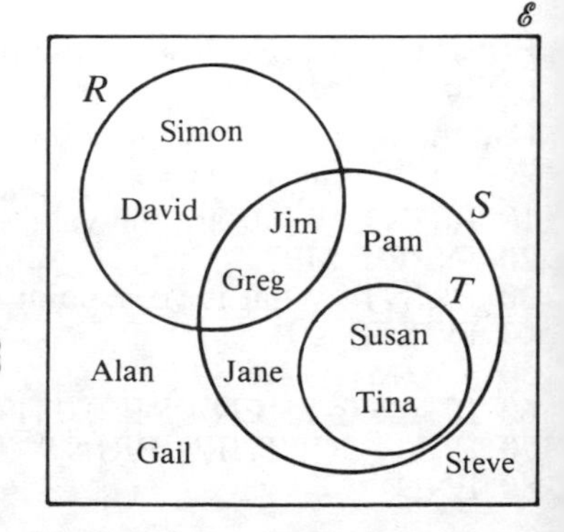

Sets

6 Match the letters on these Venn diagrams with the given sets.

a

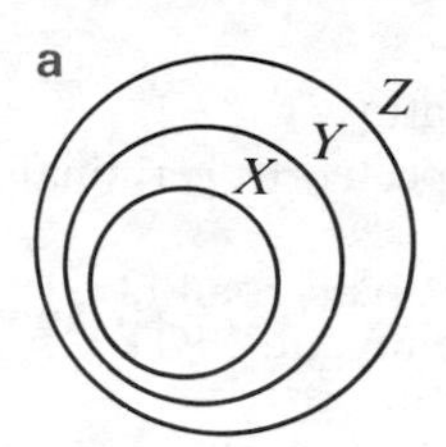

{all children}
{all babies}
{all young people}

b

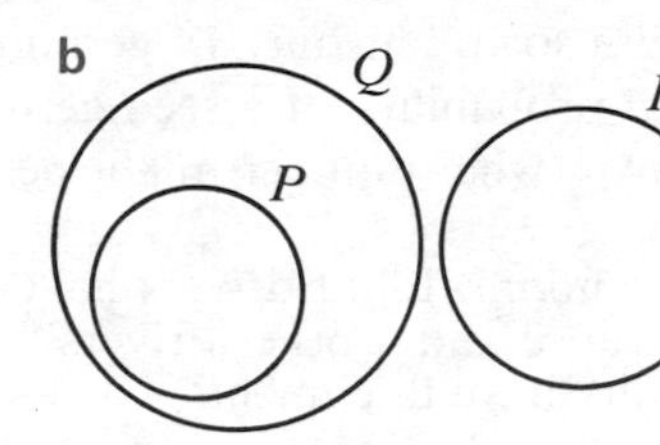

{all saloon cars}
{two-door saloon cars}
{estate cars}

c

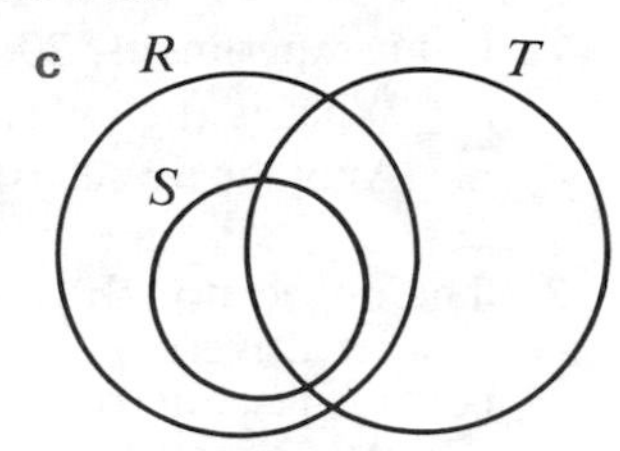

{all cats}
{all leopards}
{African animals}

7 Copy this Venn diagram eight times and shade the given sets.

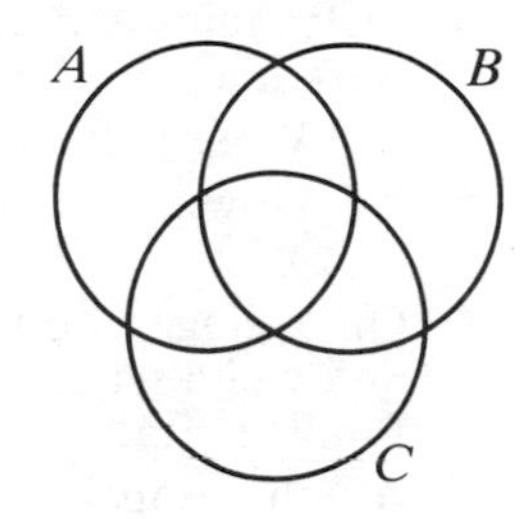

a $B \cap C$ **b** $A \cup C$ **c** $A \cap B \cap C$
d $A \cup B \cup C$ **e** $(A \cap B) \cup C$ **f** $A \cap (B \cup C)$
g $(A \cup B) \cap C$ **h** $A \cup (B \cap C)$

8 In a class of twenty nine pupils, twenty two choose to study science, eighteen choose to study art and four study neither of these subjects.
Use this diagram to find how many study both these subjects.

9 A firm advertises two jobs, one for a clerk and the other a technician. Sixty four people apply for both posts and thirteen apply for only the clerk's position. If there are ninety four applicants altogether, how many of them applied for the technician's post?

10 Sixteen of the eighty eight employees at a factory do not use public transport to get to work. Of those who do, sixty seven use a bus and seventeen use a train. How many use both bus and train?

11 A garage inspects twenty cars for their MOT tests of which twelve cars pass immediately. Five cars have faulty lights and six have faulty brakes. How many cars have faulty brakes but good lights?

12 A school of 800 pupils has 420 boys. There are 140 first-formers of whom exactly half are boys. How many girls are not in the first form?

Probability

Part 1 Experimental and theoretical

1 In an experiment, 20 seeds are sown but only 14 germinate.
- **a** What is the experimental probability of a seed germinating?
- **b** Another 60 seeds are sown, what number might be expected to germinate?

2 The police stop 100 cars at random. Eight drivers have not paid road tax, 3 other drivers have no insurance and 4 other drivers have invalid MOT test certificates; the rest of the cars have documents which are in order.
- **a** What is the probability of a driver's documents being in order?
- **b** If the police stop another 300 cars, how many might they expect to be untaxed?

3 Fifty trains arrived at Southwich station this morning. Twelve arrived early, 8 were late and the rest were on time.
- **a** What is the probability of the next train being on time?
- **b** How many of the next 15 trains would you expect to be on time?

4 On average, 3 out of every 100 households have no television, 21 have a black and white set and the rest have a colour set. If a school has pupils from 800 families, find
- **a** the probability of a pupil having a colour television at home
- **b** how many of all the families might have colour televisions.

5 In which of these trials are the given outcomes *equally likely* to occur?

	Trial	*Outcomes*
a	tossing a coin	either a head or a tail
b	rolling a dice	either an odd score or an even score
c	picking up one shoe	either for a right or a left foot
d	cutting a pack of cards	either a picture card or not
e	arriving at a bus stop	either being first in the queue or not
f	choosing any letter in the alphabet at random	a letter either before and including M or after M

6 What is the theoretical probability of
- **a** rolling an even score on one dice
- **b** selecting any club from a pack of cards
- **c** a watch stopping at between 3 and 9 o'clock
- **d** selecting BBC2 at random from all the channels on a television
- **e** cutting a pack of cards at a male picture card
- **f** choosing a vowel from the letters of the alphabet?

7
- **a** What is the probability of cutting a pack of cards at any picture card?
- **b** If you cut the same pack of cards 65 times, how many picture cards might you expect?

8
- **a** What is the probability of this spinner falling on an odd score?
- **b** How many odd scores might you expect if it is spun 60 times?

Probability

9 This circle is spun about its axis 30 times in succession and the shading of the section which comes to rest at the arrow is noted each time.
How many of the 30 spins might be expected to give

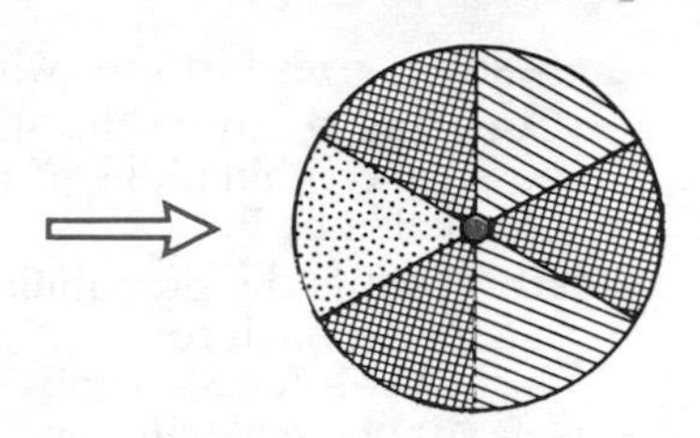

a a dotted sector **b** a striped sector?

10 0 $\frac{1}{2}$ 1 This number line from 0 to 1 represents all possible probabilities.

Indicate by arrows labelled from **a** to **f** the probabilities of

a tossing a head on a coin
b cutting a spade with a pack of cards
c May following April
d the electricity being cut off tomorrow
e scoring a 1 or 6 on a dice
f scoring 23 with one dart on a dartboard.

11 This Venn diagram shows the number of people who work in an office. Some can type; some can use a word processor.

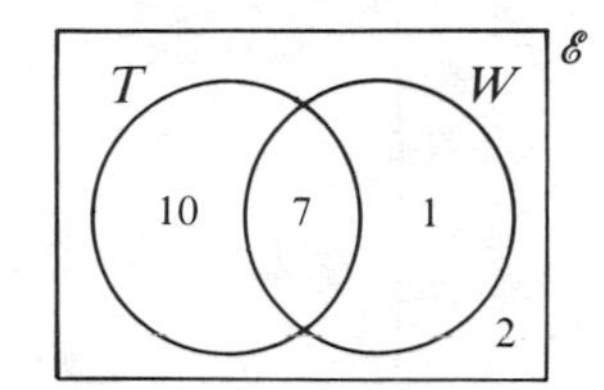

ℰ = {all office workers}
T = {those who can type}
W = {those who can use a word processor}

a How many office workers are there?
b If one of them is chosen at random, what is the probability that he or she can type but not use the word processor?

12

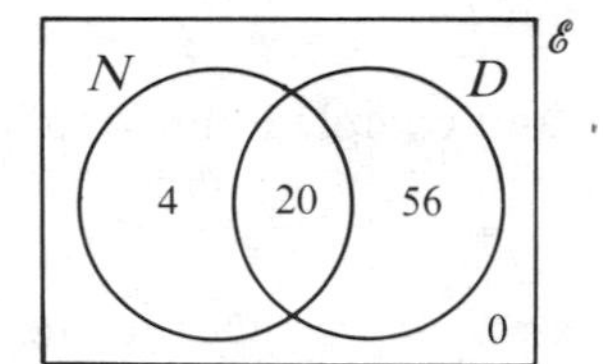

A factory employs people who work on day-shifts and night-shifts.
ℰ = {all those employed}
D = {those who have worked day-shifts}
N = {those who have worked night-shifts}

a How many employees work in this factory?
b If one of them is chosen at random, what is the probability that he or she has worked only day-shifts? Give your answer as a decimal.

Part 2 Successive events

1 Two fair dice are rolled, one after the other. Copy and complete this diagram showing all possible total scores.

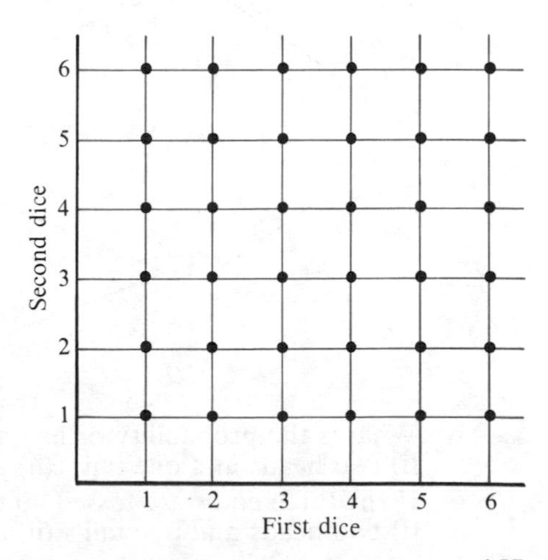

a Which total score is most likely?
b What is the probability of a total score of
(i) 9 (ii) 8 (iii) 7?
c If the two dice are rolled 180 times, how many times would you expect a total score of 5?

Probability

2 A sheep gives birth to twin lambs.
Copy and complete this diagram to show all possible combinations of male lambs M and female lambs F.

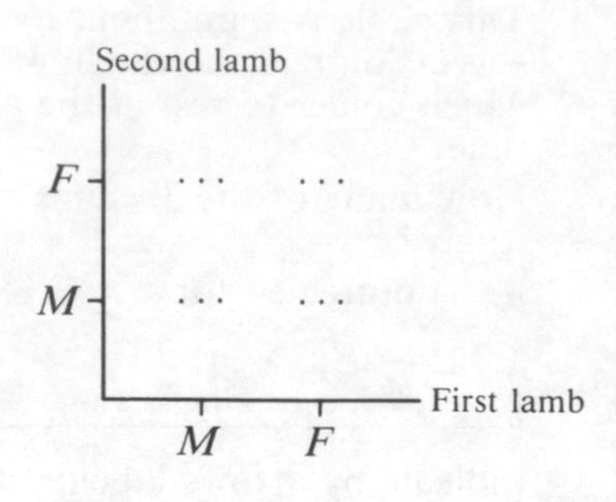

a What is the probability of the sheep giving birth to
(i) two female lambs
(ii) one male and one female lamb?

b If 60 sheep in a shepherd's flock give birth to twins in springtime, how many of them would you expect to give birth to one of each sex?

3 A machine produces nuts and bolts for children's toys. Four different colours of bolt and two different colours of nut are equally likely.
Copy and complete this diagram to show all the possible combinations of colours when a nut and bolt are used together.

Colour of nut
blue B … … … …
red R … … … …
red R green G blue B yellow Y
Colour of bolt

a What is the probability of a nut and bolt having the same colour?

b A toy needs 40 nuts and bolts. How many pairs do you expect there to be of the same colour, if they are paired together at random?

4 a Three coins are tossed, one after the other. Copy and complete this tree diagram to show all possible outcomes.

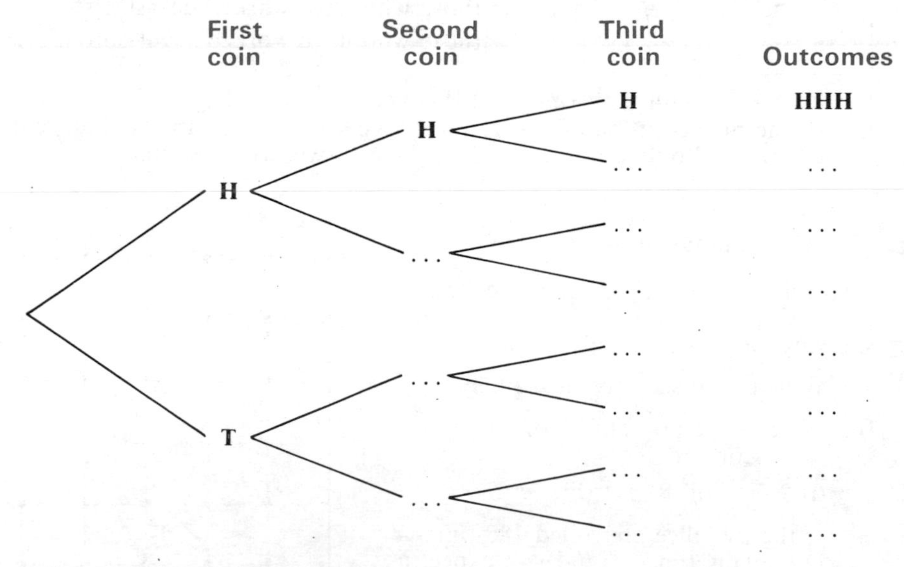

b What is the probability of having
(i) two heads and one tail (ii) all three coins showing the same side?

c If the three coins are tossed 40 times, how many times would you expect
(i) two heads and one tail (ii) all three coins showing the same side?

Probability

5

First cut	Second cut	Outcomes	Probabilities
S	...	...	...
	...	...	...
no S	...	...	...
	...	...	...

a A pack of 52 cards is cut, shuffled and cut again in the hope of cutting two spades. Copy and complete this tree diagram, list the possible outcomes and calculate their probabilities.

b In a certain game, the pack is cut twice on 32 separate occasions. How many times might you expect to see
(i) two spades (ii) just one spade?

6

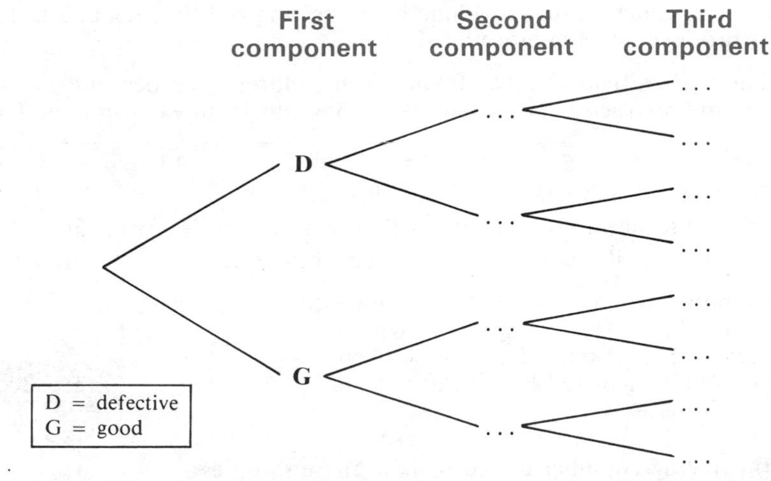

a A machine stamps out plastic components and packs them in boxes of three. One day it is discovered that a random fault is making one fifth of all components defective.
Copy and complete this tree diagram and calculate the probability of a box of three components having
(i) all three defective (ii) just one defective (iii) none defective.

b If 500 boxes were packed before the fault was rectified, how many boxes might be expected to contain
(i) three defective components (ii) just one defective component
(iii) no defective components?

7 A film crew arrive on location intending to spend twenty days filming. The probability of the weather being suitable is $\frac{3}{4}$; the probability of all those taking part being well is $\frac{4}{5}$ and the probability of the equipment breaking down is $\frac{1}{3}$. If they film only in suitable weather with everyone well, draw a tree diagram to find on how many days they might expect filming to take place.

8 There are three vacant seats on a committee which are to be filled by any three from six people, Mr Jones, Mrs Morgan, Miss Reece, Mr Price, Mr Pritchard and Mr Evans. Three people will be selected by having their names drawn from a hat. What is the probability of the vacancies being filled by

a three men **b** one woman and two men?

Statistics

Part 1 Sampling

1 Describe in what ways these methods of sampling the populations might give *biassed* samples.

- **a** A margarine salesman tests the popularity of his product by asking those people leaving a supermarket who have already bought some of the margarine.
- **b** A political party samples the voters in a town by ringing up a large number of them on the telephone.
- **c** A cross-section of first-year infants in a primary school are to be tested and those with birthdays in August and September are chosen.

2 Which of the following will produce *random samples* and which will not? Give your reasons.

- **a** List the pupils in your year at school and select every fifth name to be in your sample.
- **b** Sample the cars in a multi-storey car park by selecting those with registrations ending in the letter Y.
- **c** Check the production of a machine by inspecting the first ten and last ten articles produced on the morning shift.

3 A school has a grass field of area 700 m^2. Ten children each peg out one square metre and extract as many worms as they can from each square. The results are:

15 8 12 10 7 14 21 14 11 13

- **a** What is the total number of worms collected?
- **b** Estimate the total number of worms living in the whole of this grass area.
- **c** How could the children obtain a more accurate estimate of this total?

4 This sample bottle contains 500 balls, some white and some red. With the bottle upside-down, a sample of ten balls is taken. This is done twelve times and the number of red balls in each sample is counted with these results:

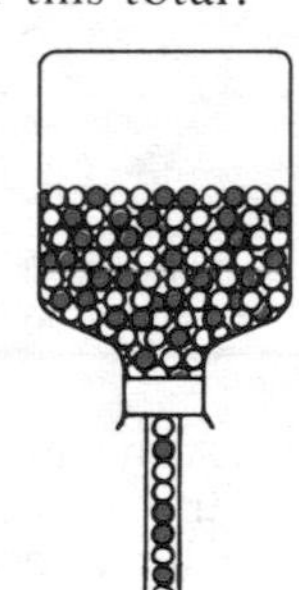

2 3 4 3 3 5 1 4 3 2 4 2

- **a** Find the average number of red balls in these samples.
- **b** Estimate the number of red balls in the bottle.
- **c** What would you do to improve this estimate?

5 A school has its children grouped into 36 classes. Eight classes are chosen at random and the numbers of pupils in each class who have had measles are counted with these results:

15 21 14 18 20 19 16 17

- **a** Find the average of these samples.
- **b** Estimate the total number of pupils in the school who have had measles.
- **c** How could this estimate be improved?

6 You want to estimate the number of bees in a hive. You catch 40 bees, mark them and release them. Some time later, you catch a second sample of 50 bees and find that 16 of them are marked.

Use this table to estimate the total number n of bees in the hive.

	In the second sample	In the whole hive
Number of marked bees	16	40
Total number of bees	50	n

7 One hundred marked fish are released into a lake. Some time later, 60 fish are caught and 12 of them are found to be marked.
Estimate the total number of fish in the lake.

Part 2 Tabulation

1 Which of these quantities are *counted* and which are *measured*?

a dots on a dice
b water in a bucket
c area of a window
d people out of work
e weight of scrap metal
f angle of rotation

2 A football team played 30 matches last season, and the number of goals scored in each match is given below.

0 2 1 3 0 1 2 4 3 0 0 1 2 2 1
4 0 1 2 1 1 3 0 2 1 3 0 0 2 3

Copy and complete this frequency table and use it to draw a bar graph illustrating your results.

Number of goals	0	1	2	3	4	Total
Frequency						

3 A commuter notes how late his morning train is for each of the 22 working days in September. His results are:

3 min 20 sec	2 min 15 sec	1 min 25 sec	5 min 45 sec	3 min 35 sec
8 min 55 sec	5 min 10 sec	0 min	2 min 30 sec	7 min 10 sec
0 min 30 sec	4 min 40 sec	3 min 50 sec	6 min 20 sec	1 min 40 sec
2 min 5 sec	4 min 30 sec	6 min 45 sec	5 min 35 sec	0 min
1 min 10 sec	3 min 15 sec.			

Copy and complete this grouped frequency table, and use it to draw a bar graph illustrating your results.

Time late, minutes	0 – 2	2 – 4	4 – 6	6 – 8	8 – 10	Total
Frequency						

4 The number of defective components produced by an automatic machine is counted every hour. During the fifteen hours of operation on one day, the number of defects are:

6 8 5 9 4 5 6 6 7 8 5 7 6 6 5

Construct a suitable frequency table and illustrate your results with a suitable diagram.

5 The power consumed by each of twenty identical resistors supposedly carrying the same electrical current is (in kilowatts).

1.25	1.43	1.32	1.14	1.47	1.52	1.32
1.37	1.24	1.18	1.48	1.36	1.16	1.38
1.42	1.35	1.29	1.37	1.48	1.27	

Construct a frequency table using classes of data 0.1 kW wide and illustrate your results with a suitable diagram.

Statistics

Part 3 Display of data

1 A large industrial firm employs young men and women on training courses. This pictogram shows the number of new recruits taken on each year.

1981	Male	
	Female	
1982	Male	
	Female	
1983	Male	
	Female	
1984	Male	
	Female	
1985	Male	
	Female	

one face = 10 trainees

a In which years did more young women than men start their training?

b How many male and how many female trainees started in 1982?

c Copy and complete this table.

Year	1981	1982	1983	1984	1985
Number of male					
Number of female					
Yearly total					

d Describe the trend in employment of trainees over this five-year period.

2

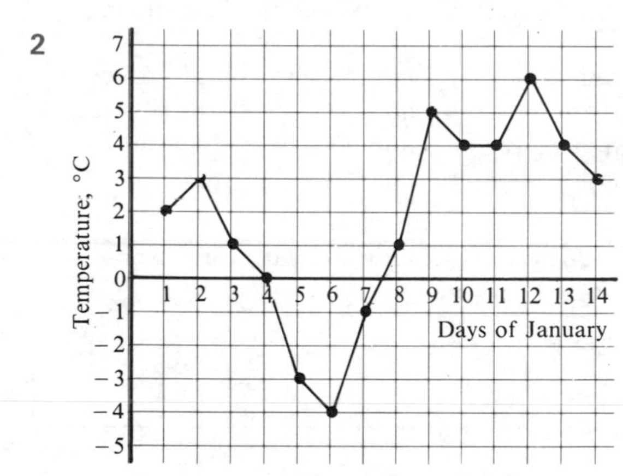

This graph shows the variation in midday temperatures for the first fortnight of January.

On which day did

a the highest temperature occur

b the lowest temperature occur?

On which two consecutive days was there

c the largest fall in temperature

d the largest rise in temperature?

3 A British manufacturing firm sells its products both in Britain and overseas. Its sales performance over five years is shown on this graph.

a In which years did exports exceed home sales?

b In which year were home sales double the export sales?

c What were the *total* sales worth in 1984?

d In which year were the *total* sales greatest?

e Describe the trend in both home and export sales over this five-year period.

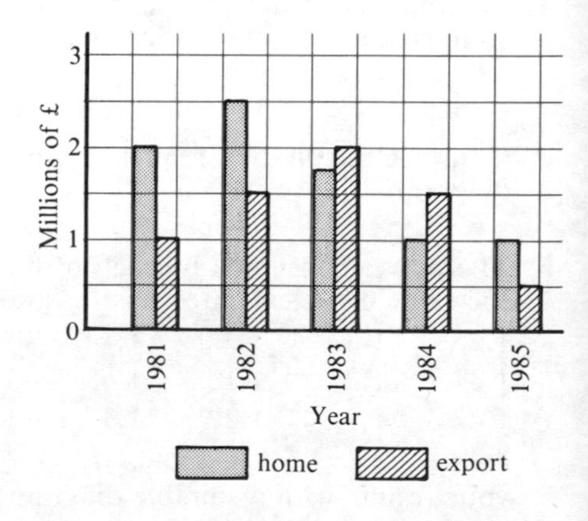

Statistics

4 A research biologist catches butterflies and measures their wing-spans. His results are shown in this diagram.

a How many butterflies have wing-spans between 32 mm and 34 mm?

b How many butterflies have wing-spans of over 36 mm?

c What is the *total* number of butterflies caught by the biologist?

d What percentage of this total number has wing-spans of over 36 mm?

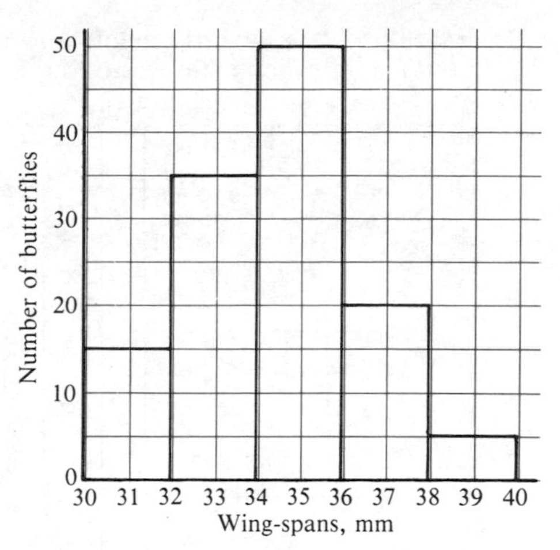

5 This pie chart indicates the occupations of 240 school-leavers in a certain town. Copy this table and use a protractor to help you complete it.

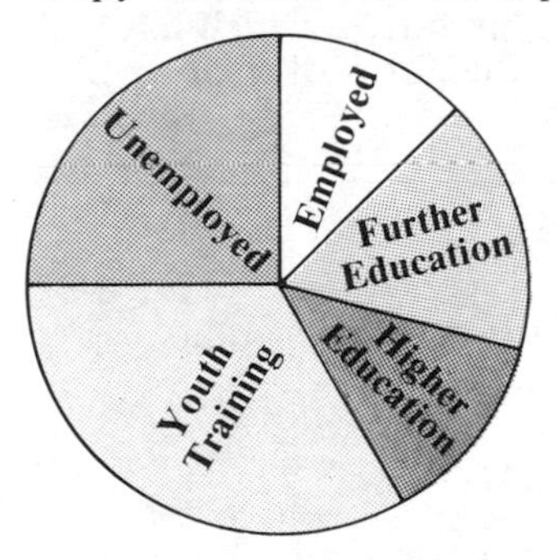

Occupation	Angle	Fraction of total	Number of school-leavers
Employed			
Further Education			
Higher Education			
Youth Training			
Unemployed			
Total			

6 This table shows the usage of land on a farm. Copy and complete the table and draw an accurate pie-chart.

Land usage	Area (hectares)	Fraction of total	Angle
Arable	80		
Pasture	60		
Woodland	60		
Waste	40		
Total			

7 A city museum surveys the 80 visitors it has during one day to see how far they have come. Draw a pie chart of the results shown in this table.

Within the city	Within 30 miles of the city	Over 30 miles from the city	Overseas
12	8	40	20

Statistics

8 Examine these advertisements and their claims, and criticise each one to say how it might be misunderstood.

a

b

c

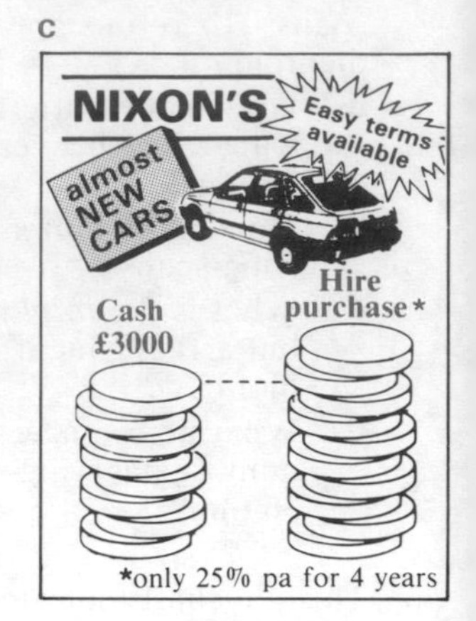

9 A groundsman measures the heights of the twelve birch trees which have recently been planted in a park. He also measures the girths of their trunks.

Height, m	3	3	4	4	$4\frac{1}{2}$	5	5	5	6	6	7	$7\frac{1}{2}$
Girth, cm	10	15	10	18	12	15	20	25	20	30	25	30

a Draw a scatter diagram and a line of best fit through your points.

b If the groundsman plants another tree with a height of $3\frac{1}{2}$ metres, what girth might you expect it to have?

10 A tourist officer at a seaside resort keeps a record of the number of weekend visitors to the resort. He also records the amount of rain which falls each weekend for the fourteen weeks of the summer season.

Rainfall, mm	0	1	2	5	5	7	10	10	13	15	17	18	20
Number of visitors, (thousands)	30	28	31	26	21	25	18	22	14	20	15	8	11

a Draw a scatter diagram and a line of best fit through your points.

b If 12 mm of rain had fallen one weekend, make an estimate of the number of weekend visitors to the resort.

11 A bank manager interviews ten customers wanting to borrow money. He notes their annual salary and the amount of the loan requested.

Annual salary, £ 000's	6	7	7	8	8	11	12	12	13	14
Loan requested, £ 000's	2	5	10	6	12	4	1	7	10	2

a Draw a scatter diagram and describe the correlation between annual salary and amount of loan.

b If the manager's next customer requests a loan of £8000, can you estimate the customer's annual salary?

Statistics

Part 4 Measures of average and of spread

1 Find the mode, the median and the mean of:

a the number of goals scored by the home team in eight football matches
2 3 1 1 0 4 0 1

b the number of passengers leaving the ten trains which stop at a small station
4 8 10 21 4 9 8 16 12 4

c the number of job interviews attended by a group of friends before each is successful
5 8 2 1 2 10 8 3 7 2.

2 A student measures the amount of rain which falls each day over a period of sixteen days. When arranged in order, his results (in millimetres) are:
0 0 0 0 0 0 0 0 0 2 6 8 10 13 16 17.

a Calculate the mode, the median and the mean of his results.

b Do you think that your answers for mode, median and mean give a reasonable measure of the average rainfall? Would you reject any of your answers as not much use? Which of the three averages do you think is the most useful in this case?

3 Find the mode, the median and the mean of each of these distributions of discrete data.

a The number of pupils arriving late at a primary school is noted over a ten-day period.

Number of late pupils	0	1	2	3	4	5	6
Frequency	2	3	0	1	2	1	1

b A police station records the number of emergency calls it answers each evening over a period of fifteen days.

Number of calls	0	1	2	3	4	5	6	7
Frequency	5	2	4	2	0	1	1	0

c A lock keeper records the number of boats per hour passing through his lock during the twenty daylight hours of a weekend.

Number of boats per hour	0	1	2	3	4	5	6	7	8
Frequency	0	1	3	4	5	2	4	1	0

4 Find the mode and calculate the mean of these distributions by using the third row of each table.

a A travelling salesman claims overnight expenses according to the number of days spent away from home on business.

Length of trip, x days	2	3	4	5	6	7	8	9	10	Totals
Frequency, f	12	8	2	10	3	6	4	3	2	
fx										

Statistics

b A record is kept of the number of charter flights leaving London for Majorca each day over a five-month period.

Number of daily flights, x	5	6	7	8	9	10	11	12	13	14	15	16	Totals
Frequency, f	2	5	6	8	10	16	23	25	31	14	8	2	
fx													

5 For each of these distributions of continuous or grouped data, write down the modal class and estimate the mean. (A copy of the table will be useful and similar tables can be constructed for the other distributions.)

a Drumburgh Castle is open to visitors all year round. The weekly numbers of visitors are recorded in the ticket office.

Number of visitors	0–500	500–1000	1000–1500	1500–2000	2000–3000	Totals
Midpoint of class, x	7	5	23	12	5	
Number of weeks, f						
fx						

b The members of a slimming club are each asked how much they would like to lose.

Amount to be lost, kg	5 – 10	10 – 15	15 – 20	20 – 25	25 – 30
Number of people	4	10	8	2	1

c The singles records released by a music company are played and the length of each song noted in minutes.

Length of song, minutes	$2-2\frac{1}{2}$	$2\frac{1}{2}-3$	$3-3\frac{1}{2}$	$3\frac{1}{2}-4$	$4-4\frac{1}{2}$
Number of songs	8	16	48	24	4

d A firm has a fleet of lorries and the amount of diesel fuel used each day for a month is tabulated.

Amount of fuel, litres	100 – 110	110 – 120	120 – 130	130 – 140	140 – 150	150 – 160
Number of days	1	5	8	10	4	2

6 The numbers of fish caught by eight men on a fishing trip are:
10 8 9 12 10 x 9 14.
If the modal number of fish is 9, find the value of x.

7 An office is lit by six light bulbs of these wattages:
60 100 100 x 150 150.
If the median wattage is 125, find the value of x.

Statistics

8 A dice is thrown five times with these results:
2 6 x 1 4
If the mean score is 3, calculate the value of x.

9 When playing cricket, a batsman is *out* in each of four innings after scoring these runs:
30 x 42 24
If his mean score is 31, calculate the value of x.

10 Mrs Willoughby is employed for six months. She earns £150 per month for three months and £185 per month for two months. In the sixth month she earns £x.
If her mean monthly earnings over the whole six months is £160, find the value of x.

11 Twenty pupils take an English examination. Their percentage marks are arranged here in order and the distribution split into quarters.

54 56 59 62 64	68 70 70 72 72	73 76 76 78 79	81 86 87 87 88

a Write down the median, the lower quartile and the upper quartile of the distribution.
b Calculate (i) the range (ii) the interquartile range.
(Note that the *range* is the difference between the highest and lowest values in the distribution.)

12 Three employees of a firm assemble electronic parts of television sets. The numbers of completed kits assembled each day are noted over two working weeks.

Mrs Kaye	9	8	7	7	14	12	7	12	9	10
Mrs Lamb	10	6	6	11	8	12	8	12	11	11
Mrs McPhee	12	8	7	9	12	9	12	13	9	11

Arrange this data in order of size and find, for each employee's figures,
a the median **b** the lower and upper quartiles
c the interquartile range **d** the range.

13 Two makes of battery, Excell and Powerite, are tested until failure. The number of hours of working is noted for each battery. Here are the results arranged in order:

Excell	21	24	25	25	27	29	32	36	38	41	42	44	45	47	48	48
Powerite	2	34	35	35	35	36	36	37	38	38	39	40	42	42	42	43

a Find the *median* of the results for each make of battery.
b Find the *range* of the results for each make of battery.
c Find the lower and upper quartiles and hence calculate the *interquartile range* for each make.
d Look at the results above. How can you tell that one of the Powerite batteries was faulty? If you ignore this faulty battery, which of the two makes gives the wider spread of results?
e Look at your answers to **b** and **c**. Is it the *range* or the *interquartile range* which ignores the freak results of the dud battery?
Which of these two measures of spread do you think gives the *fairer* measure?
f If you wanted to buy a battery which worked reliably for at least 35 hours before failure, which make of battery would you buy?

Statistics

14 A weights and measures inspector visits a greengrocer's shop to investigate bags of potatoes advertised as '2 kg'. He finds the masses of 120 bags to the nearest 0.01 kg and tabulates his results.

Mass, kg	1.81 – 1.90	1.91 – 2.00	2.01 – 2.10	2.11 – 2.20	2.21 – 2.30
Number of bags	2	18	57	38	5

a Copy and complete the following table using the inspector's results.

Mass, kg	$\leqslant 1.80$	$\leqslant 1.90$	$\leqslant 2.00$	$\leqslant 2.10$	$\leqslant 2.20$	$\leqslant 2.30$
Cumulative frequency						

b Draw a cumulative frequency graph and find
(i) the median (ii) the lower and upper quartiles
(iii) the interquartile range.

15 The words used by a popular daily newspaper are analysed by taking the first two thousand words used in one of its editions. The number of letters used in each word is counted and the results tabulated.

Number of letters	1 – 3	4 – 6	7 – 9	10 – 14	15 – 21
Number of words	480	635	490	340	55

Construct a cumulative frequency table and draw a graph of the cumulative frequencies. Find

a the median **b** the lower and upper quartiles
c the interquartile range.

16 **a** The seaside resort of Drayton Bay advertises its hours of sunshine during the summer months as in this table. (The hours are measured to the nearest tenth.)

Hours of daily sunshine	0 – 4.0	4.1 – 8.0	8.1 – 10.0	10.1 – 12.0	12.1 – 14.0
Frequency	15	17	26	22	20

Construct a cumulative frequency table and draw a graph of the cumulative frequencies. Estimate
(i) the median (ii) the lower and upper quartiles
(iii) the interquartile range.

b A second resort, Whitbeach, issues its own figures as follows:

Hours of daily sunshine	0 – 4.0	4.1 – 8.0	8.1 – 10.0	10.1 – 12.0	12.1 – 14.0
Frequency	12	23	28	25	12

Draw a graph of the cumulative frequencies from this table, and estimate
(i) the median (ii) the lower and upper quartiles
(iii) the interquartile range.

c If you were going on holiday and wanted to have, as far as possible, the same amount of sunshine each day, which of these two resorts would you go to? Give reasons for your choice.

17 An examination is taken by 200 candidates and the marks given produce this cumulative frequency curve.

a Find the 50th percentile (or *median*).

b Find the 25th percentile (or *lower quartile*).

c If the weakest 10% of the candidates fail the examination, find the mark which separates the highest failure from the lowest pass. This mark is the *10th percentile*.

d The best 20% of the candidates gain a distinction. What is the lowest mark for which a distinction is given? This mark is the *80th percentile*.

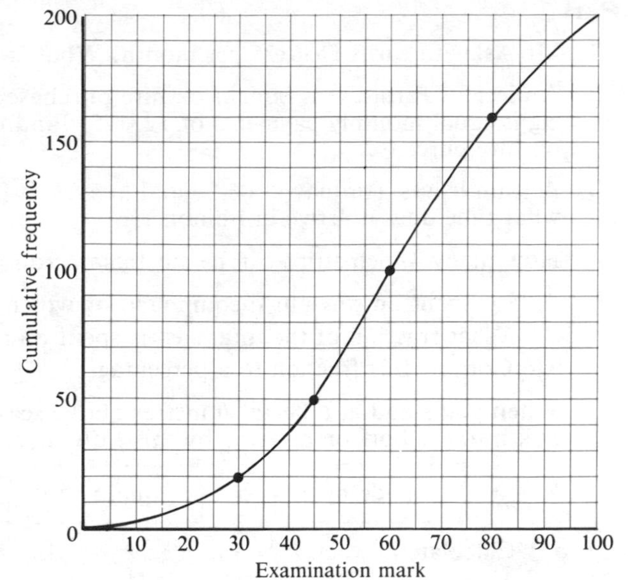

18 150 golfers enter a two-day competition. Only the top 25% of them after the first day will be allowed to continue on the second day.
The scores after the first day are tabulated as shown.

Score	61 – 70	71 – 80	81 – 90	91 – 100	101 – 110	111 – 120
Number of players	8	17	35	62	22	6

a Construct a cumulative frequency table and hence draw a graph of the cumulative frequencies.

b Use the graph to find the 25th percentile (or *lower quartile*).

c What is the highest score at the end of day 1 which a golfer can afford to have and still qualify for day 2?

19 An automatic machine is monitored and the number of faulty items which it produces every hour is noted for the whole of a 40-hour working week.

Number of faulty items per hour	0 – 100	101 – 200	201 – 300	301 – 400	401 – 600	601 – 1000
Number of hours	8	12	9	6	4	1

a Construct a cumulative frequency table and hence draw a graph of the cumulative frequencies.

b Find the 90th percentile.

c If, for 90% of the time which the machine is working, it is producing no more than 500 faulty items per hour, then it continues working and does not need maintenance. Is this particular machine allowed to continue working or is it stopped for maintenance?

A mixture of short problems

Part 1

1 Mr Askwith earns £586·25 per month. What is his annual salary?

2 A piece of furniture is bought on hire purchase by paying a deposit of £40 and eight equal monthly payments of £23·75. Find the total cost of the furniture on hire purchase.

3 A train leaves Torquay at 06.20 and arrives in Edinburgh 9 h 45 mins later. At what time does it arrive in Edinburgh?

4 How many 4-inch strips can be cut from a roll of plastic 5 feet long?

5 £375 is spent on some hi-fi equipment of which £225 is spent on the turntable.

a What fraction of the total cost is spent on the turntable?
b Convert this fraction to a percentage.

6 When you stand at a height h metres above sea-level, you can see a distance of D km to the horizon as given by this flow diagram.

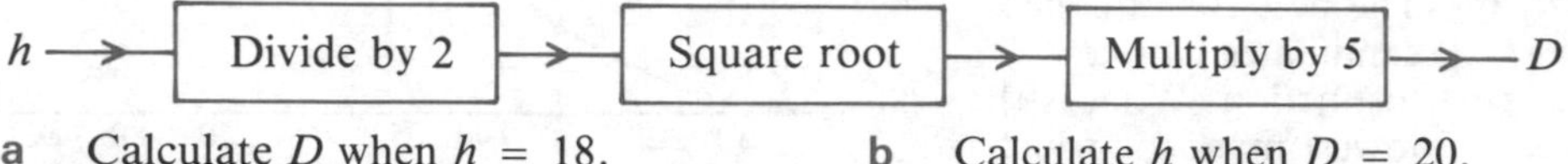

a Calculate D when $h = 18$. **b** Calculate h when $D = 20$.

7 A bag of flour is weighed and found to have a mass of 463.7 kg. Write its mass

a correct to the nearest whole kilogram **b** correct to two significant figures.

8 The six faces of a dice are marked, as if they were playing cards, with the nine, ten, Jack, Queen, King and ace of hearts. What is the probability of rolling

a the nine of hearts **b** either the Queen or the King of hearts?

9 Find the size of angles p, q and r.

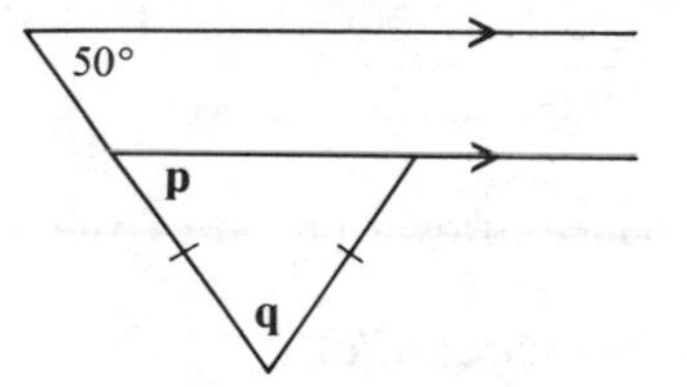

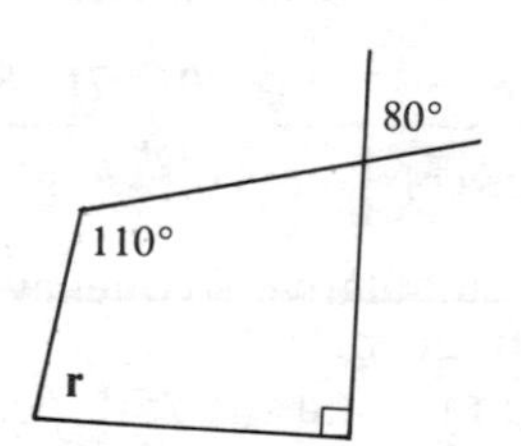

10

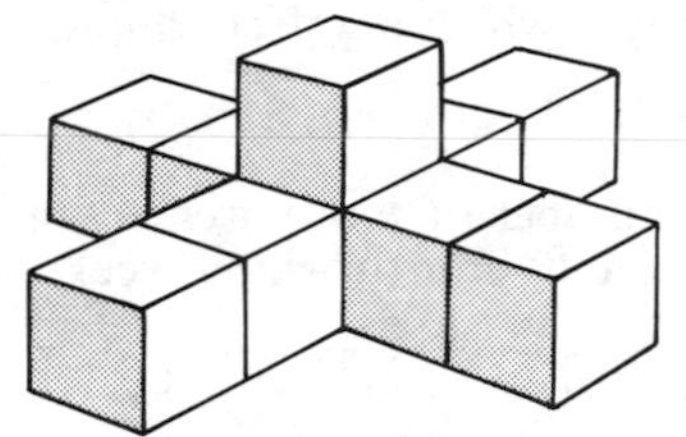

a Several white cubes are fixed together to make this star shape. How many cubes are used?
b If all the faces of the star shape (including its base) are painted red, how many of the cubes will have exactly four red faces?

11 **a** Two pieces of metal, X and Y, are welded together. Calculate
(i) the total distance from A to B
(ii) the area of piece X, as a fraction in its lowest terms.

b A rectangle has an area of 0.64 m^2 and a width of 0.04 m. Calculate its length l.

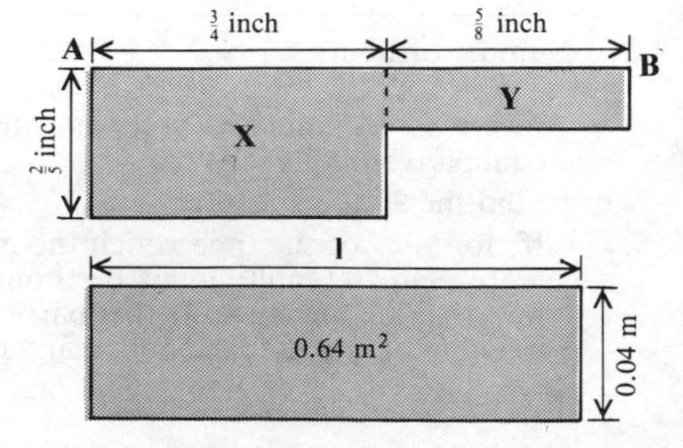

A mixture of short problems

12 Miss Eskrick is going on holiday and she orders some films for her two cameras by post.
She decides to buy two 36-exposure Kodicolor films (size 135) for one camera.
She also has a small pocket camera which takes size 126.
She decides to buy one Kodicolor film and one of the cheaper Farbesnap films for this camera.
How much should she pay in total for her order?

Direct Snaps plc	PO Box 99 Banbury · Oxon		
	Size	Exposures	Price per film
KODICOLOR FILMS for colour prints (200 ASA)	126	24	£2·00
	110	24	£1·90
	135	24	£2·40
	135	36	£2·95
	disc	15	£1·90
FARBESNAP FILMS for high resolution colour prints	126	24	£1·50
	110	24	£1·50
	135	24	£1·60
	135	36	£2·00
	disc	15	£1·40

PRICES INCLUDE V.A.T.
RETURN POSTAGE AND PACKAGE – add 40p per order

13 Solve these equations for x.

a $2(3x - 6) = 15$

b $\frac{x}{2} - 3 = 5$

14 **a** The distance around one lap of a race track is 400 m. If I run three laps, how far have I gone in *kilometres*?

b If 2.1 litres of washing-up liquid is divided equally amongst 3 empty containers, how many cubic centimetres are put in each bottle?

c A sheet of glass has an area of 60 000 cm^2. What is this area in *square metres*?

15 A crate is protected by metal strips along all of its edges. It has a square base of edge x metres and a height h metres.
If $x = 3$ and $h = 2$, calculate

a the volume of the crate V where $V = x^2h$

b the length L of the metal strips where $L = 4(2x + h)$.

16 Mr Wilkins bought a new car for £8200 in March 1984.
In its first year, the value of the car depreciated by 10%.
In its second year, the value of the car depreciated by 5% of its value at the end of the first year.
Calculate the value of the car in March 1986 (i.e. two years after purchase).

17 In her will, Miss Annabel Wood left her three friends Laura, Mary and Nicola a total of £15 000 to be divided amongst them in the ratio 2:3:7 respectively.
How much does each of the three friends receive?

A mixture of short problems

18

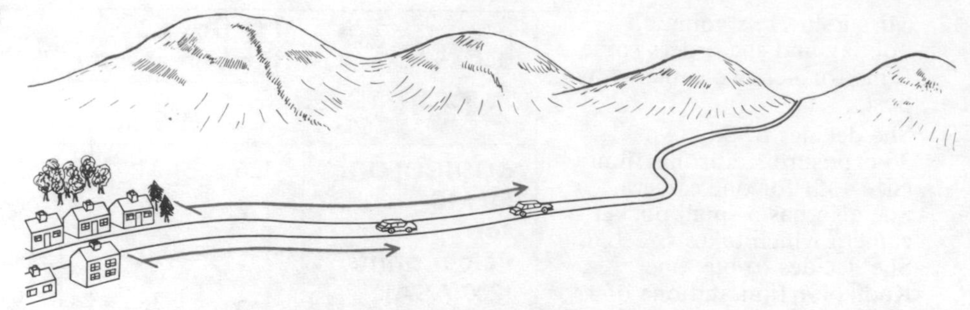

Two cars set off in the same direction from different points on the same road. Their distances, y km from the nearest village are given by

$$y = 2t + 1 \quad \text{and} \quad y = \tfrac{1}{2}t + 7,$$

where t is the time of travel in minutes.

a Copy the table and complete it for the equation $y = 2t + 1$.

t	0	1	2	3	4	5
y						

b Complete a second copy of the table for the equation $y = \frac{1}{2}t + 7$.

c Use your results to draw two straight lines on the same axes.

d How long does it take for one car to overtake the other?

19 This diagram shows the plan of a garden in which the shaded area indicates the lawn.

Calculate

a the area of the vegetable plot

b the area of the drive

c the area of the lawn

d the cost of covering the lawn with turf costing £1·40 per square metre.

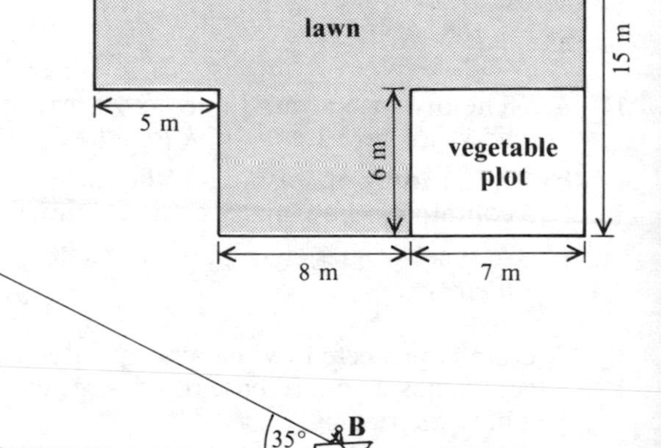

20

The angle of elevation of the top of a lighthouse is 35° from a rowing boat B when it is 30 metres from the lighthouse. A man is standing at 10 metres from the lighthouse at point A. Calculate

a the height of the lighthouse

b the angle of elevation α of its top from A.

21 An open cylindrical container of depth 30 cm has a top of diameter 20 cm. It is filled to half its depth with liquid, 1 cm^3 of which has a mass of 1.4 grams. Calculate

a the volume of the liquid

b its mass to the nearest 100 grams.

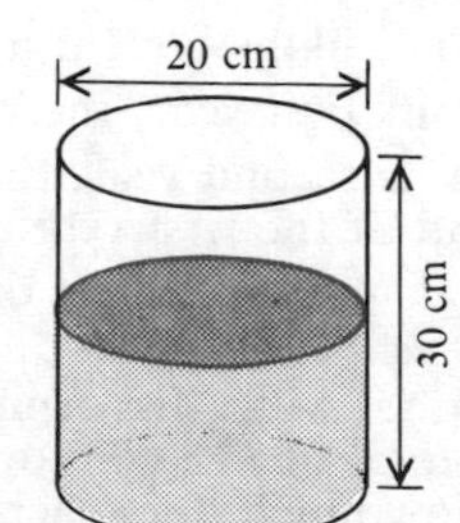

A mixture of short problems

22

	Petrol, litres		
	2✱	3✱	4✱
Rentet	60	50	200
Quickar	50	0	250

Cost of petrol, pence

Garage A	Garage B	
40	40	Each litre of 2✱
42	40	Each litre of 3✱
44	46	Each litre of 4✱

➡

Total cost of petrol, £

Garage A	Garage B	
...	...	Rentet
...	...	Quickar

The first table gives the number of litres of petrol to be bought by two car-hire firms, Rentet and Quickar, in one week. The amounts of 2-star, 3-star and 4-star petrol are shown separately.

The second table gives the cost in pence of each grade of petrol at the two garages A and B which the car-hire firms use.

Calculate the total cost (in £s) for both firms at each of the two garages. Copy the third table and enter your total costs.

23 a On axes labelled from 0 to 12, plot the points (1, 3), (3, 3), (2, 5), (1, 5). Join them and label the trapezium T.

b A reflection in the line $y = x$ maps the trapezium T onto its image T'. An enlargement with centre the origin and length scale factor 2 maps T' onto T''. On the same diagram, draw the positions of T, T' and T''.

24 **London → Slough → Maidenhead → Reading**

Mondays to Fridays (Local service only)

The times shown here are valid only for this exercise and may not represent actual train times.

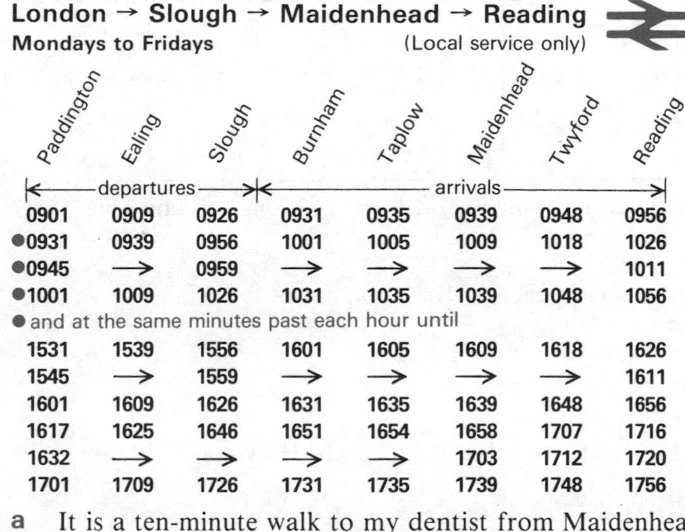

Paddington	Ealing	Slough	Burnham	Taplow	Maidenhead	Twyford	Reading
departures			arrivals				
0901	0909	0926	0931	0935	0939	0948	0956
●0931	0939	0956	1001	1005	1009	1018	1026
●0945	→	0959	→	→	→	→	1011
●1001	1009	1026	1031	1035	1039	1048	1056
● and at the same minutes past each hour until							
1531	1539	1556	1601	1605	1609	1618	1626
1545	→	1559	→	→	→	→	1611
1601	1609	1626	1631	1635	1639	1648	1656
1617	1625	1646	1651	1654	1658	1707	1716
1632	→	→	→	→	1703	1712	1720
1701	1709	1726	1731	1735	1739	1748	1756

a It is a ten-minute walk to my dentist from Maidenhead Station. If my appointment is for 10.30 a.m. and I live in Slough, what is the time of the latest train I can catch from Slough?

b At what time does the latest possible train leave Paddington to get you to Reading for 2 p.m.?

c Alison lives a twenty-minute walk away from Ealing Station. If she leaves home at 5 minutes to 4 o'clock one afternoon, how long will she have to wait before catching the next train to Reading?

A mixture of short problems

25 An office clerk records the temperatures at 10 a.m. both inside and outside the office building. This graph shows his results for one working week.

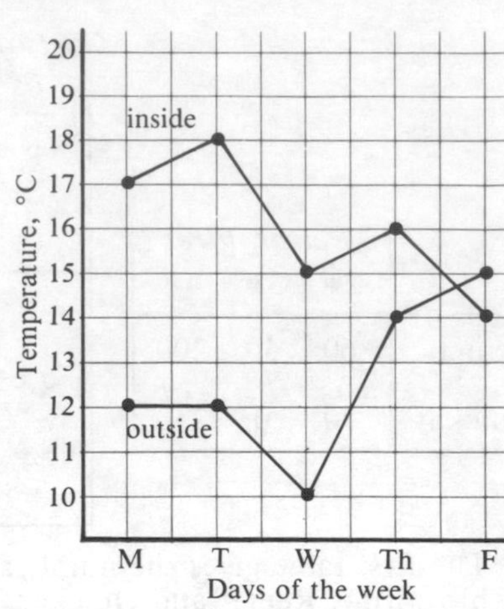

- **a** On which two consecutive days is the outside temperature the same?
- **b** Write down the outside temperatures on the five days of this week, and calculate the mean outside temperature.
- **c** On which day is there the greatest difference between the outside and inside temperatures?
- **d** On which day do you suspect the heating system in the building is not working properly?

26 **a** **b** **c** **d** **e**

Copy and complete this table for each of these solids.

	Name of shape	Number of faces F	Number of vertices V	Number of edges E	F + V − E
a					
b					
c					
d					
e					

From your results, decide which solid is the *odd-one-out* and give your reasons.

27 Find the next two numbers in each of these sequences.

a	1	2	4	7	11	...	...
b	2	5	10	17	26	...	...
c	1	8	27	64	125	...	...

28 A regular polygon has exterior angles of 45° each. How many sides has the polygon?

29 Standing on top of Midsummer Hill, Worcester is 11 miles away on a bearing of 030°, Gloucester is 12 miles away on a bearing of 160° and Ross is 13 miles away on a bearing of 230°.
Use a scale of 1 cm = 2 miles and draw an accurate diagram showing the positions of these four places.

- **a** What is the shortest distance between Worcester and Gloucester?
- **b** What is the bearing of Gloucester from Ross?

A mixture of short problems

30 Square cakes are cut into pieces by cuts which are at right angles or parallel to each other.
These diagrams show the cakes being cut into 4, 9 and 16 pieces.

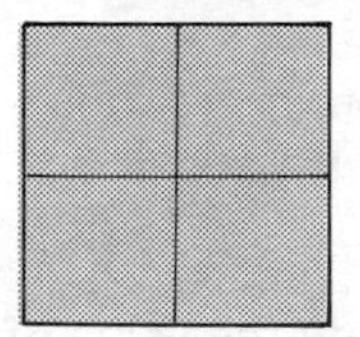
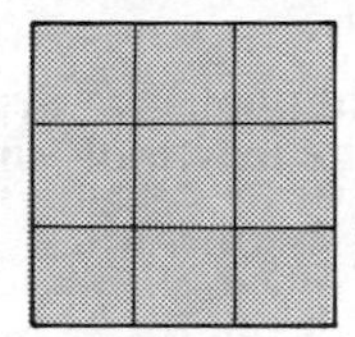
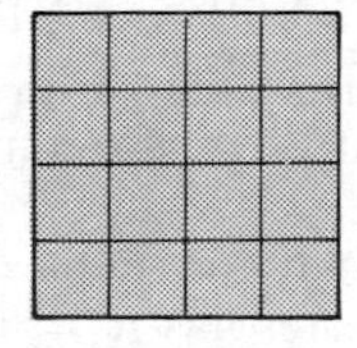

a Copy and complete this table.

Number of pieces	4	9	16	25	36
Number of cuts					

b If this pattern of cutting is continued with other similar cakes, how many pieces would be formed if 24 cuts are made?

Part 2

1 A workman 'clocks-in' at 07.30 and 'clocks-out' at 17.00.
a How many hours has he worked?
b If he is paid £4 per hour, how much has he earned in this time?

2 A builder receives a bill for £67·40 and a 10% discount is given if it is paid within thirty days.
a Calculate 10% of £67·40.
b How much will the builder pay if the discount is deducted from the bill?

3 Eighty people apply for the same job and 44 of them have identical qualifications.
a What fraction of the applications have identical qualifications?
b Express this fraction as a percentage.

4

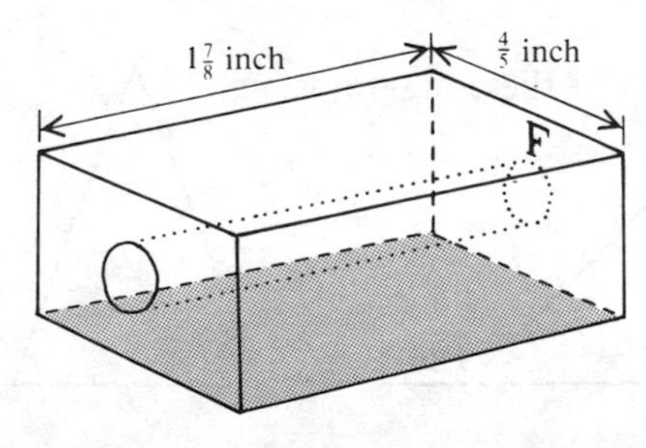

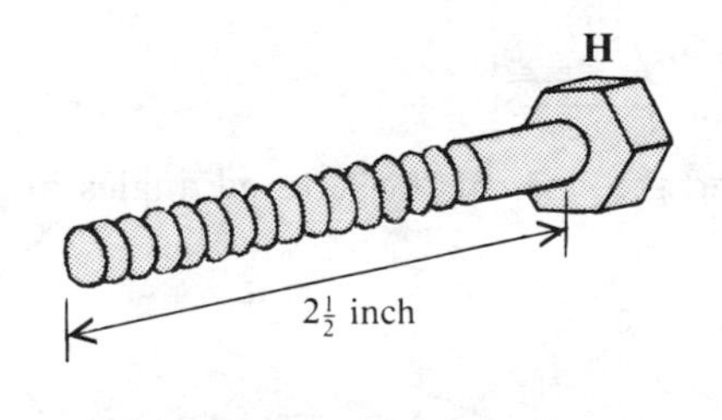

A bolt, $2\frac{1}{2}$ inch long is passed through a hole $1\frac{7}{8}$ inch long in a block of metal, so that the head of the bolt H comes into contact with the face F of the block.
a What length of bolt will protrude from the other end of the hole?
b If the block is $\frac{4}{5}$ inch across as shown in the diagram, what is the area of the shaded base of the block?

A mixture of short problems

5 a A family returns from a holiday in France with 11 litres of milk. How many gallons of milk is this?

b If they pour it into pint milk bottles, how many bottles will they need? (1 gallon = 4.5 litres = 8 pints)

6 The energy E joules used by an electrical appliance depends on its resistance R ohms, the current flowing i amps and the length of time t sec for which it is switched on.

E is then given by the formula $E = Ri^2t$.

An immersion heater has $R = 100$, $i = 3$, and $t = 2$. Calculate the value of E.

7 Samantha Payne goes for a run on three mornings every week. Last week she ran 0.8 miles, 1.75 miles and 3 miles.

a Calculate the total distance she ran.

b How far did she run to the nearest whole mile?

8 A man is charged £5·80 for a piece of wood 2.5 metres long. What is the cost of a one-metre length?

9 a A cylindrical tin of orange juice is 15 cm tall with a diameter of 8 cm.

What volume of juice is in a full tin (to the nearest cm^3)?

b If a full tin is emptied into a plastic box with a rectangular base 12 cm by 9 cm, what depth of juice will there be (to the nearest cm)?

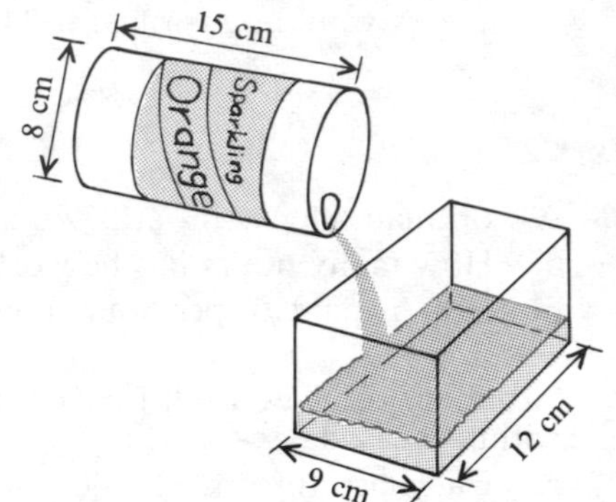

10 These four rectangles and two trapeziums form the net of a prism. The lengths of some edges are given in centimetres.

Write the lengths of the edges lettered a, b, c, d, e and f.

11 Find the sizes of the lettered angles in each of these diagrams.

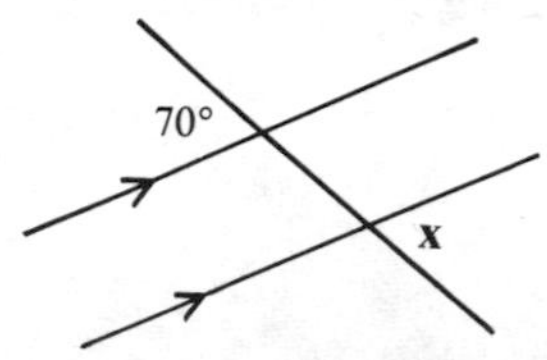

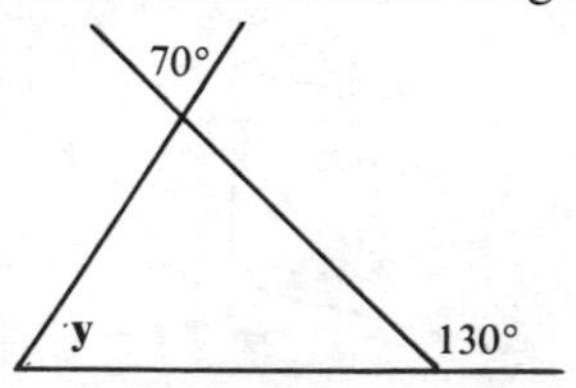

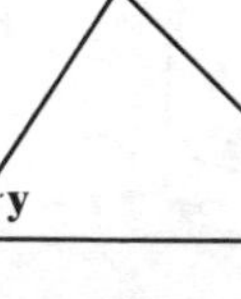

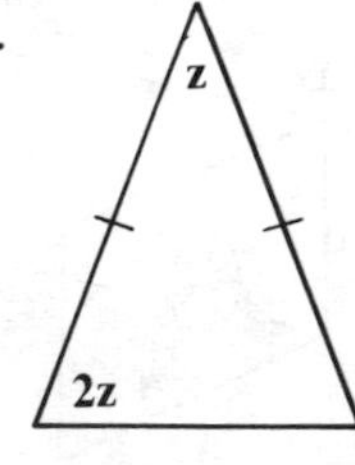

12 Solve these equations for x.

a $8x - 3 = 2(3x + 1)$

b $\frac{x}{2} + 3 = 7$

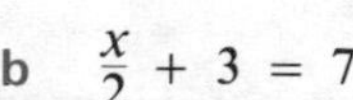

13 A donkey is tethered to a stake in a field by a rope 10 metres long.

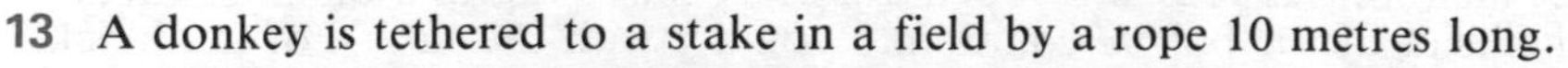

a Take $\pi = 3$ and calculate the area which the donkey can graze.

b If the length of the rope is reduced to 5 metres, what fraction of the original area can the donkey now graze?

A mixture of short problems

14 The thirty pupils in a class at school are asked how many children there are in their families. The results are given in this diagram.

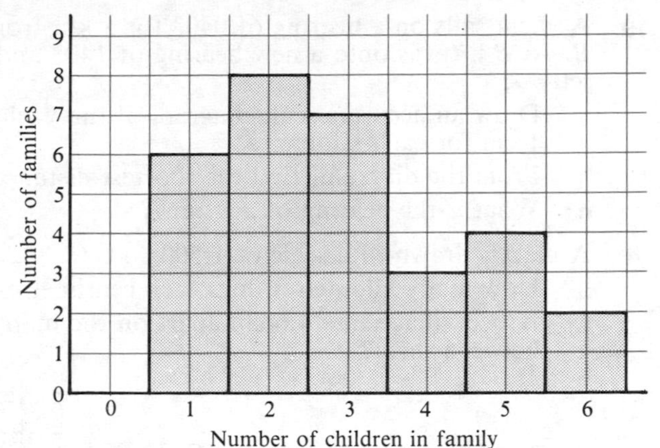

a How many children are in the most common size of family?

b If you choose a pupil at random from this class, what is the probability that he or she comes from a family of four children?

c What percentage of the children in this class have no brothers or sisters?

d If no child has a brother or sister in the class, what is the total number of children in all the thirty families from which this class of pupils came, and what is the mean number of children per family?

15 This diagram shows an L-shaped metal casting with measurements given in centimetres. Calculate

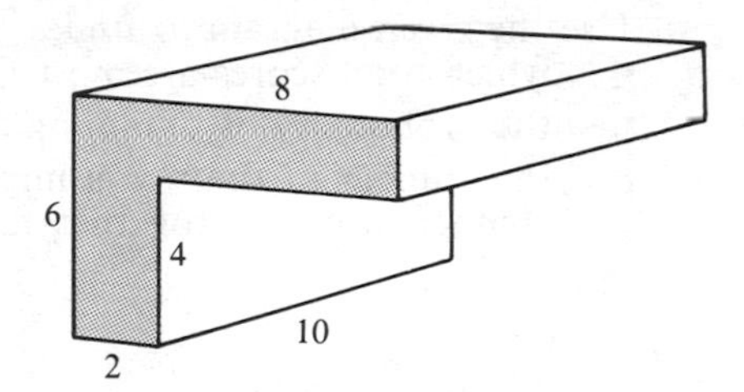

a the shaded area of cross-section

b the volume of the casting

c its mass, in kilograms, if 1 cm^3 has a mass of 7 grams.

16 A tap is left running into an empty water tank. Every minute the water level rises by $2\frac{1}{2}$ cm.

Copy and complete this table.

Time, min	1	2	5	10	25
Height of water, cm	$2\frac{1}{2}$				

Draw a graph of height against time.

How long does the tap have to run for the water level to rise 30 cm?

17 Three women win £9000 on the football pools. They share this sum in the same ratio as their stakes.

If Jean's stake was 50 pence, Kate's stake was £1 and Laura also paid £1, find how much of the winnings they each receive.

18 Draw and label both axes from 0 to 6.

Plot the points $P(1, 1)$, $Q(6, 3)$, $R(6, 6)$ and $S(3, 6)$ and draw the quadrilateral $PQRS$.

a What is the name of this type of quadrilateral?

b Draw the line of symmetry of the quadrilateral and write its equation.

c Calculate the area of the quadrilateral.

A mixture of short problems

19 A yacht sails on a bearing of 040° for 5 km from the harbour H to the buoy B. At B it turns onto a new bearing of 140° and sails a further 8 km to reach a jetty J.

a Draw an accurate scale diagram of the yacht's journey using a scale of 1 cm for each kilometre.

b From the diagram, find the shortest distance between H and J.

c What is the bearing of J from H?

20 A map is drawn to a scale of 1:500 000.

a How many kilometres does each centimetre on the map represent?

b If two villages are 4.6 cm apart on the map, what is the actual distance between them?

21

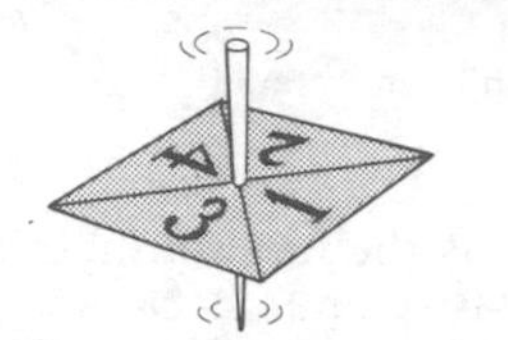

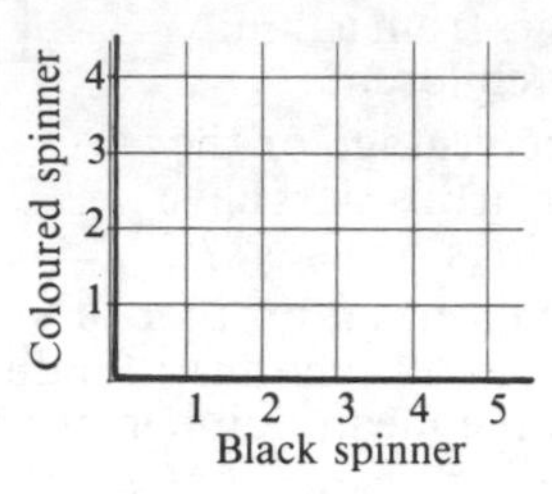

In a certain game, these two spinners are spun and their scores added together. Use the given diagram to find

a which total scores are most likely

b the probability of obtaining a total score of 3

c the number of times you might expect a total score of 8 when both spinners are spun together 120 times.

22 Triangle T is rotated onto triangle T'.

a What point is the centre of the rotation?

b Find the angle and direction of the rotation.

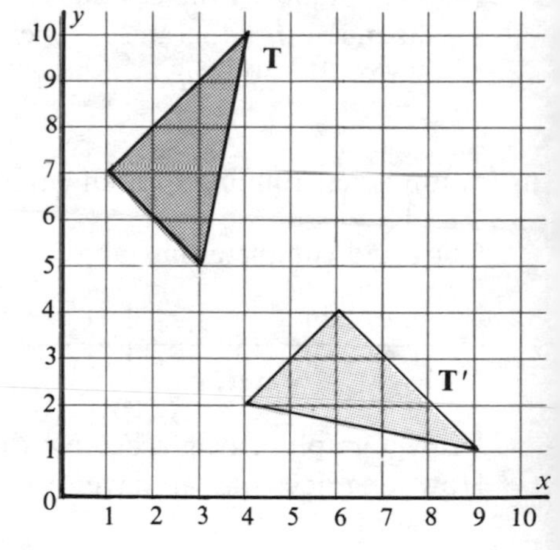

23 a The distance from the Earth to the Sun is 93 000 000 miles. If this distance is written as 9.3×10^m miles, write the value of m.

b A molecule of oxygen has a diameter of 0.000 000 034 cm. If this measurement is written as 3.4×10^n cm, write the value of n.

24 A factory has two similar tanks for storing petrol. One is $1\frac{1}{2}$ metres long and the other is $4\frac{1}{2}$ metres long.

a What is the length scale factor relating to the sizes of these two tanks?

b What is the corresponding volume scale factor?

c If the smaller tank holds 120 litres of petrol, what is the capacity of the larger tank?

A mixture of short problems

25 **Freshfield – Waterloo – Liverpool**

Monday to Saturday

Merseyside Transport

	NS				⊕ NS									
Freshfield	0630	0700	0730	0800	■	0830		00	30		1700	1735	1805	1830
Royal Hotel	0639	0709	0739	0809	0814	0839		09	39		1709	1744	1814	1839
Ince Blundell	0645	0715	0745	0815	0820	0845	then	15	45		1715	1750	1820	1845
Thornton	0652	0722	0752	0822	0827	0852	at	22	52		1722	1757	1827	1852
Crosby, St. Luke's	0657	0727	0757	0827	0832	0857	these minutes	27	57	until	1727	1802	1832	1857
Waterloo Interchange ≠	0707	0737	0807	0837	▲	0907	past	37	07		1737	1812	1842	1907
Litherland Station ≠	0715	0745	0815	0845		0915	each	45	15		1745	1820	1850	1915
Linacre Lane	0720	0750	0820	0850		0920	hour	50	20		1750	1825	1855	1920
Bootle New Strand ≠	0723	0753	0823	0853		0923		53	23		1753	1828	1858	1923
Liverpool Bus Station ≠	0740	0810	0840	0910		0940		10	40		1810	1845	1915	1940

■ From Southport Bus Station, departs 0736
▲ To Waterloo Boys' School, arrives 0837
⊕ School service – liable to alteration or suspension if not required.
NS Not Saturday
≠ Near to railway station

This timetable shows the bus service on one route into Liverpool.

a How long does it take the quarter-past-seven bus from Ince Blundell to reach Linacre Lane?

b Mr Kelly has never worked on a Saturday but his firm asks him to do so this week. He usually goes to work on the 7.07 am from Waterloo Interchange. Why will he be disappointed if he tries to catch this bus to go to work on this particular Saturday?

c Annette Hardy has a dentist's appointment in Crosby for ten past ten in the morning. At what time is the last possible bus she can catch from the Royal Hotel?

d You want to get to Liverpool Bus Station before 7 pm and you live in Thornton. You catch the latest possible bus from Thornton. If it leaves two minutes early and arrives at Liverpool bus station five minutes late, how long was the journey?

26 Eighty students on a Commercial Studies course are entered for
either R.S.A. examinations,
or Pitman examinations
or a mixture of both R.S.A. and Pitman examinations.

If eight students take only R.S.A. exams and forty eight are entered for both R.S.A and Pitman exams, find the percentage of the whole group which is entered for *only* Pitman examinations.

27 The scale of this map is 1 cm = 500 metres.
From the crossroads C a footpath climbs directly to the top T of the nearby hill.

a What is the horizontal distance in kilometres from C to T?

b How much higher is T than C?

c Calculate the angle to the horizontal at which the footpath climbs.

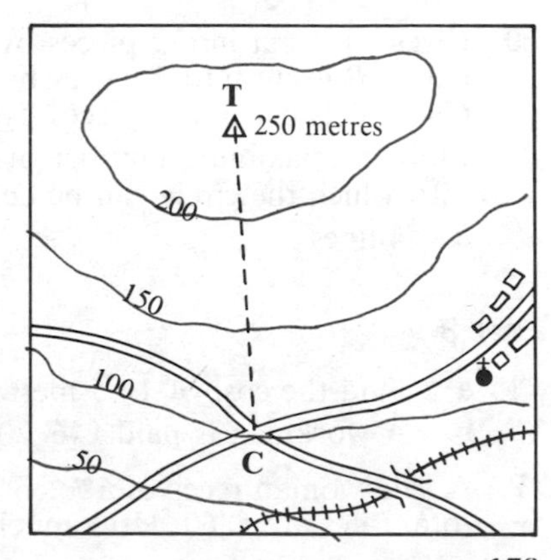

A mixture of short problems

28

Distance from home, km: 0, 10, 20, 30, 40, 50, 60, 70

Time of day: 1000, 1100, 1200 noon, 1300, 1400, 1500, 1600

Oma Venkat went shopping to Birmingham yesterday by car. This graph shows her journey there and back. She left home at 10.00 a.m. After driving 40 km she stopped at a garage for petrol and oil. She arrived in Birmingham at 12 noon. Later in the afternoon she set off home.

a What was her average speed during the first hour of her journey?

b How long did she stop at the garage for petrol and oil?

c How far away is Birmingham from her home and how long did she spend in Birmingham?

d What was her average speed on her return journey?

e How long did she spend driving the car during the whole day?

29

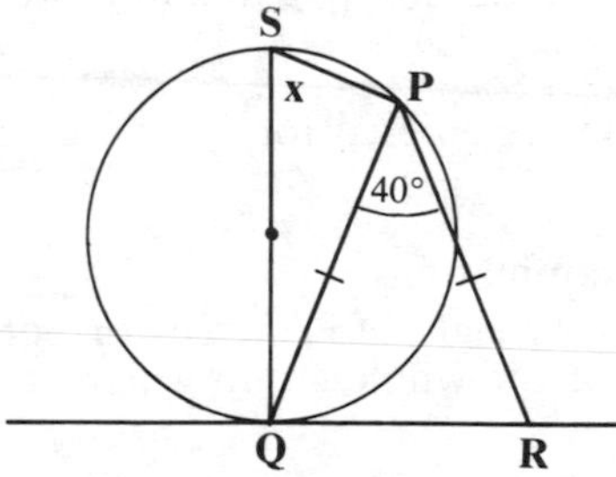

Triangle PQR is isosceles. QR is a tangent to the circle and QS is a diameter of the circle.

If angle QPR is 40°, calculate the value of x.

30 Circle A is cut into 2 pieces by 1 line. Circle B is cut into 4 pieces by 2 lines. Circle C is cut into 7 pieces by 3 lines.

Find the maximum number of pieces into which the circle can be cut by

a 4 lines **b** 5 lines.

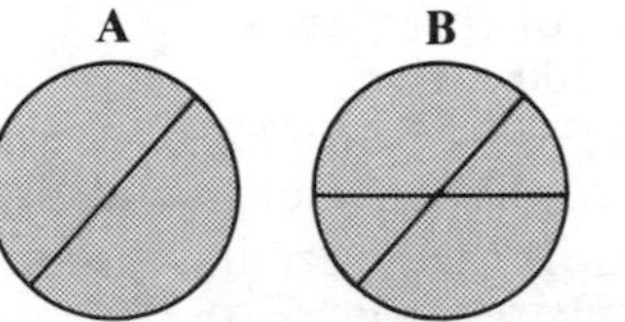

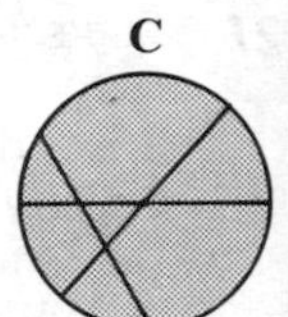

Part 3

1 a Find the cost of 14.5 metres of cloth if it costs £4·20 per metre.

b A workman is paid £46·20 for 15 hours work. Find his hourly rate of pay.

2 A saleswoman receives 6% commission on her total sales. In one month, her total sales are £860. How much commission will she receive?

A mixture of short problems

3 A savings certificate is bought for £150 and one year later it is worth £159. Calculate the percentage rate of interest which it has earned.

4 **a** If you buy a *Sony 36G* colour television with five £50 notes and three £20 notes, what change will you receive?

b (i) What saving do you make on a *Panasonic 420* hi-fi system by buying during the sale?

(ii) What percentage saving is this?

c You buy a *Bush 14B* colour television in the sale. The shop will deliver it to your house and supply and install an aerial for an extra 8% of the sale price. How much extra do you have to pay for this service?

SALE NOW ON

PATEL'S BAZAAR
8 Tythe Barn St.
Pontebridge

HI-FI	WAS	NOW
SANYO 6XL system	£345	**£299**
NYTECH AL7 system	£412	**£380**
PANASONIC 420 system	£250	**£234**
NAIM NK76 system	£474	**£415**
AKAI TX3A system	£246	**£190**
COLOUR TV's		
PANASONIC (20")	£242	**£215**
SONY 36G (24")	£312	**£298**
NATIONAL P47 (20")	£317	**£299**
BUSH 14B (16")	£237	**£205**
PIONEER 4Z (26")	£475	**£390**

5

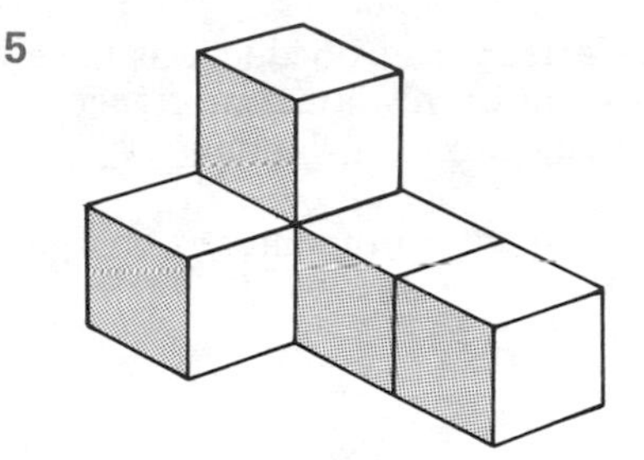

This solid is built from interlocking cubes of edge length 2 cm.

a How many cubes are used to make the solid?

b Find the volume of the solid in cm^3.

c How many faces has the solid, including the base?

d Calculate its total surface area.

6 Miss Heyes received this electricity bill, but she notices that there are several mistakes in the calculations. She knows that the four meter readings are correct and she also knows that the standing charges of £6·45 and £2·50 are correct. She makes enquiries and finds that each unit of electricity costs 6.24 pence and 2.25 pence for the two tariffs as shown on the bill.

Check all the calculations and find the correct total amount which she has to pay.

W·E·B
WESTLANDS ELECTRICITY

MISS S. HEYES,
34 THE BUTTS,
LUCKWITH,
WORCS.

ACCOUNT DATE	CONSUMER REFERENCE NO.	VAT REGISTRATION NO.
29 SEPT	9876/Z/5432	000 5079 00

METER READINGS PREVIOUS	METER READINGS PRESENT			AMOUNT £
		DOMESTIC SUPPLY TARIFF		
		Standing charge		6·45
46823	47715	Units	992 at 6.24p	61·90
		OFF-PEAK SUPPLY TARIFF		
		Standing charge		2·50
21023	21335	Units	312 at 2.25p	7·20

DATE OF READING OR ESTIMATE 29 SEPT

E INDICATES ESTIMATED READING
C INDICATES CONSUMERS OWN READING THE DATE SHOWN ON THE LEFT IS THE DATE WE RECEIVED YOUR CARD
R INDICATES METER REMOVED OR PREMISES RE-ASSESSED

TOTAL AMOUNT NOW DUE FOR PAYMENT £ 78·05

A mixture of short problems

7 Two companies, Etronco and Mitushi, both made electronic parts for television sets. This bar chart shows how many they produced over several years.

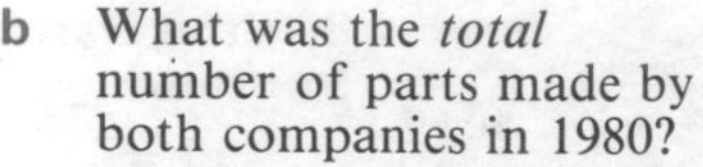

a In which year did Etronco produce more parts than Mitushi?

b What was the *total* number of parts made by both companies in 1980?

c In which year did Mitushi make *four* times as many parts as Etronco?

d One company went bankrupt. In which year did it happen, and can you suggest a possible reason?

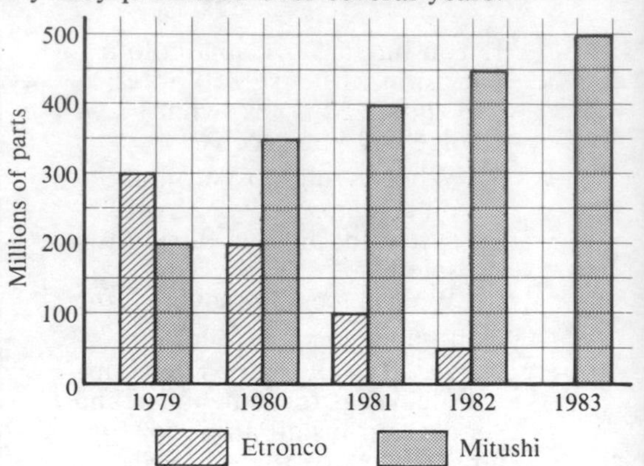

8 A train leaves Cardiff at 10.40 and arrives at Paddington Station in London $2\frac{3}{4}$ hours later. Give its time of arrival at Paddington as on a 24-hour clock.

9 Mr Brace receives two thirds of his gross salary after various deductions have been made. One quarter of what he receives is spent on his house mortgage, and three fifths of what he spends on this mortgage pays off the interest charged.

Calculate the fraction of his gross salary which is spent on the interest charged on his mortgage.

10 Calculate the angles x, y and z.

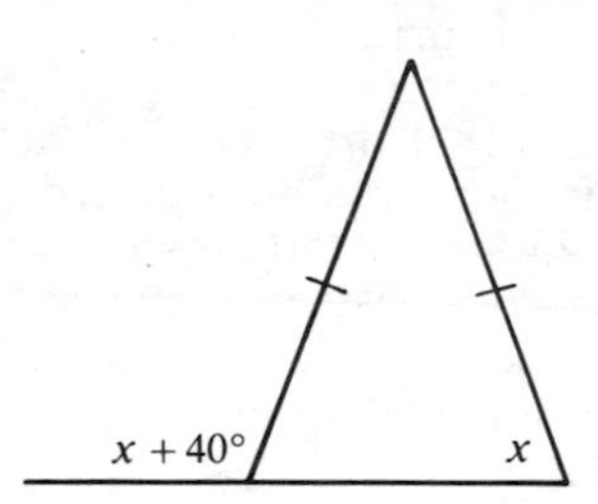

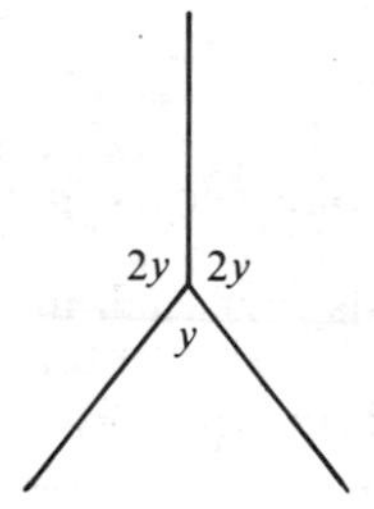

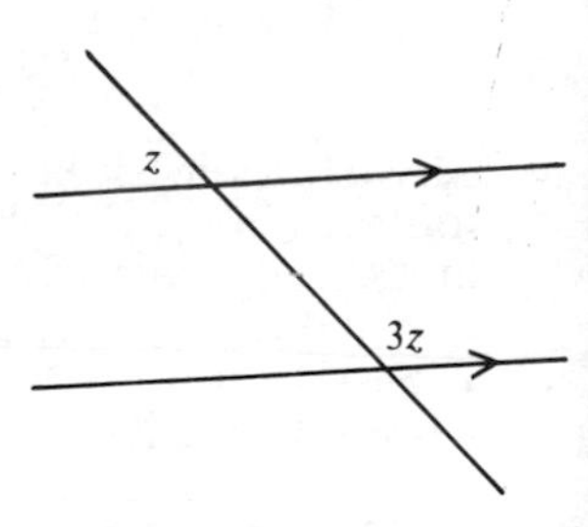

11 The pupils at a large school are collected by ten buses at the end of the day. The number of buses arriving late is noted over a period of two weeks with these results:

0 0 3 2 1 4 0 1 0 2

Find **a** the mode **b** the median **c** the mean of these ten numbers.

12 The time t taken for a car to travel a distance x from rest depends on its acceleration a, where

$$t = \sqrt{\frac{2x}{a}}.$$

a Find the value of t when $x = 50$ and $a = 4$.

b Find the value of t when $x = 1.2$ and $a = 0.15$.

c Rewrite the formula with x as subject.

13 A pack of 52 playing cards is shuffled and a card is selected at random. Find the probability of selecting

a an ace **b** a red King **c** either a queen or a diamond.

A mixture of short problems

14 Over a certain period, the Riley family uses 360 units of electricity and 200 units of gas. In the same period, the Wood family uses 150 units of electricity and 380 units of gas.

Each unit of electricity costs 10 pence and each unit of gas costs 14 pence.

a Copy these two tables and enter the above information.

	Units of Electricity	Gas
Riley	...	...
Wood	...	...

Cost per unit in pence	
...	Electricity
...	Gas

b Calculate the total amount spent by each family on these two fuels.

15 Draw and label both axes from -5 to 5.

A trapezium T has vertices $(-2, 1)$, $(-2\frac{1}{2}, 3)$, $(-2, 4)$, $(-1, 3)$. It is reflected in the x-axis onto its image T'. T' is then rotated onto T'' through 90° anticlockwise about the origin.

a Draw T and its images T' and T'' on one diagram.

b Describe in detail the single transformation which maps T onto T''.

16 A wooden cylindrical peg has a radius of 4 cm and height of 8 cm. Calculate

a the circular area of cross-section of the peg

b the volume of the peg, to the nearest whole cm^3

c the volume of wood wasted if the peg has been made from a cube of wood of edge 8 cm.

17 A drain-pipe slopes down the wall of a house 5 metres high so that it covers a horizontal distance of 4 metres.

Calculate

a the length of the drain-pipe

b the angle which the drain-pipe makes with the horizontal.

18 The formula for calculating the stopping distance D metres of a car travelling at a speed S km/h is

$$D = \frac{S}{10}\left(\frac{S}{20} + 2\right).$$

The flow diagram given by this formula is:

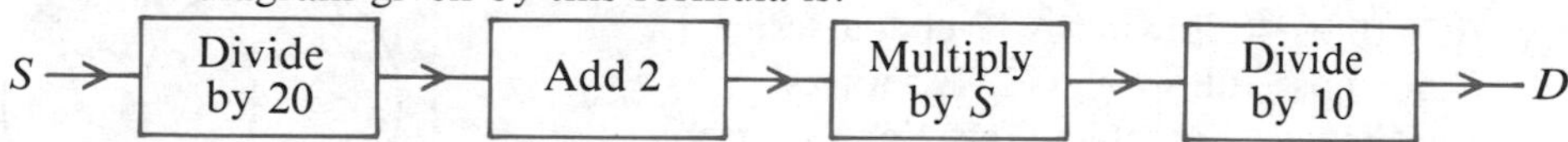

Use either the formula or the flow diagram to calculate the stopping distance of a car travelling at 130 km/h.

19 This map shows the main roads between several towns and cities.

a Is it possible to start from Tewkesbury and travel on each of these roads only once?

b Where would you have to start a journey to travel on each of these roads only once? Where would the journey end?

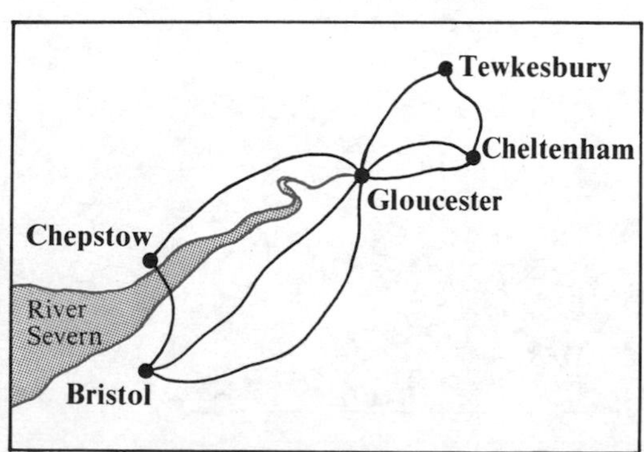

A mixture of short problems

• **20** a Trains travel the 120 miles between Birmingham and London.
Copy and complete this table to show the times taken by trains travelling at various average speeds.

Average speed, mph	20	30	50	60	80
Time taken, h					

b Using scales of 1 cm for 1 hour and 1 cm for 10 mph, draw a graph of your results on squared paper.

c Estimate the time taken at an average speed of 34 mph.

21 A straight path is made by tesselating paving stones in the shape of kites and triangles. The triangles are used to fill the gaps between two kites laid next to each other.

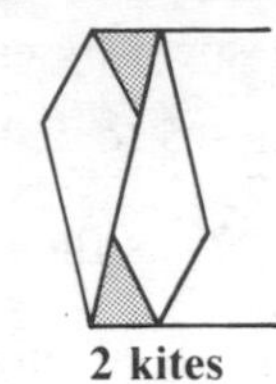
2 kites

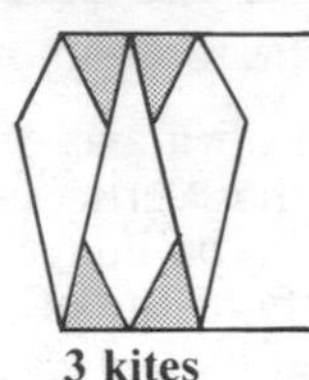
3 kites

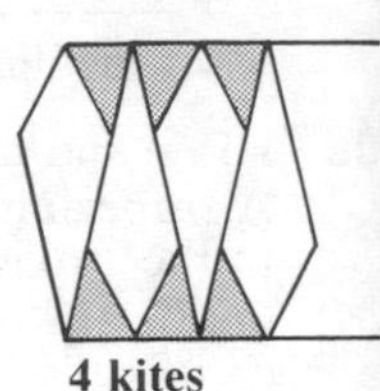
4 kites

a Copy and complete this table.

Number of kites	2	3	4	5	6	7	8
Number of triangles	2	4					

b If a long path is made using 50 kites, how many triangular paving stones will be needed?

c How many triangles are needed if n kites are used?

d If a total of 130 paving stones are used, how many of them will be kites?

22 A dairy herd is fed a ration of 10% hay, 40% silage and 50% kale.

a Write the ratio of hay to silage to kale in its simplest terms.

b If each cow in the herd is given $3\frac{1}{2}$ kg of hay each day, how much silage and how much kale is each cow given daily?

c If the herd eats 1750 kg of hay in 20 days, how many cows are in the herd?

23 In the sixth form of a school there are 120 students, all of whom are following either A-level or GCSE courses (or both). 84 are following A-level courses and 56 are following GCSE courses.

$\mathscr{E}$ = {members of the sixth form}
A = {those following A-level courses}
G = {those following GCSE courses}

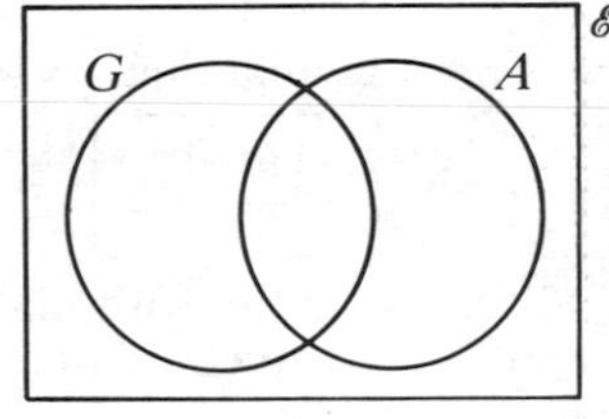

a Copy and complete this Venn diagram.

b Find the percentage of the sixth-form students who are following only GCSE courses.

24

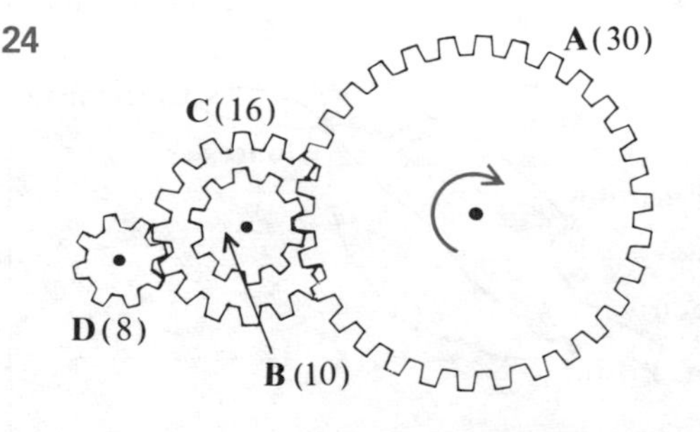

Four cogs A, B, C, D are mounted on three axles, with B and C on the same axle. The numbers in brackets give the number of teeth on the cogs. If cog A rotates clockwise at a speed of 2 revolutions per second, find the direction and speed of rotation of

a cog B b cog D.

A mixture of short problems

25 Draw and label both axes from 0 to 10.
On the axes, draw the three straight lines with the equations
$y = 2x + 1$, $y = 7 - x$, and $x = 3$.

a Shade the region where $y < 2x + 1$, $y > 7 - x$ and $x < 3$.

b Calculate the area of the shaded region.

26 A map has a scale of 1:500 000.
A forestry plantation has the shape of a square which has sides 3 cm long on the map. Find

a the number of kilometres which 1 cm on the map represents

b the length of the sides of the actual square, in kilometres

c the area of the plantation.

27 This net (with measurements given in centimetres) is used to make a prism with the shaded rectangle as the base.
Calculate

a the height h of each triangular face

b the total surface area of the prism

c the volume of the prism.

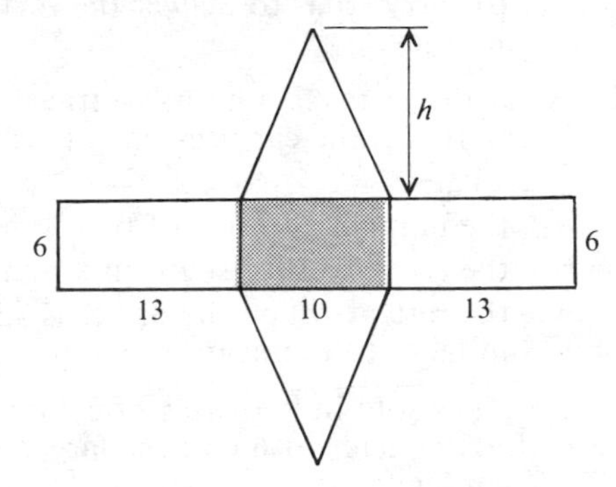

28

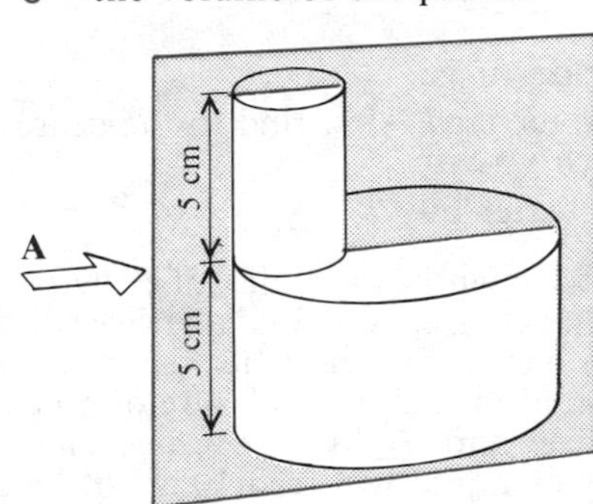

This metal casting takes the shape of two solid cylinders, both 5 cm high, joined as shown. The radii of the cylinders are 2 cm and 6 cm.

a Draw the plan of the casting and its elevation in the direction of arrow A.

b Calculate the area of the cross-section of the casting indicated by the shaded plane.

29 This pie chart represents the cost of imports to the country of Dalriada for the year 1985.

a What is the value of x?

b What fraction (in its simplest form) of the total imports is given to 'food'?

c If the total cost of all imports is £240 million, find the cost of the food imports.

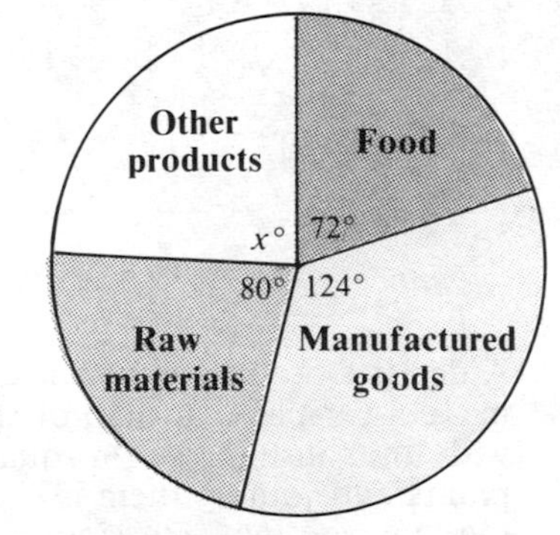

30

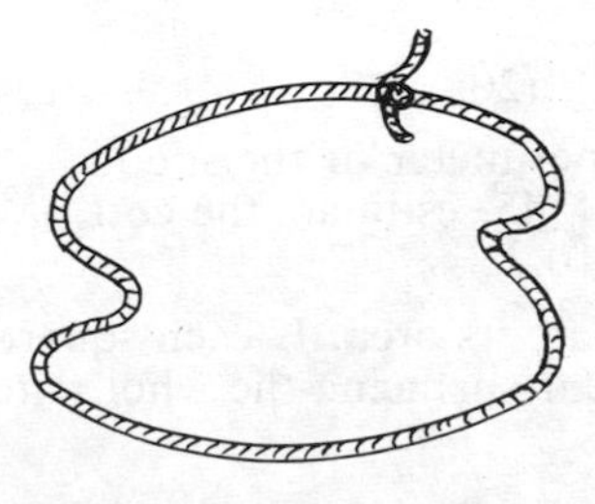

A piece of string is tied to form a loop of 36 cm. Calculate the area inside the string when it takes the shape of

a a square

b a rectangle with the longer sides of 12 cm

c a circle.

Costing problems

Part 1

1 A house-owner decides to make an L-shaped lawn by laying turf in the shape shown. The turf costs 85 pence per square metre and is cut to a depth of 5 centimetres.

Calculate

- **a** the area of turf required
- **b** the cost of the turf
- **c** the volume which a lorry must be able to carry so as to deliver the turf in one journey.

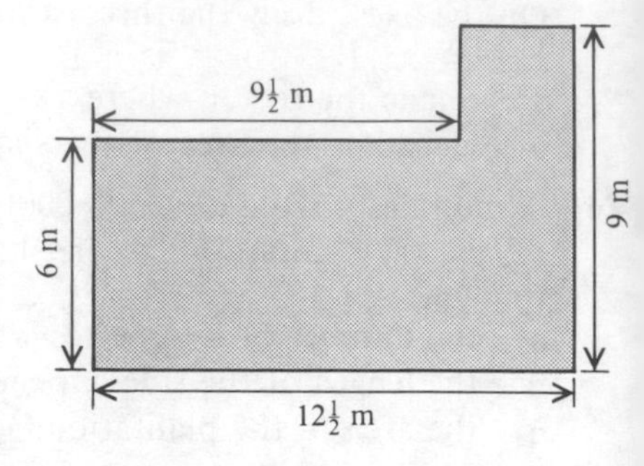

2 A car driver takes 4 hours to travel 336 km from London to Stockport. Every 16 km the car uses one litre of petrol, costing 46 pence per litre.

Calculate

- **a** the number of litres of petrol needed and its cost
- **b** the car's average speed in km/h
- **c** the rate at which the driver is spending money on petrol in pence per minute, to the nearest penny.

3 Coffee is sold in two sizes of cylindrical jars. The smaller jar is 14 cm high with a diameter of 6 cm; the larger jar has a diameter of 8 cm and stands $17\frac{1}{2}$ cm high.

- **a** Calculate the volume of coffee in each size of jar.
- **b** If the smaller jar costs £1·85 and the larger jar £3·03, find to the nearest penny the cost of 1 cm^3 of coffee in each jar.
- **c** Which size of jar would you say is the 'better buy'?

4

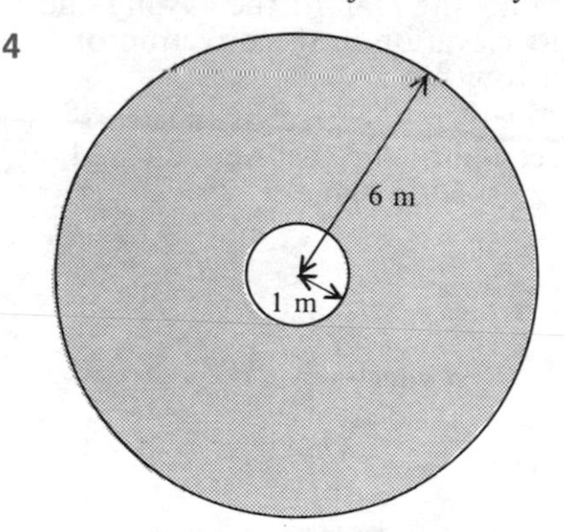

A circular lawn has a radius of 6 metres and, at its centre, is a circular flower-bed of radius 1 metre. The gardener feeds the lawn with a liquid which is a mixture of 50 cm^3 of fertilizer and 10 litres of water. This quantity of liquid will feed 10 m^2 of lawn.

Calculate

- **a** the area of the lawn, to the nearest whole m^2
- **b** the number of times the gardener mixes fertilizer with 10 litres of water
- **c** the cost of the fertilizer required if one litre of it costs £2·40.

5 a A derelict site in a city centre is to be fenced, levelled and surfaced to make a car-park. A map of the site has both axes labelled from 0 to 120 with units in metres. On squared paper draw the map by plotting these points and joining them in order:

(20, 20), (20, 80), (60, 120), (120, 60), (100, 20), (20, 20).

- **b** Taking measurements from the map, find the perimeter of the site in metres. If each metre length of fencing costs £1·45, estimate the cost of fencing around the whole site, to the nearest £10.
- **c** By dividing the site into suitable shapes, calculate its area. If each square metre of tarmac costs £2·84, estimate the cost of surfacing the whole site with tarmac, to the nearest £100.

Costing problems

d It is estimated that 450 cars will use the car-park each day, paying an average of 30 pence per car. How many days will it take for these payments to cover the cost of fencing and surfacing the site?

6 Mr Barraclough is a bus conductor. His hours of duty for a particular week are given in this table. He works during the weekend but has Friday free. His basic rate of pay is £4·60 per hour but on weekend duty he earns 'time and a quarter'. If he works a 'split shift' where there is a break in his duty of more than three hours, then he earns an extra £5 for that day.

Day	Hours of duty
Mon.	06.00 – 14.00
Tues.	07.30 – 09.30 and 15.00 – 22.00
Wed.	15.00 – 23.30
Thurs.	13.30 – 21.30
Fri.	–
Sat.	08.30 – 12.00 and 14.00 – 17.30
Sun.	12.30 – 18.30

Calculate

a how many hours he works from Monday to Friday and his total earnings for these days, including any extra for split shifts

b his weekend rate of pay and his earnings at the weekend

c his total earnings for the whole week.

7 Mrs Whitman paid £154 for a single ticket on a flight from London Heathrow to New York which is due to take $7\frac{1}{2}$ hours. The plane left at 08.40 (London time) and the distance travelled is 3400 miles. Time in New York is five hours behind British time, and the plane was thirty minutes late landing at New York.

a If Mrs Whitman's watch keeps good time and was correct on leaving London, what time did it read on landing at New York and to what time should she then alter it?

b Calculate, to 3 significant figures, the average speed of the plane
(i) during the actual flight
(ii) if it had arrived on time in New York.

c The only reason for the delay was a head-wind which slowed down the aircraft. What was the average speed of this head-wind?

d How much did Mrs Whitman's ticket cost her
(i) in £ per minute (ii) in pence per mile?

8 A new house has the three sides of its back garden enclosed with wooden fencing 1.75 metres high. A metre length of fencing costs £2·25 and 25 posts are needed at 80 pence each.

Both sides of the fence are painted with creosote oil costing £2·40 for a 5-litre tin and one litre of creosote will treat a surface area of 3 m^2.

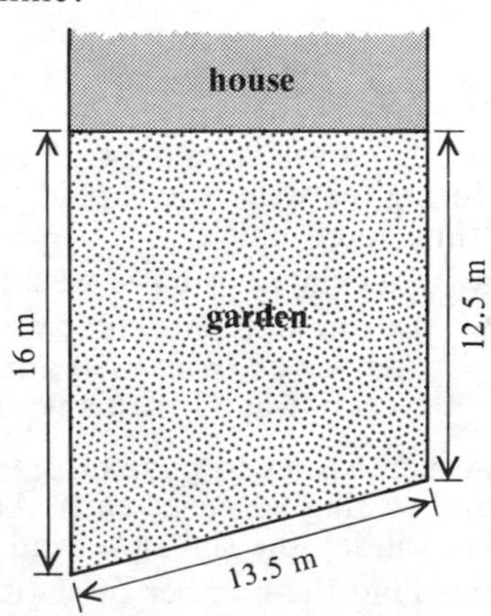

Calculate

a the total cost of the fencing and posts

b the total surface area to be treated

c the number of 5-litre tins required and their cost.

Costing problems

9 A market gardener has borrowed £13 000 on a mortgage to buy his house. The building society charge him 12% interest per annum on this mortgage. He has also borrowed £9000 from his bank to buy materials and equipment for his business. The bank charges him 18% interest per annum on the loan, but he can claim back one third of this interest from the Government. His monthly income from the sale of his produce averages £812·50 and the monthly cost of running his equipment is £117·50.

a In one year, what is
(i) his income from the sale of produce
(ii) his expenditure on running his equipment?

b How much interest does he pay in one year
(i) to the building society (ii) to the bank?

c How much can he claim back from the Government after paying interest to the bank?

d What is the total amount he spends in one year? Include the cost of running his equipment and the interest he pays on his two loans (after allowing for the amount he reclaims from the Government).

e What is his annual profit (i.e. the difference between income and total amount spent)?

10 An open barn is made from four posts, each 8 metres high, standing at the corners of a concrete floor 6 metres by 15 metres. There are no walls but the roof has two semi-circular ends as shown.

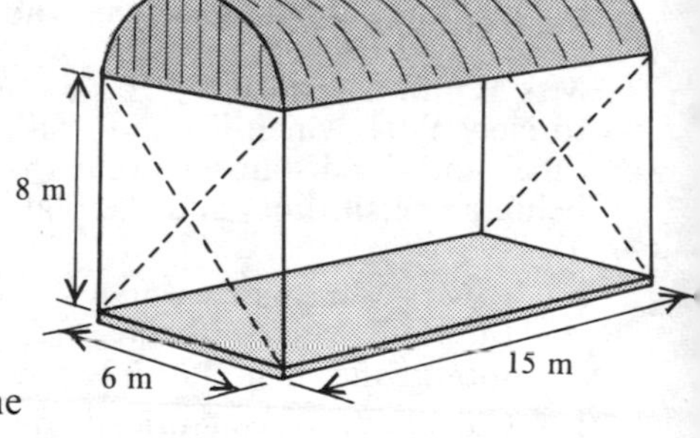

a If the concrete floor is 10 cm thick and concrete costs £72 for a cubic metre, calculate the volume of concrete required to make the floor and its cost.

b Calculate the total area of sheet metal required to make the whole roof (i.e. the curved section and its two ends).

c If the sheet metal costs £3·24 per m^2, what is the cost of the metal required for the roof, to the nearest £10?

d Four metal rods are connected diagonally across the ends to strengthen the barn. They are indicated by dotted lines on the diagram. Each metre length of these rods costs £1·25. What is the cost of the four of them?

Part 2

1 An employee is paid £4·28 per hour for a 40-hour week. If she works overtime during the week she is paid 'time and a quarter', and Saturday work gives 'time and a half'.

a Calculate her basic weekly wage for a 40-hour week.

b During one particular week she works four hours overtime on Wednesday and works from 8.30 a.m. to midday on Saturday. Calculate her total weekly wage for this week.

2 A man wishes to pave his garden path with square concrete slabs of side 50 cm and costing 95 pence each. The path is to be $1\frac{1}{2}$ metres wide and 12 metres long. He will lay the slabs on sand 2 cm deep. One cubic metre of sand costs £8·50.

a Find the number of slabs required and their cost.

Costing problems

b Find the volume of sand required and its cost.

c If a delivery charge of 5% of the total cost is made, calculate the price paid for slabs, sand and delivery.

3 A water tank with internal dimensions of 3 metres and 2.5 metres as shown and 1.4 metres deep, is made from concrete. The walls and base of the tank are to be 10 cm thick.

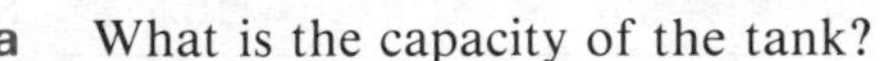

a What is the capacity of the tank?

b What volume of concrete is required to make the tank?

c If twenty bags of cement at £1·26 each are needed for each cubic metre, how many bags are needed to make the tank and at what cost, to the nearest £?

4

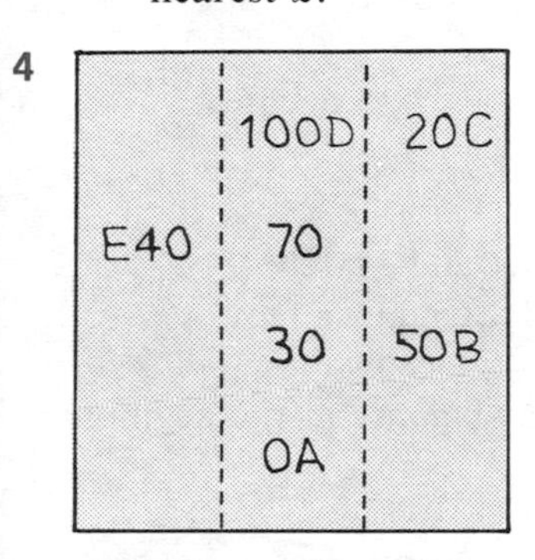

A builder buys a plot of land on which to build twenty five identical houses. An offset survey in metres produces the results shown here.

a Draw a plan of the site using a scale of 1 cm for 10 metres and calculate its area.

b The land cost the builder £120 000 per hectare and each house will cost £15 500 to build. Find the total amount spent by the builder in buying the land and building the houses.

c If the builder wants to make a profit of 25% of his total cost price, at what price should he advertise each house (to the nearest £100)?

5 Mrs Marsden hires a car at a rate of £8 per day plus five pence for every mile travelled. There is an additional charge of £2 per day for insurance and she must pay for her own petrol.
She uses the car to travel 406 miles from Leeds to Bristol and back in one day of 8 hours of driving, using 58 litres of petrol at a cost of 43.6 pence per litre.
Find

a how much she pays for the hire of the car

b the cost of the petrol she uses

c the total cost of her journey

d her average speed during the time she is driving

e the fuel consumption of the car in miles per litre

f the average rate (in pence per minute) at which she is spending money on petrol while driving. Give your answer to three significant figures.

6 Anne Barnet makes dresses for sale in a small shop. From a roll of material 24 metres long she can make eight identical dresses and wastes only 5% of the material.
One particular roll of material costs her £5·20 per metre length and she charges the shop 80% more than the cost of the material for each dress.
Calculate

a the cost of one roll of this material and the cost of the material lost in wastage

b the cost of the material (including the wastage) used in one dress and the price at which she sells each dress to the shop.

Costing problems

7 John Kenworthy is the treasurer of a church youth club. He collects the membership fees and pays various bills. The club has thirty two members who each pay an annual subscription of £2·50 and 20 pence each time they attend. The club opens from October until March and the monthly attendance figures are shown in this table.

Month	October	November	December	January	February	March
Number of members attending	110	112	86	94	102	88

During this time, he pays the following bills:

Table tennis equipment	£23·46
Snooker equipment	£8·75
Badminton equipment	£19·25
Other sports equipment	£12·34
Christmas party expenses	£62·70
Hire of minibus	£24·33
Postage	£6·72
Miscellaneous expenses	£5·73

Calculate

a the total number of attendances between October and March

b the total income from annual subscriptions and attendance fees

c the total expenditure

d the 'balance in hand' at the end of March, if the balance at the beginning of October was £37·42.

8 A garage has the shape and size shown where the sides and roof are made from rectangles. The front and back of the garage are part rectangle and part isosceles triangles.

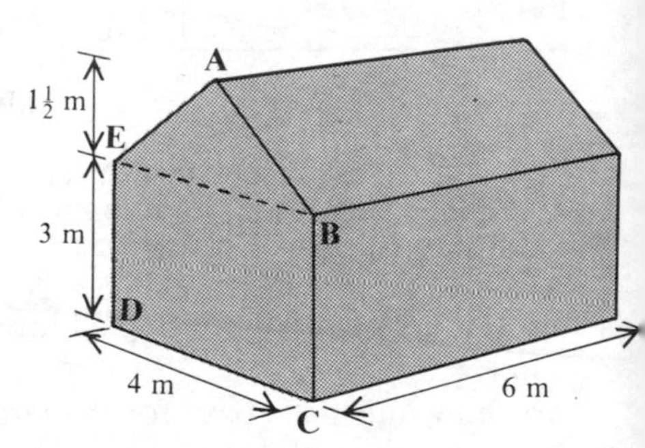

a Calculate the area of one end of the garage.

b Use triangle ABE and Pythagoras' Theorem to calculate the length AB.

c Find the total surface area of the garage, excluding the floor.

d The owner decides to paint the sides, the front and the back, but not the roof, with two coats of paint. If one litre of paint will cover 12 m^2 and ignoring any reduction for window space, calculate how many litres are needed.

e If paint is sold in one-litre tins at £4·70 each and two-litre tins at £8·90 each, estimate how much he will need to spend on paint.

9 A wall is built 12 metres long to a height of 2 metres above ground level and to a depth of 25 cm below the ground. It is 10 cm wide. Every square metre of the side of the wall needs 70 bricks and a load of one thousand bricks costs £94.

a Calculate the area of the side of the wall and the number of bricks required to build it.

b Only full loads of one thousand bricks each are delivered. How many of these loads are needed and how many bricks will not be used in the wall?

c Value Added Tax at 15% is added to the cost of the bricks. Calculate the total amount which is paid for the bricks delivered.

Costing problems

10 A package holiday to Italy is advertised at a cost of £224 per person. This does not include a fuel surcharge of 3% of the advertised cost, nor does it include airport taxes of £2·50 per person, nor medical insurance of £7·60 per person.

a If Mr & Mrs Shah decide to go on this holiday, what will be the *total* cost?

b If they pay for their holiday at least three months in advance, they will receive a reduction of 4%. How much is this reduction, to the nearest penny?

c Before leaving for Italy, they change £150 into lire. If the exchange rate is 2430 lire for each £1 sterling, how many lire will they receive?

d At the end of the holiday, they wish to change 58 450 lire back into sterling at the same rate of exchange. How much will they receive to the nearest penny?

Part 3

1 Janie Hellon has the offer of two jobs. The first provides for a 38-hour week and pays £2·45 per hour with no possibility of overtime. The second offers £2·16 per hour for a 32-hour week and in addition, if Janie wishes, it guarantees at least five hours overtime each week at 'time and a quarter'.

a Calculate the basic weekly earnings for both jobs, excluding any overtime.

b If Janie works five hours overtime, what would be her total weekly earnings from the second job?

c How many hours of overtime would she have to work for the income from the second job to equal that of the first?

d Which job do you think Janie accepted? Give your reasons.

2 a A garden swimming pool is 8 metres long and 5 metres wide. It is 2 metres deep at all points. Find the total area of its sides and floor.

b It is completely covered in square ceramic tiles of side 20 cm. How many of these tiles cover
(i) each square metre (ii) the sides and floor of the pool?

c Each tile costs 16 pence but a discount of 8% is given on a large order. Calculate the cost of the tiles used for the pool if the discount is given.

3 An oil drum, used to store petrol, has a diameter of 1.2 metres and stands 2.3 metres high. It costs £1170 to fill with petrol. A dipstick is used to find the level of petrol in the drum and in one week the level fell by 45 cm.
Calculate

a the volume of petrol in the drum when full

b the cost per litre of the petrol

c the amount of petrol used in the week and its cost.

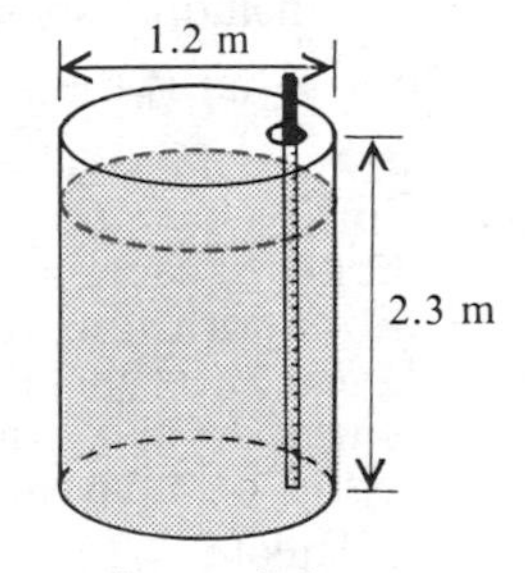

4

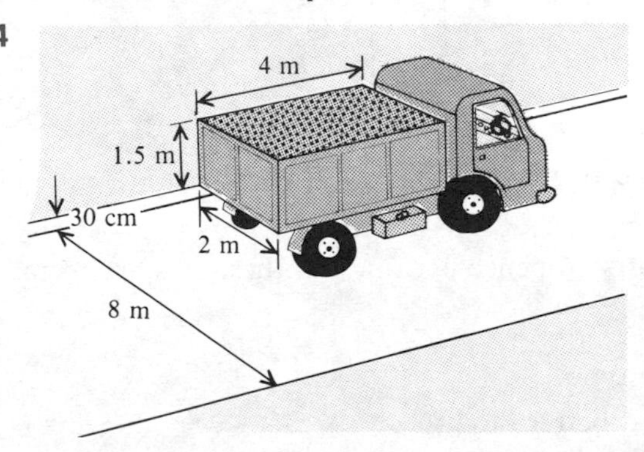

To build a new road, earth has to be removed to make the road-bed. The land which is excavated is 8 metres wide and 300 metres long, and a depth of 30 cm of earth is removed by lorry. The back of each lorry is 1.5 m high, 2 m across and 4 m long, and is filled so that the earth is level with the top of the back. Each journey which a lorry makes costs an average of £12 but the earth is sold as top soil at 90 pence per cubic metre.

Costing problems

Calculate

a the volume of earth to be removed

b the volume of earth which fills one lorry

c the number of lorry loads required to remove the earth

d the net cost of using the lorries.

5 In one year Mr Patel's car costs him £80 to insure, £100 to tax and £280 to service. He also pays for petrol at an average of 46 pence per litre and his car consumes petrol at an average rate of 16 km per litre. In this particular year he travels 32 000 km.

a How much does he spend altogether on his car in the year?

b If his car depreciates by £1200 during the year, what is the total cost incurred by having the car
(i) over the whole year (ii) per week, to the nearest £.

If he were not to own a car, he would have to travel the same distance by public transport; three-quarters by rail and the rest by bus. Rail travel costs eight pence per kilometre on average and bus fares six pence per kilometre.

c Calculate how much he would spend on train and bus fares.

d Which is the cheaper method of transport, by car or by train and bus; and by how much?

6 Mr and Mrs Willoughby take their two children on a package holiday to Spain, flying from Manchester Airport. The holiday is advertised at £145 per adult with children at half-price. Spanish airport taxes are an extra £2·50 per person and there is an extra charge of £5 per person for holidays from Manchester. It will cost £12 return for the family to reach Manchester Airport, and they have allowed themselves a total of £8 per day spending money during their ten-day holiday.

a What is the total cost they are expecting to pay for the whole holiday, including spending money?

b What is the average total cost per person per day, to the nearest penny?

7 Tins of baked beans, costing the customer 32 pence each, are packed in three layers in a box with four rows of six tins to each layer. Sheet metal, costing 14 pence per m^2, is used to make the tins so that they have a radius of 4 cm and a height of 10 cm.

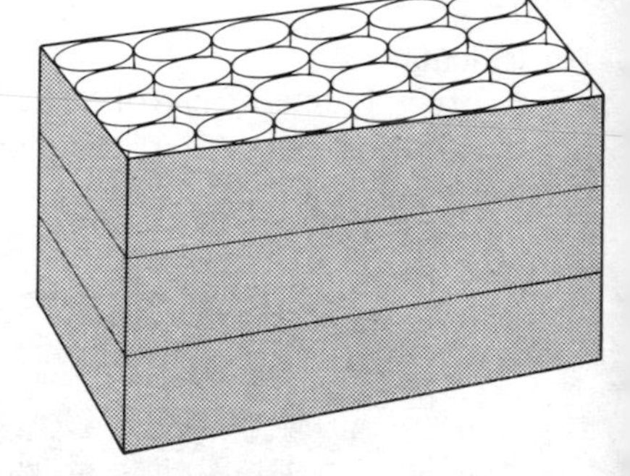

Calculate

a the number of tins in a full box and their total cost to a customer

b the total surface area of one tin

c the cost of the sheet metal used to make all the tins in one full box.

Another brand of baked beans costs 16 pence per tin, and these tins are 8 cm high with a radius of 3 cm.

Find

d the volume of one tin of each type

e which tin gives the customer a 'better buy'.

Costing problems

8 The tarmac playing area of a small primary school has the shape shown here. It is to be resurfaced to a depth of 5 cm with tarmac costing £24 per cubic metre. The job takes three days and employs two men for eight hours per day, at a rate of £2·35 per hour per man. They are using machinery which has to be hired at a fee of £18·70 per day.

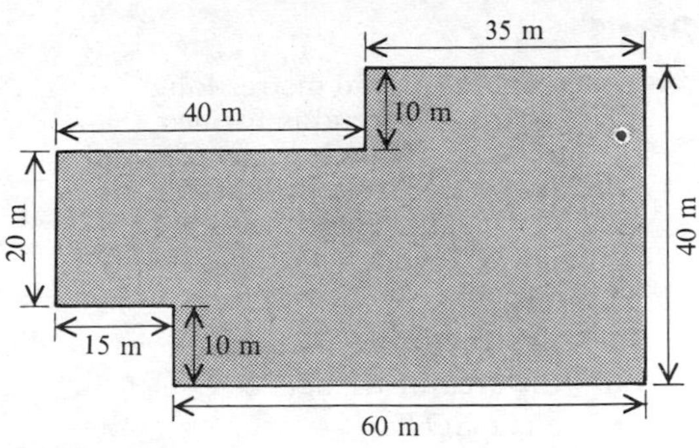

a Calculate the cost over the three days of
(i) the hire of the machinery (ii) the wages of the two men.

b Calculate the surface area of the playground, the volume of tarmac needed and its cost.

c Find, to the nearest £10, the total cost of tarmac, men and machinery.

9

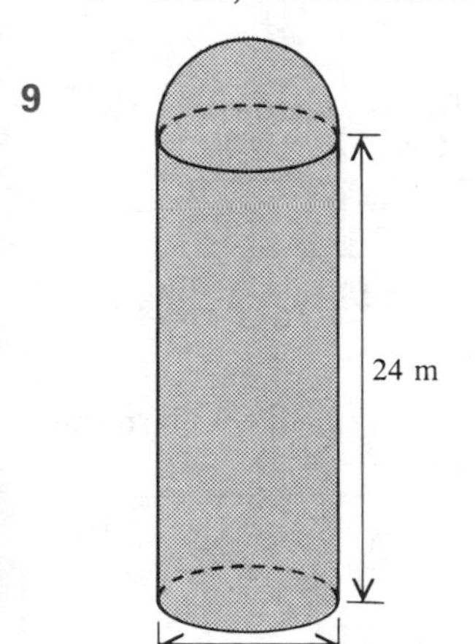

A farmer's silo holds cattle food. It has a cylindrical shape 24 metres high topped by a hemisphere of diameter 8 metres. The farmer is to paint the outside of the silo with paint, one litre of which covers 32 m^2.

a Find the total surface area of the silo, excluding the base, to the nearest square metre.

b How many litres of paint will be needed for two coats of paint?

c If paint is available in 10-litre tins costing £25 each or in 2-litre tins costing £5·60 each, what is the cheapest combination of tins which the farmer could buy?

10 Mr and Mrs Cooper use electricity in their home but no gas. At the start of the year, their meter reading was 38510 and at the end of the first quarter of the year it read 41436. Each unit of electricity costs 5.6 pence and there is a standing charge of £7·45.

a Find the number of units of electricity used and their cost to the nearest penny.

b What is their total electricity bill for this quarter?

They think the bill is too high and decide to convert some of their appliances to use gas, though they cannot avoid using electricity altogether.
Their next quarter's bill shows that they now use only 937 units of electricity but they now have to pay for 127.5 therms of gas. The charges for electricity have not altered, and gas costs 42.3 pence for each therm with a standing charge of £10·60.

c Find the cost, to the nearest penny, of their new electricity bill, including the standing charge.

d Find the cost, to the nearest penny, of their first gas bill, including the standing charge.

e Estimate how much they have saved on fuel by making the change.

A mixture of longer problems

Part 1

1 A greenhouse is 50 metres long. The shape of its end is formed from two right-angled triangles and the sector QST of a circle with centre Q and radius 5 metres. Angles SQP and TQR are both 30°.

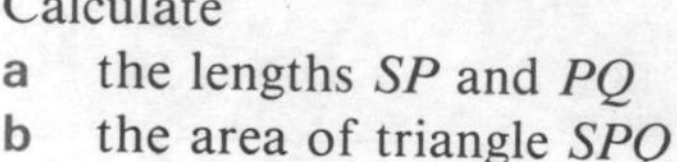

Calculate

- **a** the lengths SP and PQ
- **b** the area of triangle SPQ
- **c** angle SQT
- **d** the area of sector SQT
- **e** the total area of the end of the greenhouse
- **f** the volume of the greenhouse, to the nearest m^3.

2 A farmer grows wheat on 100 hectares of land. When harvesting the crop, he notes the amount of wheat which each hectare yields.
His results are tabulated as shown.

Yield, tonnes	0-2	from 2-4	from 4-6	from 6-8	from 8-10
Number of hectares	5	11	42	36	6

a Approximate each group of yields by the mid-value and hence estimate
(i) the total yield
(ii) the mean yield per hectare, to one decimal place.

b Copy and complete the following table of cumulative frequencies. Use the table to draw a graph of cumulative frequency against yield.

Yield, tonnes	$\leqslant 2$	$\leqslant 4$	$\leqslant 6$	$\leqslant 8$	$\leqslant 10$
Cumulative frequency					

Estimate the median yield per hectare.

3 a Solve these equations.
(i) $3(2x - 1) = 4x + 2$ (ii) $5x = \dfrac{2x + 9}{2}$

• **b** One angle is 15° bigger than twice the size of another angle $x°$. Together they make a straight line.
Find the value of x.

c The floor $PQRS$ of a room is partly covered by a square carpet of side x metres.
The uncovered part of the floor is 2 metres wide as shown.
Use the letter x to write down expressions for

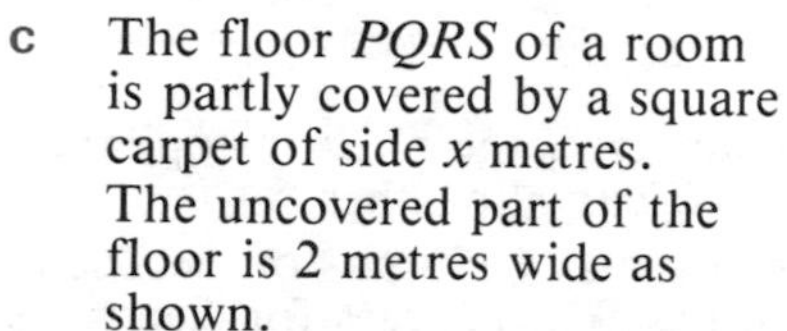

(i) the area of the carpet
(ii) the area of the floor not covered by the carpet
(iii) the total area of the floor.
(iv) Given that the total area of the floor is 24 m^2, form a quadratic equation and solve it to find the value of x.

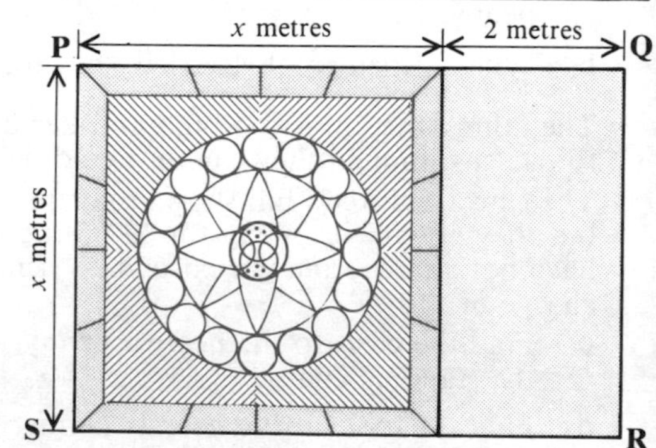

A mixture of longer problems

4 The radio mast FT on a remote Scottish island is 13 metres high and is held vertical by four sloping wires. The wires are fixed to the top T of the mast and are anchored to the ground at the corners J, K, L and M of a square. The square has centre F and sides 5.66 metres long.

Calculate

a the lengths JL and JF

b the length of the wire JT

c the angle TJF between the wire JT and the ground

d the angle between the two wires JT and TL.

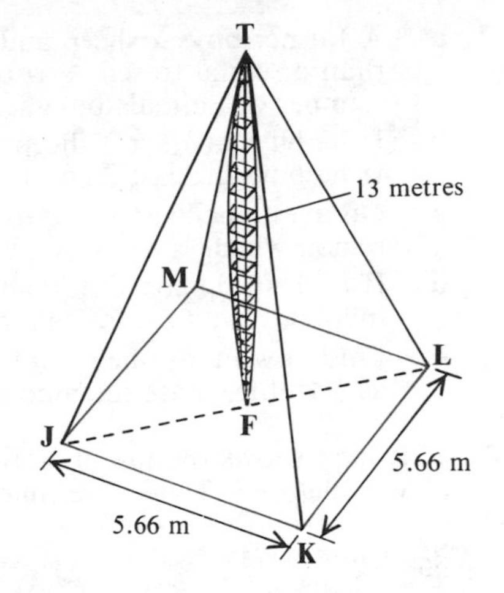

5 a Copy and complete this table for the equation $y = x^2 - 2x + 2$.
The values for $x = -\frac{1}{2}$ and $x = 2\frac{1}{2}$ are already completed.

x	-1	$-\frac{1}{2}$	0	$\frac{1}{2}$	1	2	$2\frac{1}{2}$	3
x^2		$\frac{1}{4}$					$6\frac{1}{4}$	
$-2x$		$+1$					-5	
$+2$		$+2$					$+2$	
y		$3\frac{1}{4}$					$3\frac{1}{4}$	

Draw the graph of $y = x^2 - 2x + 2$ on squared paper, using 4 cm for each unit on the x-axis and 2 cm for each unit on the y-axis.

b Use your graph to explain why the quadratic equation $x^2 - 2x + 2 = 0$ has no solutions.

c Draw the straight line $y = 2x - 2$ on the same diagram and hence write down the solution of the equation $x^2 - 2x + 2 = 2x - 2$.

d Check your answer to **c** by an algebraic method.

6 a Draw and label both axes from -5 to 10 using a scale of 1 cm for each unit on both axes.
Draw the rectangle R with corners (1, 4), (3, 4), (3, 1), (1, 1).

b R' is the image of R after a transformation T given by the matrix $\begin{pmatrix} 2 & 1 \\ 0 & 1 \end{pmatrix}$
Calculate the co-ordinates of the corners of R'. Plot its position on the same diagram and label it.

c R'' is the image of R' after a transformation U given by the matrix
$\begin{pmatrix} 0 & -1 \\ -\frac{1}{2} & \frac{1}{2} \end{pmatrix}$
Calculate the co-ordinates of the corners of R''. Plot its position on the same diagram and label it.

d Describe, giving full details, the single transformation which maps R directly onto R'' in one step.
Calculate the matrix UT of this transformation.

A mixture of longer problems

7 **a** A farmer buys x sheep and y cows. The total number of animals is greater than or equal to ten. Write an inequality involving x and y for the total number of animals bought.

b If he buys at least 3 sheep, write an inequality in x.
If he buys at least 2 cows, write an inequality in y.

c On axes labelled from 0 to 15, draw three boundary lines and shade the region which is *not* given by your three inequalities.

d If each sheep costs £150 and each cow costs £400, write an expression involving x and y for the total amount which the farmer spends.

e Find how many sheep and how many cows he must buy to spend as little as possible. State the amount spent in this case.

8 This map shows the major stations on the Metro serving the area around Newcastle-upon-Tyne. The timetable gives part of the daily service.

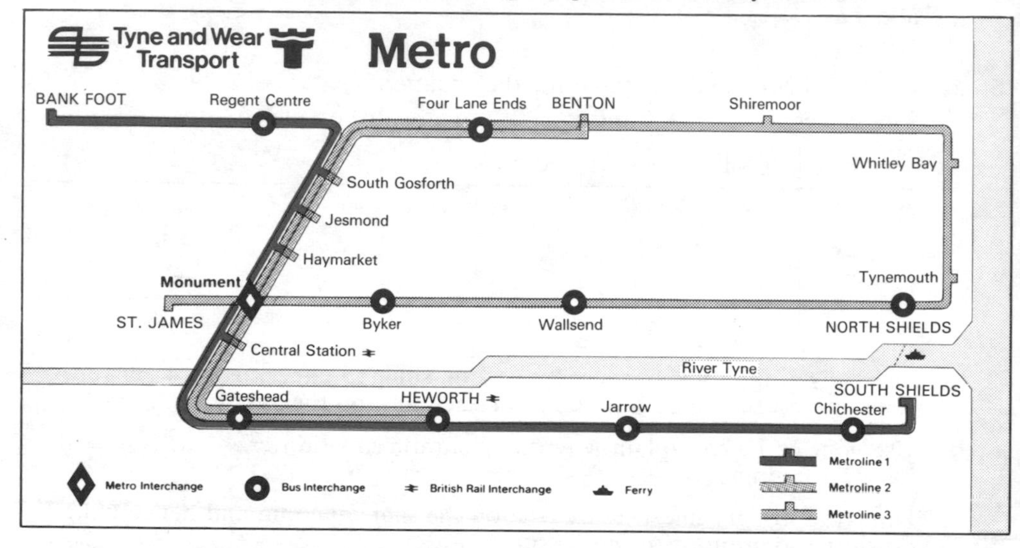

Metroline	
SOUTH SHIELDS – HEWORTH – BANK FOOT	1
HEWORTH – BENTON	2
HEWORTH – BENTON – THE COAST – NORTH SHIELDS – ST. JAMES	3

MONDAY TO SATURDAY

Station	1	2	3	1	2	3	1	2	3	1	2	3	1	2	3
SOUTH SHIELDS	1633	...	...	1643	...	...	1653	...	...	1703	...	...	1713	...	...
Chichester	1635	...	...	1645	...	...	1655	...	...	1705	...	...	1715	...	...
Jarrow	1642	...	...	1652	...	...	1702	...	...	1712	...	...	1722	...	...
HEWORTH ⇌	1649	1652	1656	1659	1702	1706	1709	1712	1716	1719	1722	1726	1729	1732	1736
Gateshead	1654	1657	1701	1704	1707	1711	1714	1717	1721	1724	1727	1731	1734	1737	1741
Central Station ⇌	1656	1659	1703	1706	1709	1713	1716	1719	1723	1726	1729	1733	1736	1739	1743
Monument	1657	1700	1704	1707	1710	1714	1717	1720	1724	1727	1730	1734	1737	1740	1744
Haymarket	1658	1701	1705	1708	1711	1715	1718	1721	1725	1728	1731	1735	1738	1741	1745
Jesmond	1659	1702	1706	1709	1712	1716	1719	1722	1726	1729	1732	1736	1739	1742	1746
South Gosforth	1705	1708	1712	1715	1718	1722	1725	1728	1732	1735	1738	1742	1745	1748	1752
Regent Centre	1707	...	...	1717	...	...	1727	...	...	1737	...	...	1747	...	...
BANK FOOT	1713	...	...	1723	...	...	1733	...	...	1743	...	...	1753	...	...
Four Lane Ends	...	1712	1716	...	1722	1726	...	1732	1736	...	1742	1746	...	1752	1756
BENTON	...	1713	1717	...	1723	1727	...	1733	1737	...	1743	1747	...	1753	1757
Shiremoor	...	...	1721	...	...	1731	...	...	1741	...	...	1751	...	...	1801
Whitley Bay	...	...	1728	...	...	1738	...	...	1748	...	...	1758	...	...	1808
Tynemouth	...	...	1733	...	...	1743	...	...	1753	...	...	1803	...	...	1813
NORTH SHIELDS	...	...	1736	...	...	1746	...	...	1756	...	...	1806	...	...	1816
Wallsend	...	...	1745	...	...	1755	...	...	1805	...	...	1815	...	...	1825
Byker	...	...	1751	...	...	1801	...	...	1811	...	...	1821	...	...	1831
Monument	...	...	1755	...	...	1805	...	...	1815	...	...	1825	...	...	1835
ST. JAMES	...	...	1756	...	...	1806	...	...	1816	...	...	1826	...	...	1836

⇌ Adjoining or near a Rail Station.

A mixture of longer problems

a You catch the 16.43 from South Shields. What time do you arrive at Monument?

b If you catch the 16.52 from Heworth, which Metroline are you using and how long does it take to reach Jesmond?

c You are in Jarrow and *just* miss the 17.02 to Monument. When is the next train and how long do you have to wait?

d You are in Gateshead and want to get to Regent Centre before 5.30 p.m. Which Metroline will you use and what is the latest possible time you can leave Gateshead?

e Two youngsters leave Monument on a round trip, arriving back at Monument without changing trains. Which Metroline do they use and how long will the journey take?

f A family, returning from London by British Rail, arrive at Central Station at 17.25. If they live in Whitley Bay, which Metroline will they use to get home? When does their earliest Metro train leave Central Station and when does it arrive in Whitley Bay?

g Mrs Foggo lives in Jarrow and she wants to visit her sister living in Shiremoor. Which two Metrolines will she use? List the possible stations at which she could change trains. If she arrives at Jarrow station at 5 p.m. exactly,

(i) how long has she to wait before her first train

(ii) how long has she to wait when she changes trains

(iii) what time does she arrive in Shiremoor?

h Mr Stokoe works in Chichester and lives in Benton. Will he need to change trains when travelling between work and home? If he is at Chichester station at 16.40, when does he catch the first possible train and when does he arrive in Benton? Which Metrolines will he have used and how long will the whole journey have taken?

9 **a** Draw and label axes from 0 to 9 using 2 cm for each unit on both axes. Triangle T has vertices $P(0, 8)$, $Q(3, 7)$ and $R(1, 4)$. Its image T' has vertices $P'(5.7, 5.9)$, $Q'(6.3, 2.8)$ and $R'(2.7, 3.0)$. Draw and label both T and T' on your axes.

b T maps onto T' under a clockwise rotation. Find the centre of the rotation by construction using ruler and compasses only. Show all construction lines clearly and label the centre of rotation C.

c Draw the angle of rotation, measure it and write its value. Use compasses to draw the path of point P as it rotates onto its image P'. Measure CP in centimetres and calculate the length of the path from P to P'.

10 **a** John Cox works a basic 36-hour week, earning £3·80 per hour. In one particular week he also worked six hours overtime at 'double-time'. Calculate his total gross earnings for this week.

b The weekly deductions from his total gross earnings are:

National Insurance	£14·25
Income tax	£36·48
Pension fund	5% of total gross earnings.

Calculate

(i) how much he pays into his pension fund

(ii) his total weekly deductions

(iii) his net weekly wage after these deductions have been made.

A mixture of longer problems

c

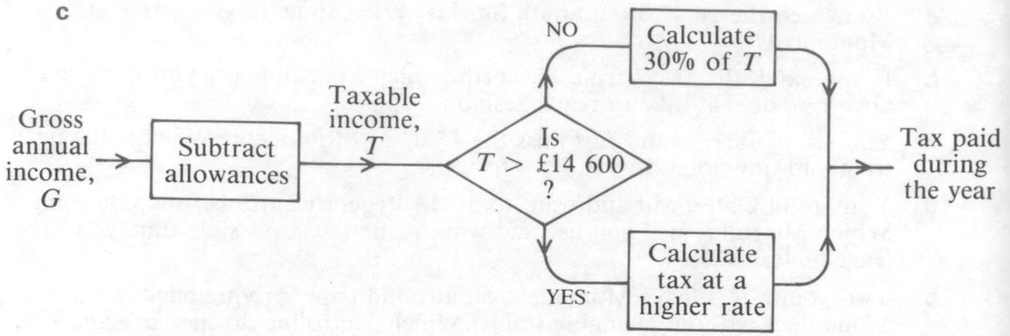

The income tax which John pays in a year can be calculated using this flow diagram.
He is not married and has the single person's tax-free allowance of £1846.
Assume he earns £548 on average each month, and ignore any overtime payments.
Calculate
(i) his gross annual income, G
(ii) his taxable income, T for the year
(iii) the income tax which he pays
(iv) his annual income after income tax is deducted.

Part 2

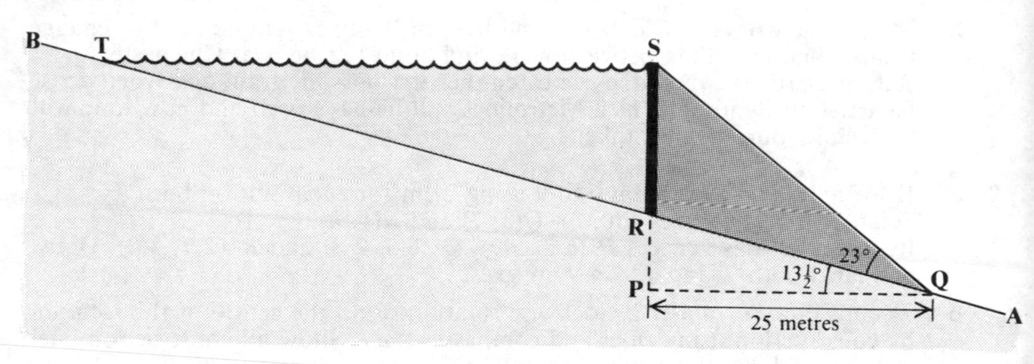

• **1** A hillside AB slopes upwards at an angle PQR of $13\frac{1}{2}°$ to the horizontal. A dam is constructed by building a vertical wall RS and earth is then tipped on the downhill side as shown by the shaded triangle RSQ. Water is stored behind the wall and ST is the distance across its surface.
Given that PQ = 25 metres, find
- **a** the vertical distances PR and PS
- **b** the height of the wall RS
- **c** angles PRQ and TRS
- **d** the distance ST across the surface of the water.

• **2 a** A manufacturing company makes parts for a certain machine. When the number of parts they make is x hundreds, the cost y of making them is given by

$$y = \frac{6}{x - 2}$$

where y is in thousands of pounds.

A mixture of longer problems

Copy and complete this table and draw the graph of y against x for $3 \leqslant x \leqslant 8$, labelling both axes from 0 to 8.

Number of parts, (in hundreds) x	3	4	5	6	7	8
$x - 2$						
Cost (in £000s), y						

b On selling these parts, the income which the company receives is given by $y = \frac{1}{2}x$, where x is in hundreds and y is in thousands of pounds.
Copy and complete this table and draw the graph of y against x on the same axes.

Number of parts sold, (in hundreds) x	2	4	6	8
Income (in £000s), y				

c How many parts must the company make and sell for it to 'break even' (i.e. for income and cost to be equal)?

d What profit does the company have when it makes and sells 650 parts?

3 a An electricity bill can be calculated using this flow diagram where N is the number of units used and A is the amount to be paid.

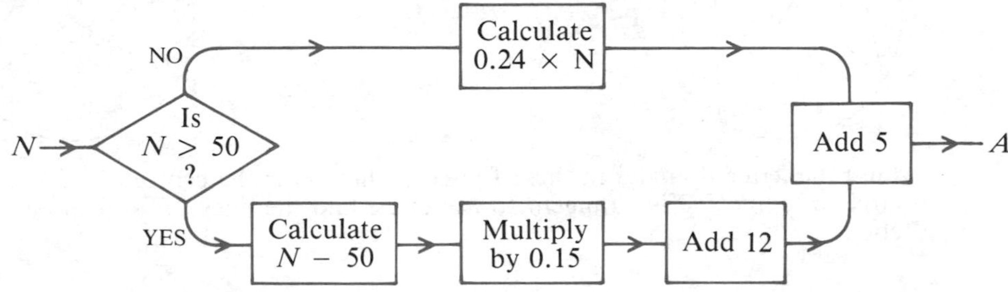

Find the amount to be paid, A when (i) $N = 32$ (ii) $N = 264$.

b $x \rightarrow$ [ ] $\rightarrow$ [ ] $\rightarrow$ [ ] $\rightarrow$ [ ] $\rightarrow y$

This flow diagram calculates values of y given by the formula

$$y = \frac{(3x - 2)^2}{4}.$$

The instructions missing from the four boxes are:

Subtract 2	Divide by 4	Square	Multiply by 3

Copy the flow diagram and insert these instructions into the boxes in the correct order.
Calculate the value of y when $x = 4$.

c When two electrical resistors of values p and q are connected in parallel, their total effect T is given by the formula

$$T = \frac{pq}{p + q}.$$

Find the value of T when
(i) $p = 6$ ohms, $q = 4$ ohms (ii) $p = 1.6$ ohms, $q = 0.8$ ohms.

A mixture of longer problems

4 a Draw and label both axes from 0 to 12. Use a scale of 1 cm for each unit on both axes. Plot and label the points A to F, and join them in order to make a hexagon.

$A(3, 2)$, $B(3, 9)$, $C(6, 12)$, $D(8, 11)$, $E(11, 8)$, $F(11, 2)$

b Imagine that the diagram you have drawn is a map of an area of open moorland purchased by the Forestry Commission. The scale of the map is 1 cm = 1 km.

Use a ruler to find the perimeter of this land in kilometres.

c Divide the land into strips and so find its area in square kilometres. If the purchase price was £625 per km^2, find the total amount paid by the Forestry Commission, to the nearest thousand pounds.

d The land is to be drained by a straight pipe running from B to D. What will be the length of this pipe?
If B is 525 metres above sea-level and D is 75 metres above sea-level, find the difference in height between B and D and calculate the angle which the pipe will make with the horizontal.

5 a

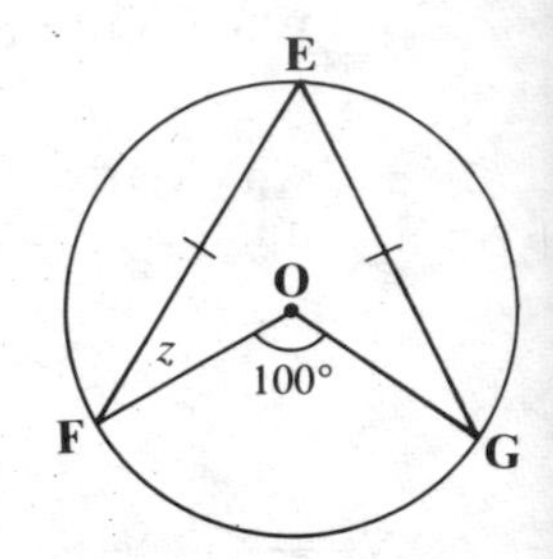

Find the lettered angles in these figures, where O is the centre of each circle, the line PQ is a tangent to the circle and the lines EF and EG are equal.

b A jib of a crane overhangs the edge of a dock. Three struts of the jib, AE, BE and BC are equal in length. Two struts AB and CD are parallel. If angle $ABE = 25°$, calculate angle EDC.

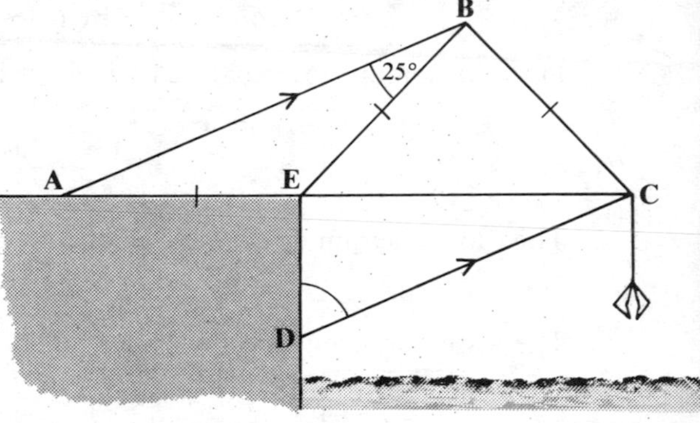

c An aeroplane leaves Heathrow Airport at 08.30 GMT on a bearing of 100° for Frankfurt, Germany. It averages a speed of 300 miles per hour and arrives in Frankfurt 1 hour 20 minutes later. It stays $1\frac{1}{4}$ hours at Frankfurt to refuel, before it continues its journey on a new bearing of 045°. It now averages a speed of 200 miles per hour and lands in Berlin after a further $1\frac{1}{4}$ hours.

(i) At what time GMT does it land in Berlin?

(ii) Make an accurate drawing of the flight to a scale of 1 cm to 50 miles.

(iii) If the plane returns *directly* from Berlin to London, how many miles does it cover on its return flight and on what bearing does it travel?

A mixture of longer problems

6 a Draw and label both axes from -5 to 5.
The square S with vertices (2, 1), (2, 2), (1, 2), (1, 1) is transformed onto its image S' by the matrix $M = \begin{pmatrix} 1 & 1 \\ 0 & -1 \end{pmatrix}$.
Calculate the co-ordinates of the vertices of S' and draw both S and S' on the same diagram.

b S' now maps onto S'' under a transformation given by the matrix $N = \begin{pmatrix} 2 & 2 \\ 0 & -2 \end{pmatrix}$.
Calculate the position of S'' and draw it on the same diagram.

c Find the areas of S' and S'' and hence find the area scale factor of the transformation which maps S' and S''.

d Describe in detail the single transformation which maps S directly onto S'' in one step, and calculate the matrix NM which represents this transformation.

• **7 a** A sheep gives birth to twins. The probability of any lamb being male or female is $\frac{1}{2}$.
Copy and complete this tree diagram to show the possible combinations of female lambs, F and male lambs, M.

What is the probability of both lambs being male?

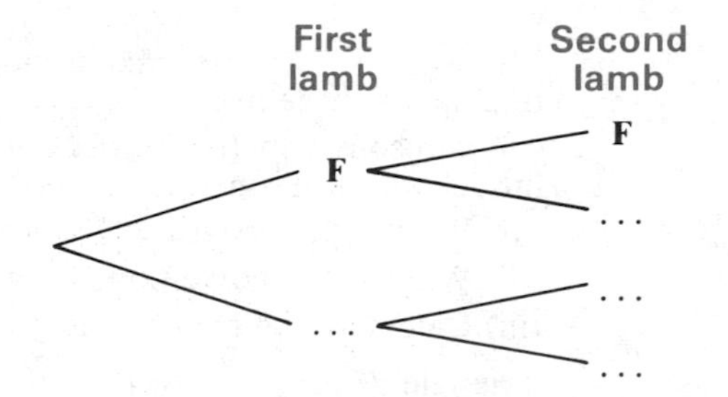

b A pigeon fancier has twenty birds. He times each bird over the same distance to the nearest tenth of a minute with the following results.

12.8	14.5	11.7	12.5	10.8
10.3	12.1	12.0	11.3	12.9
12.6	10.7	13.8	12.4	13.4
13.2	13.6	14.7	12.5	10.9

Copy and complete this table using these results.

Time, minutes	10.0 – 10.9	11.0 – 11.9	12.0 – 12.9	13.0 – 13.9	14.0 – 14.9	Total
Midpoint of class	10.45					
Number of pigeons						

(i) State the modal class.

(ii) Use the table (*not* the individual results) to estimate the total of the times for all the birds. Hence, estimate the *mean time* of the birds.

(iii) If the owner selects a pigeon at random, what is the probability that it had a time of more than 13.0 minutes?

A mixture of longer problems

8 A garden path is made of concrete so that the three straight sections, each 5 metres long and 1 metre wide, are joined by two corner sections. Each corner section has the shape of a sector of a circle, with centre at P or Q and with angles of 60° as shown.

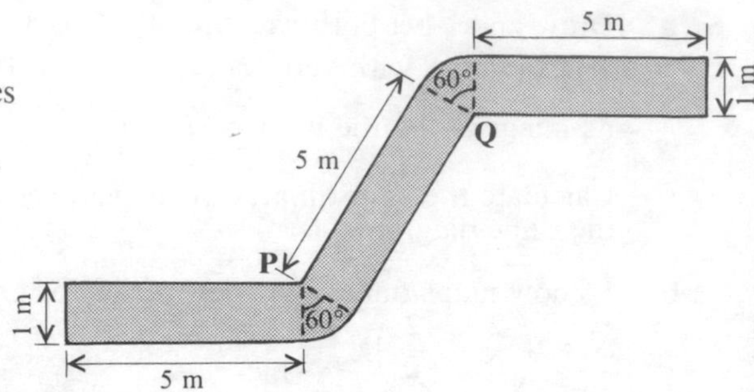

a Calculate the area in m^2 of one of the corner sections.

b Calculate the total area of the path in m^2, correct to two decimal places.

c If the concrete is laid 5 cm thick, calculate the volume of concrete required to make the path. Give your answer in m^3 correct to two significant figures.

d The concrete is made from a mixture of four parts sand, three parts cement, two parts water and one part gravel. How much cement was used to make the path?

9 a Mr Brace is going to make a model house for his grandson, similar to the one in which he lives.

The house is 6 metres across and has a ground floor area of 80 m^2. The model will be $1\frac{1}{2}$ metres across.

(i) What is the length scale factor involved?

(ii) What is the corresponding area scale factor?

(iii) Calculate the area of the ground floor of the model.

b Triangle T(2, 6), (5, 9), (7, 7) is rotated onto triangle T'(5, 5), (8, 2), (6, 0). Draw T and T' on suitable axes and find, by any method, the centre of the rotation and the angle and direction of rotation.

c The point O is the centre of the regular hexagon $UVWXYZ$ of side 2 cm. M and N are the midpoints of the sides UV and XY respectively.

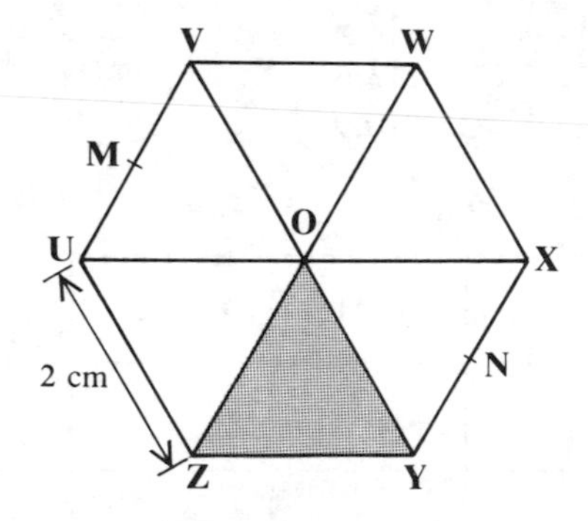

Write the triangle which is the image of the shaded triangle OYZ after each one of the following transformations:

(i) a reflection in the line WZ

(ii) a reflection in the line UX

(iii) a reflection in the line joining M and N

(iv) a clockwise rotation of 60° about O

(v) a clockwise rotation of 60° about Y

(vi) a clockwise rotation of 60° about X

(vii) a translation of 2 cm parallel to XW.

10 a Mr Cuthbertson has an annual gross income of £7200. His annual taxable income is £4800 and he pays income tax at a rate of 30% of his taxable income.

Calculate

(i) the amount of income tax he pays in the year

(ii) his net annual income after tax is deducted

(iii) his net monthly income.

A mixture of longer problems

b A newly-married couple apply for a mortgage to buy a house. The building society will give them a loan of no more than

either four times the husband's gross annual income
or two and a half times their gross annual incomes added together.

They have already managed to save a deposit of £2500.

The husband earns £5400 and his wife earns £6200 per annum.

(i) Which of the two schemes offered by the building society will provide them with the bigger loan?

(ii) What is the maximum price they can afford for a house?

The house which they buy requires a mortgage of £18 000 on which they have to pay interest of 12% in the first year. Their repayment to the building society is £240 each month.

(iii) How much do they repay to the building society in one year?

(iv) How much interest do they pay on their mortgage in the first year?

(v) How much do they still owe to the building society at the end of their first year?

Part 3

1 Water flows a distance of 160 cm every second along a cylindrical pipe of radius 1.5 cm.

1.5 cm

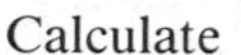

Calculate

a the area of cross-section of the pipe, to 1 decimal place

b the volume of water flowing through the pipe each second, giving your answer in both cm^3 and litres.

The water flows at constant speed for twelve hours into an empty metal tank which has the shape of a cuboid with a base 2.8 metres long and 2.5 metres wide.

Calculate

c the volume of water entering the tank to the nearest thousand litres

d the depth of the water in the tank at the end of the twelve-hour period. ($1\ m^3 = 1000$ litres)

2 A simple shed is made from wood with the roof $WXYZ$ leaning against an existing wall. The shed has two triangular ends one of which, VXY, forms the door.

Using the measurements given, calculate

a the length XY

b the total area of wood used for the roof and the two triangular ends

c the length XU

d the length XZ of the longest possible rod which can be stored in the shed

e the angle which this rod XZ makes with the horizontal floor.

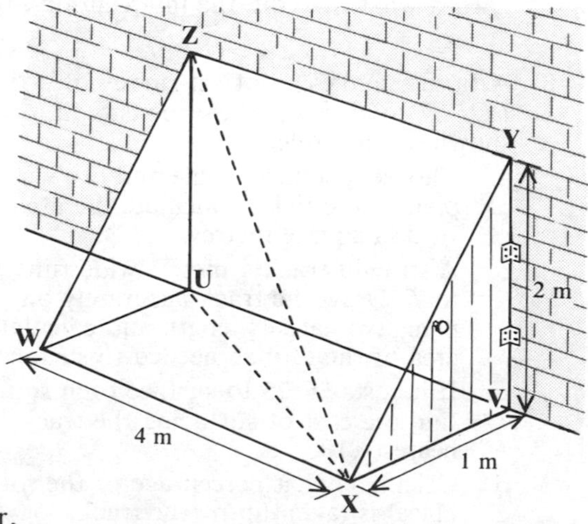

A mixture of longer problems

3 A public hall is to be used for a music concert and tickets are to be sold at two prices.

a The total number of seats must be less than or equal to 500. If there are x seats at the lower price and y seats at the higher price, write an inequality involving x and y for the total number of seats in the hall.

b The number of higher-priced seats must be less than or equal to the number of lower-priced seats. Write another inequality involving x and y.

c The number of lower-priced seats must be no greater than 350. Write an inequality involving only x.

d On axes labelled from 0 to 600, draw three straight lines which are the boundaries of a region defined by your three inequalities. Shade this region and list the co-ordinates of points at its corners.

e If tickets cost either £2 or £3, find how many of each kind must be sold to give the greatest possible income. Calculate this maximum income.

4 This question requires polar graph paper with nine circles 1 cm apart and radii at intervals of 10°.

a The Fisheries Protection Authority uses radar to scan a triangular area of sea in which no fishing is allowed. On a radar screen with a scale of 1 cm for 10 km, the corners of the triangle RST inside which fishing is forbidden are $R(80, 030°)$, $S(90, 100°)$ and $T(80, 190°)$.

Label your polar graph paper. Draw and label the forbidden region.

b At 11.40 hours, the radar screen indicates a fishing boat at point $A(60, 330°)$. It sails a straight course at a steady speed and at 14.40 hours it enters the forbidden region at point $B(40, 040°)$. Draw the path of this boat, labelling points A and B. Find the distance AB to the nearest kilometre and the speed of the boat in km/h.

c The boat continues across the forbidden region at the same speed in the same direction. Label the point C where it leaves the region. Find the distance BC to the nearest kilometre and the time taken to cross the forbidden region.

d At 14.40 when the fishing boat was at point B, a Fisheries Protection launch left the radar station at $(0, 0°)$ and sailed in a straight line to intercept the fishing boat at point C.
How far and at what speed did the launch have to travel to reach point C? At what time did the interception take place?

5 A farmer makes an offset survey of a field $PQRST$ to give the measurements in metres as shown in this table.

	S80	
	60	50R
T40	50	
	30	50Q
	P0	

a Choose a suitable scale and draw an accurate plan of the field. Calculate the area of the field in square metres.

b A straight track 4 metres wide runs from Q to T. Draw the track accurately on your plan, estimate its length and calculate the area of the tarmac needed to surface it.

c If it costs £6·75 to surface each square metre, find the cost of surfacing the track to the nearest £10.

d Calculate what percentage of the total area of the field (to one decimal place) is taken up by the track.

A mixture of longer problems

6 a Draw and label the x-axis from 0 to 16 and the y-axis from -8 to 8. The rectangle $R(1, 1)$, $(1, 3)$, $(2, 3)$, $(2, 1)$ is transformed onto its image R_1 by the matrix $M = \begin{pmatrix} 2 & 4 \\ 0 & -2 \end{pmatrix}$

Calculate the vertices of R_1 and draw both R and R_1 on the same diagram.

b Find the areas of both R and R_1 and hence find the area scale factor of the transformation given by the matrix M.

c R_1 is now transformed onto R_2 by the matrix $N = \begin{pmatrix} \frac{1}{2} & 1 \\ 0 & -\frac{1}{2} \end{pmatrix}$

Calculate the vertices of R_2. Draw and label R_2 on the same diagram.

d Find the one matrix NM which will map R directly onto R_2 in one step. Copy and complete this sentence with one word:

"The matrix N is the .... of matrix M."

7 A tennis ball is thrown from a point O at the bottom of a slope. The horizontal and vertical distances, x and y, (in metres) of the ball from O during its flight are related by the equation $y = 8x - x^2$.

The height of the slope above O is given by the equation $y = \frac{1}{2}x$.

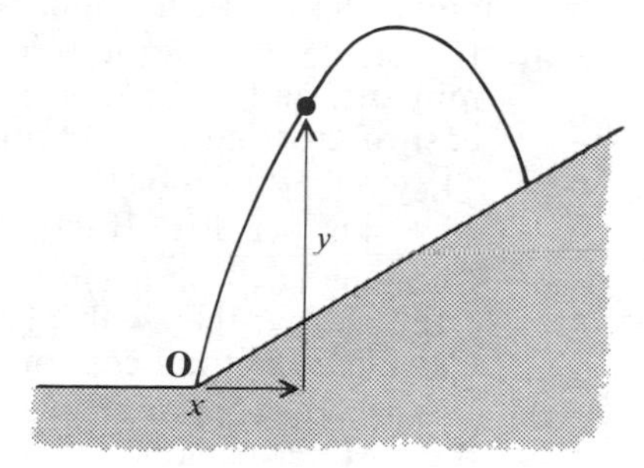

a Copy and complete these two tables:

For the ball, $y = 8x - x^2$

x	0	1	2	3	4	5	6	7	8
$8x$ $-x^2$									
y									

For the slope, $y = \frac{1}{2}x$

x	0	1	2	3	4	5	6	7	8
y									

Using a scale of 1 cm = 1 metre, label the x-axis from 0 to 10 and the y-axis from 0 to 20. Draw the path of the ball and the incline of the slope. Label the point P where the ball hits the hillside. How high is P vertically above O and how far is P horizontally from O?

b Label the highest point Q which the ball reaches.

Calculate the angle of elevation of Q from O and check your answer by measurement with a protractor.

8 a **R** denotes a clockwise rotation of 90° about the point (3, 3).

S denotes a reflection in the line $y = 3$.

Triangle T has vertices (2, 5), (1, 5), and (1, 7).

On axes labelled from 0 to 10, draw the position of T, **R**(T) and **SR**(T).

Give a full description of the single transformation which maps T directly onto **SR**(T) in one step.

b Two maps have different scales. The distance between Norwich and Ipswich on map A is 4 cm, and between the same two places on map B is 12 cm. If the county of Suffolk on map A has an area of 17 cm^2, find its area on map B.

A mixture of longer problems

c The rectangle $OPQR$ has sides of 5 cm and 10 cm and is rotated clockwise about O until its image rectangle $OP'Q'R'$ has point R on side $P'Q'$.
Find
(i) the length OP' and the angle $P'OR$
(ii) the angle of rotation
(iii) the area of the trapezium $ORQ'R'$, given that the area of triangle $OP'R$ is 22 cm^2.

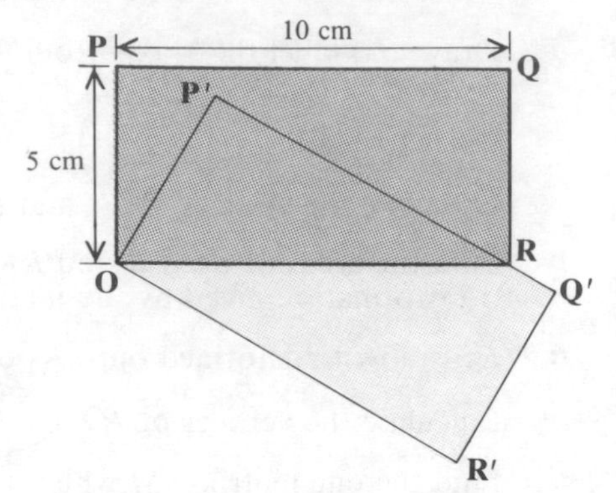

9 A group of eight young people decide to holiday for two weeks in France. The basic cost of the holiday falls into four categories; crossing the English Channel, fuel for their minibus, overnight camping charges and food.

a If a single fare on the ferry costs £17·50 per person and £47·50 per minibus, and if the return fare is double the single fare, calculate the total cost of crossing there and back.

b They intend to travel about 3000 km with an average petrol consumption of 12 km per litre. If each litre costs 4.80 francs, calculate their petrol costs.

c Camping costs an average of 20 francs per person per night and they estimate that they will spend 34 francs per person per day on food. Calculate the total cost of camping and food.

d What is the total cost of all the four categories together, and how much is the cost per person, to the nearest £. Name two other items of expenditure *not* included in the above calculations. (Take £1 = 12 francs)

10 Jane Naylor lives at 57 Walton Road and she drives her car to her friend's house at 21 Cleeve Avenue. She travels along Walton Road and Beeston Road before joining the main A62 dual-carriageway trunk road. At the first roundabout she leaves the A62, travels along Lees Road and then turns into Cleeve Avenue.

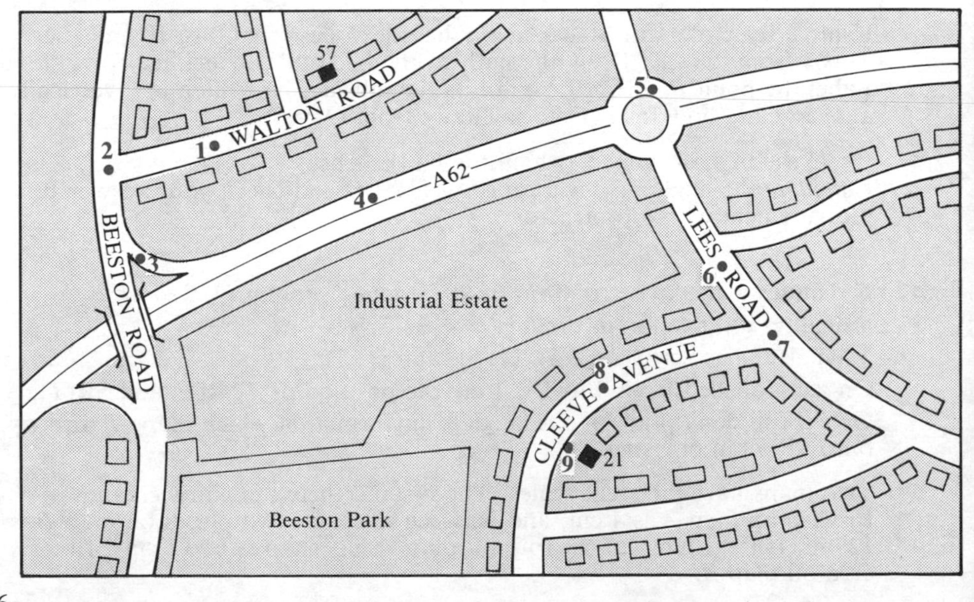

A mixture of longer problems

a When joining Beeston Road from Walton Road, does she turn to the left or to the right?

b Is the Industrial Estate on her left or right when driving along the A62?

c Which way does she turn from Lees Road into Cleeve Avenue?

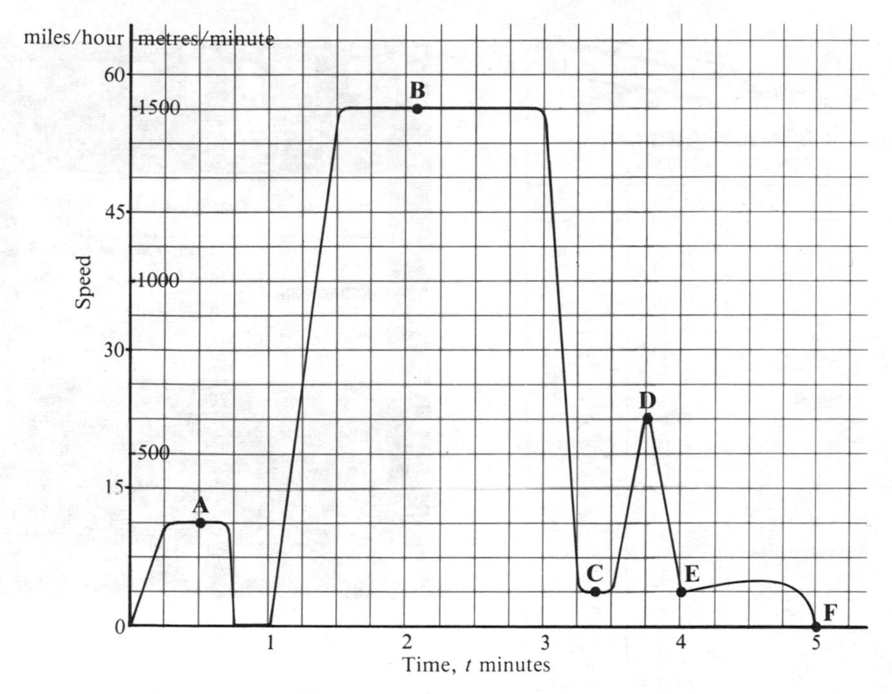

Her speed during the whole journey is shown on the graph where the vertical axis is labelled in both metres per minute and in miles per hour, and the horizontal axis gives the time t in minutes from leaving 57 Walton Road.

d She stops only once during the journey. How long does she stop for and which road is she waiting to enter?

e What is her steady speed as she drives along the A62 trunk road? Give your answer accurately in metres per minute and also approximately in miles per hour.

f What is the maximum speed (in metres/minute) which she reaches as she drives along Lees Road?

g Nine points, numbered from *1* to *9*, are marked on the map. Six points, labelled *A* to *F*, are marked on the graph. Match the letters *A* to *F* with their six corresponding numbered points on the map.

h Estimate the acceleration of the car (in metres/min^2) during the time interval from $t = 1$ to $t = 1\frac{1}{2}$.

i Estimate the total distance (in metres) which Jane travels and her average speed in metres per minute.

Aural exercises

These tables and diagrams are to be used in conjunction with the questions given in the teachers' answer-book.

Table A

NORTH BRITISH AIRLINES
PASSENGER TARIFFS

ROUTE	SINGLE £	RETURN £
Barra – Benbecula	12·60	17·10
Barra – Glasgow	46·20	60·50
Barra – London Heathrow	95·50	157·50
Barra – Tiree	14·20	20·00
Belfast – Blackpool	42·50	76·00
Belfast – Edinburgh	47·50	67·50
Belfast – Manchester	51·00	94·50
Benbecula – Stornoway	25·60	33·50
Campbeltown – Glasgow	24·70	32·20
Glasgow – Belfast	40·50	63·00
Inverness – Edinburgh	38·00	49·50
Inverness – Glasgow	37·70	47·00
Islay – Glasgow	29·50	39·50
Islay – London Heathrow	82·00	134·50
Kirkwall – Edinburgh	72·00	94·00
Kirkwall – Wick	17·80	23·60
Kirkwall – London Heathrow	121·00	191·00
Lerwick – Edinburgh	91·00	119·00
Londonderry – Blackpool	54·50	90·00
Londonderry – Glasgow	41·00	61·00
Londonderry – Isle of Man	46·20	75·50
Manchester – Edinburgh	51·00	94·50
Manchester – Glasgow	51·00	82·00
Manchester – Inverness	85·00	166·50
Skye – Glasgow	41·00	53·50
Skye – London Heathrow	91·50	150·00

Table B

SUPERFILMS

Order your new films – sizes below		Price Each
KODACOLOR Films for Colour prints	126 – 24 exposures	£2·00
	110 – 24 exposures	£1·90
	135 – 24 exposures	£2·30
	135 – 36 exposures	£2·90
	DISC – 15 exposures	£1·90
QUICKSNAP Film for Colour Prints	126 – 24 exposures	£1·40
	110 – 24 exposures	£1·40
	135 – 24 exposures	£1·40
	135 – 36 exposures	£2·00
	DISC – 15 exposures	£1·40

Diagram C

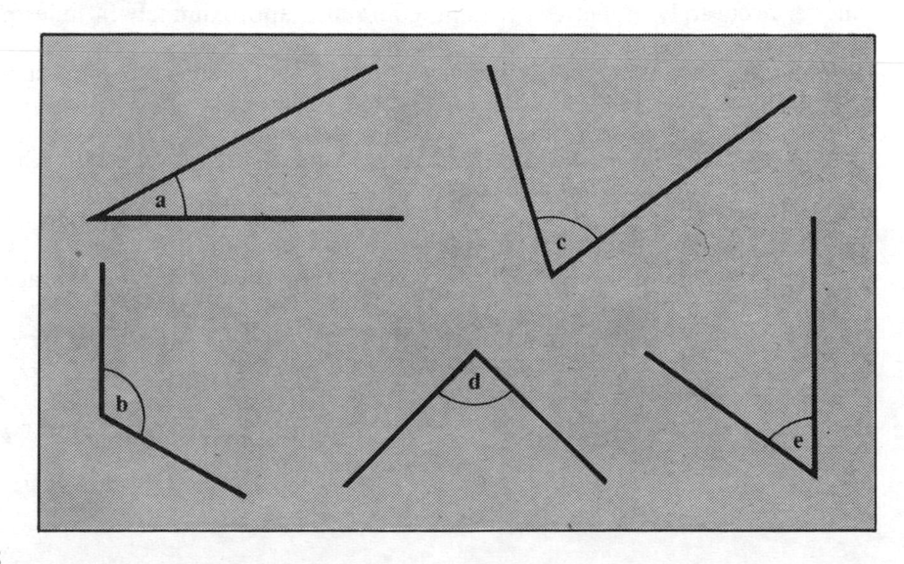

Aural exercises

Diagram D

Diagram E

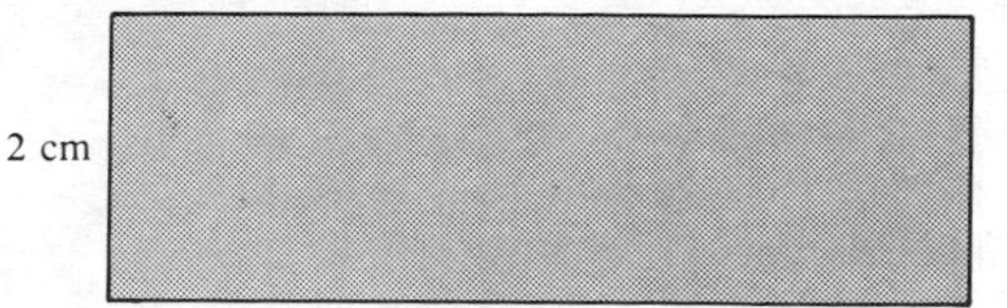

Diagram F

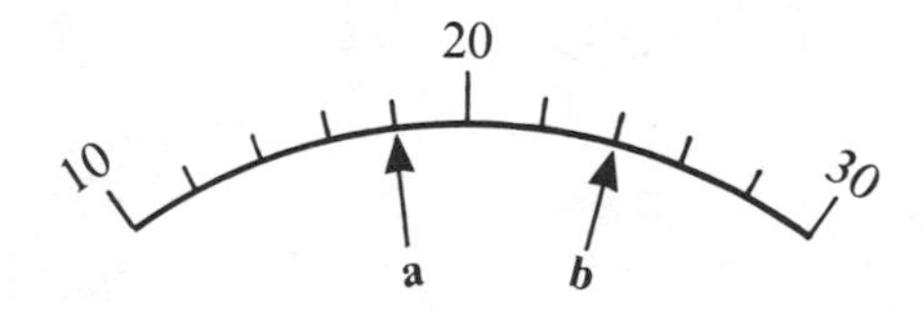

Table G

TOURIST RATES

Rates indicate foreign currency obtained for one pound sterling.

Australia	2·37 dollars
Austria	23·00 schillings
Belgium	68·00 francs
Canada	2·10 dollars
Cyprus	0·75 pounds
Denmark	12·33 kroner
Finland	7·79 markkaa
France	10·58 francs
Germany	3·30 marks
Greece	203 drachmae
Holland	3·72 guilders
Iceland	60 kronur
Ireland	1·095 punts
Israel	2·30 shekels
Italy	2,278 lire
Japan	249 yen
Malta	0·5770 lire
New Zealand	2·82 dollars
Norway	11·33 kroner
Portugal	221 escudos
South Africa	4·25 rand
Spain	210 pesetas
Sweden	10·76 kronor
Switzerland	2·67 francs
Turkey	960 lire
United States	1·53 dollars
Yugoslavia	588 dinars

Aural exercises

Diagram H

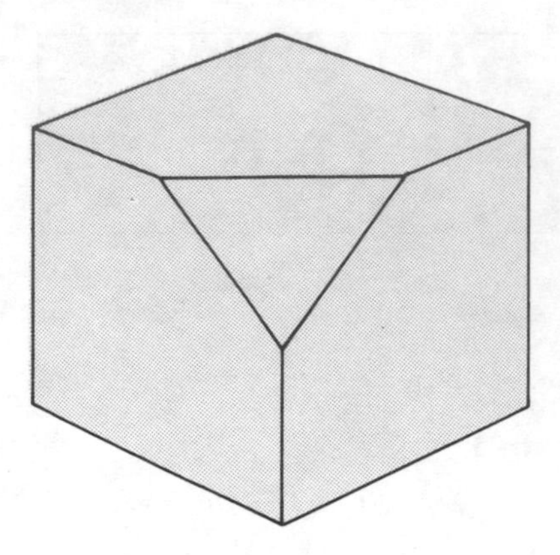

Table I

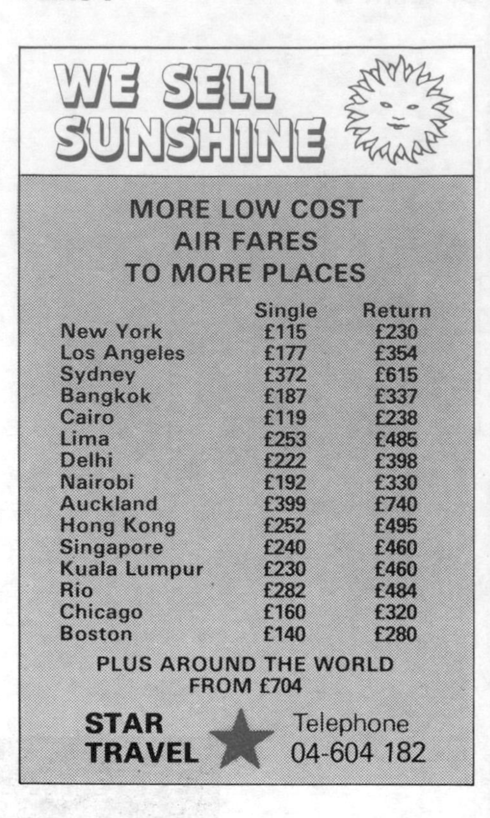

WE SELL SUNSHINE

MORE LOW COST
AIR FARES
TO MORE PLACES

	Single	Return
New York	£115	£230
Los Angeles	£177	£354
Sydney	£372	£615
Bangkok	£187	£337
Cairo	£119	£238
Lima	£253	£485
Delhi	£222	£398
Nairobi	£192	£330
Auckland	£399	£740
Hong Kong	£252	£495
Singapore	£240	£460
Kuala Lumpur	£230	£460
Rio	£282	£484
Chicago	£160	£320
Boston	£140	£280

PLUS AROUND THE WORLD
FROM £704

STAR TRAVEL
Telephone
04-604 182

Table J

Royal Mail Letters

Rates for letters within the UK

Weight not over	First class	Second class
60g	18p	13p
100g	26p	20p
150g	32p	24p
200g	40p	30p
250g	48p	37p
300g	56p	43p
350g	64p	49p
400g	72p	55p
450g	82p	62p
500g	92p	70p
750g	£1·45	£1·05
1000g	£1·85	Not admissible over 750g

Aural exercises

Table K

Superdeal Holidays

Resort	Airport	Full Board or Half Board	Date of departure	7 nights Price	14 nights Price
Tunisia	Gatwick	FB	21 July	£139	£189
Rhodes	Gatwick/Luton	HB	22 July	–	£235
	Manchester	HB	22 July	£199	£245
Ibiza	Gatwick/Luton	HB	22 July	–	£185
	Gatwick/Luton	HB	22 July	–	£199
Kos	Gatwick	FB	23 July	–	£289
Tenerife	Gatwick	HB	23 July	–	£239
Majorca	Gatwick/Luton	HB	24 July	£169	£229
	Manchester	HB	24 July	£179	–
Costa Brava	Gatwick	FB	25 July	–	£179
Ibiza	Gatwick/Luton	HB	26 July	£145	–
	Birmingham	HB	26 July	£155	£219
	Manchester	HB	26 July	£155	–
Minorca	Gatwick	HB	26 July	£185	£249
Majorca	Birmingham	HB	27 July	£185	–
	Manchester	HB	27 July	£185	–
Algarve	Gatwick/Luton	HB	27 July	–	£389
Costa del Sol	Gatwick	HB	27 July	£159	£179
	Gatwick	HB	27 July	£185	£219

WINTER BROCHURE NOW AVAILABLE

Mon-Fri 9.00am – 5.30pm: Sat 9.00am-1.00pm

08-138 6114 or 081-329 4180

Access/Visa Welcome

Diagram L

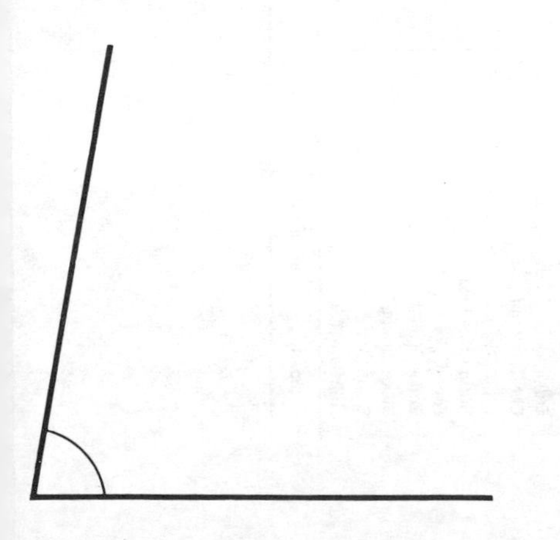

Diagram M

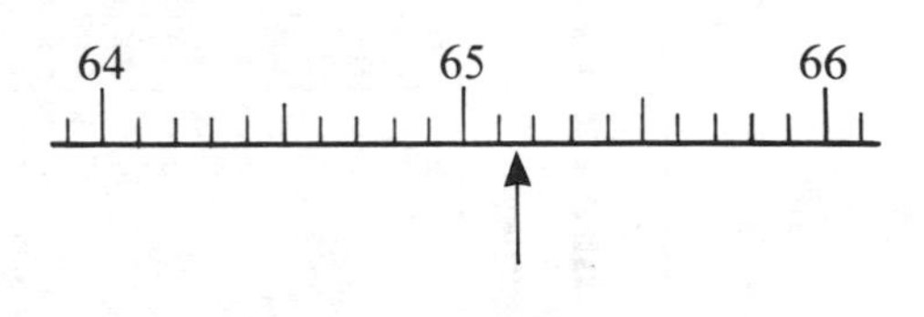

Aural exercises

Table N

Redditch → Birmingham → Lichfield

Mondays to Saturdays

Redditch	d		0757		0812					0922					1022
Alvechurch	d		0804		0819					0929					1029
Barnt Green	d	0800	0810		0825		0853			0935					1035
Longbridge	d	0807	0815	0823	0830	0843	0858	0913	0930	0943	0958	1013		1028	1043
Northfield	d	0810	0818	0826	0833	0846	0901	0916	0933	0946	1001	1016		1031	1046
King's Norton	d	0813	0821	0829	0836	0849	0904	0919	0936	0949	1004	1019		1034	1049
Bournville	d	0816	0824	0832	0839	0852	0907	0922	0939	0952	1007	1022		1037	1052
Selly Oak	d	0819	0827	0835	0842	0855	0910	0925	0942	0955	1010	1025		1040	1055
University	d	0822	0830	0838	0845	0858	0913	0928	0945	0958	1013	1028		1043	1058
Five Ways	d	0826	0834	0842	0849	0902	0917	0932	0949	1002	1017	1032		1047	1102
Birmingham New Street	a	0830	0838	0846	0853	0906	0921	0936	0953	1008	1023	1037		1053	1106
	d		0846		0857	0910	0924	0939	0954	1009	1024	1039		1054	1109
Duddeston	d		0850		0901	0914	0928	0943	0958	1013	1028	1043		1058	1113
Aston	d		0853		0904	0917	0931	0946	1001	1016	1031	1046		1101	1116
Gravelly Hill	d		0856		0907	0920	0934	0949	1004	1019	1034	1049		1104	1119
Erdington	d		0859		0910	0923	0937	0952	1007	1022	1037	1052		1107	1122
Chester Road	d		0901		0912	0925	0939	0954	1009	1024	1039	1054		1109	1124
Wylde Green	d		0904		0915	0928	0942	0957	1012	1027	1042	1057		1112	1127
Sutton Coldfield	d		0907		0918	0931	0945	1000	1015	1030	1045	1100		1115	1130
Four Oaks	d		0911		0922a	0935	0949a	1004	1021a	1034	1049a	1104		1119a	1134
Butlers Lane	d		0914			0938		1007		1037		1107			1137
Blake Street	d		0916			0941a		1009		1040a		1109			1140a
Shenstone	d		0921					1014				1114			
Lichfield City	a		0927					1020				1120			

Redditch	d					1122						1625			
Alvechurch	d					1129						1632			
Barnt Green	d					1135						1638			
Longbridge	d	1058		1113	1128	1143	1158		1613	1628		1643	1658	1705	1713
Northfield	d	1101		1116	1131	1146	1201		1616	1631		1646	1701	1708	1716
King's Norton	d	1104		1119	1134	1149	1204		1619	1634		1649	1704	1712	1719
Bournville	d	1107		1122	1137	1152	1207	and at	1622	1637	1645	1652	1707	1715	1722
Selly Oak	d	1110		1125	1140	1155	1210		1625	1640	1648	1655	1710	1718	1725
University	d	1113		1128	1143	1158	1213	the same	1628	1643	1650	1658	1713	1720	1728
Five Ways	d	1117		1132	1147	1202	1217		1632	1647	1654	1702	1717	1724	1732
Birmingham New Street	a	1122		1137	1151	1206	1221	minutes	1636	1651	1659	1706	1721	1729	1736
	d	1124		1139	1154	1209	1224		1641	1654	1704	1715	1724	1735	1743
Duddeston	d	1128		1143	1158	1213	1228	past	1645		1708	1719	1728	1739	1747
Aston	d	1131		1146	1201	1216	1231		1648	1659	1713	1722	1731	1743	1750
Gravelly Hill	d	1134		1149	1204	1219	1234	each	1651	1703	1716	1725	1734	1746	1753
Erdington	d	1137		1152	1207	1222	1237		1654	1706	1719	1728	1737	1749	1756
Chester Road	d	1139		1154	1209	1224	1239	hour until	1656	1708	1721	1730	1739	1751	1758
Wylde Green	d	1142		1157	1212	1227	1242		1659	1710	1724	1733	1742	1754	1801
Sutton Coldfield	d	1145		1200	1215	1230	1245		1702	1714	1728	1736	1746	1758	1804
Four Oaks	d	1149a		1204	1219a	1234	1249a		1706	1717	1731	1740	1750	1802	1813
Butlers Lane	d			1207		1237			1709	1720	1734	1744	1754	1806	1816
Blake Street	d			1209		1240a			1715a	1723	1737a	1748	1800a	1812	1819a
Shenstone	d			1214						1727		1752		1817	
Lichfield City	a			1220						1733		1758		1823	

d = depart a = arrive

The times shown here are valid only for this exercise and may not represent actual train times.

Aural exercises

Table O

Dialled Calls International

Approximate **Cost** to the customer including VAT

		1 min	2 mins	3 mins	4 mins	5 mins
Charge band A	**Cheap**	**40p**	**75p**	**£1·09**	**£1·44**	**£1·78**
	Standard	46p	86p	£1·32	£1·73	£2·19
Charge band B	**Cheap**	**46p**	**92p**	**£1·38**	**£1·84**	**£2·30**
	Standard	58p	£1·15	£1·73	£2·30	£2·82
Charge band C	**Cheap**	**63p**	**£1·21**	**£1·78**	**£2·36**	**£2·93**
	Standard	69p	£1·38	£2·07	£2·76	£3·45
	Peak	75p	£1·50	£2·24	£2·99	£3·74
Charge band D	**Cheap**	**86p**	**£1·67**	**£2·47**	**£3·28**	**£4·08**
	Standard	£1·04	£2·01	£3·05	£4·03	£5·00
Charge band E	**Cheap**	**86p**	**£1·67**	**£2·47**	**£3·28**	**£4·08**
	Standard	£1·04	£2·01	£3·05	£4·03	£5·00

Diagram P

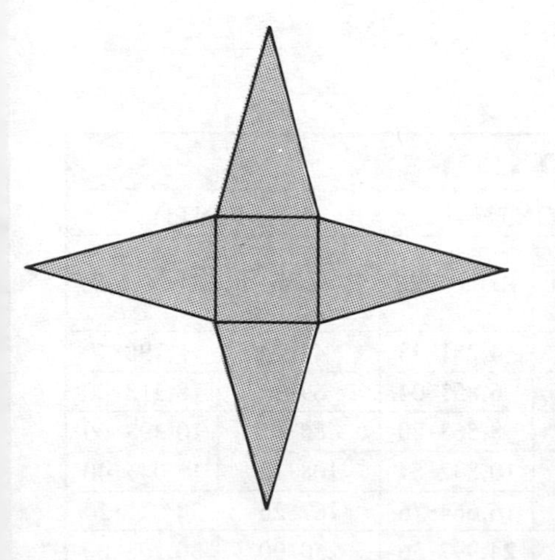

Diagram Q

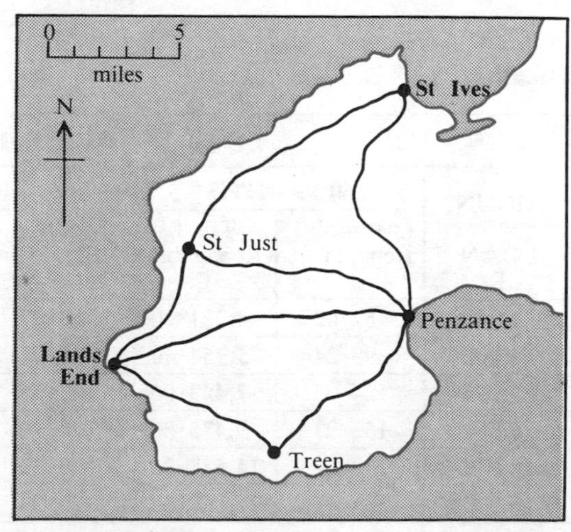

Aural exercises

Diagram R

Diagram S

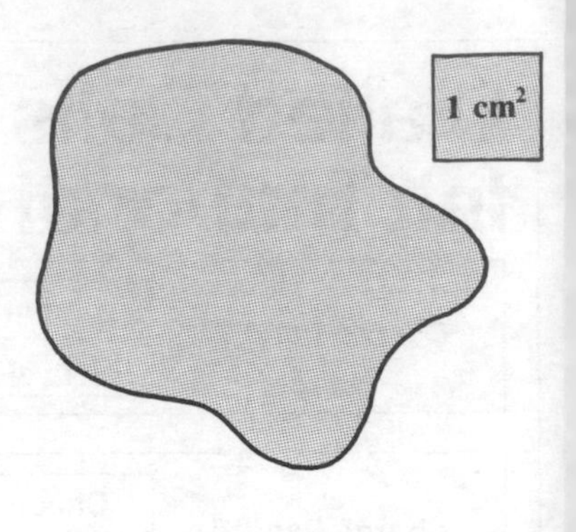

Diagram T

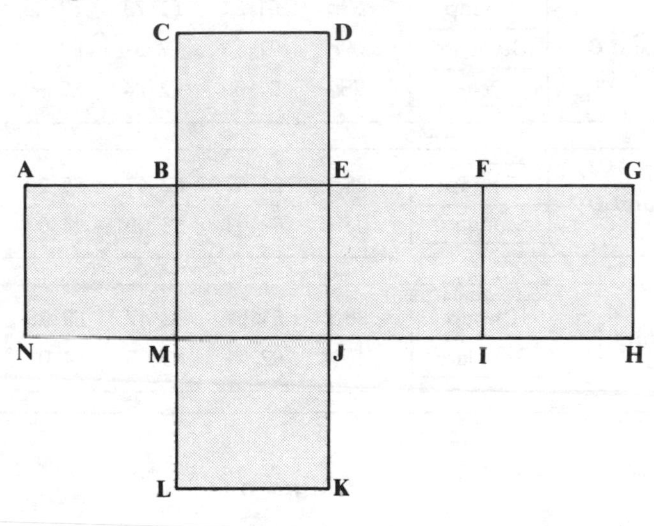

Table U

LOW REPAYMENT LOANS						
AMOUNT OF LOAN £	60 MONTHS		84 MONTHS		120 MONTHS	
	Monthly Repayment £	Total Repayable £	Monthly Repayment £	Total Repayable £	Monthly Repayment £	Total Repayable £
2,500	62·02	3,721·20	50·97	4,281·48	43·32	5,198·40
4,000	99·24	5,954·40	81·56	6,851·04	69·32	8,318·40
5,000	124·05	7,443·00	101·95	8,563·80	86·65	10,398·00
6,500	157·93	9,475·80	128·96	10,832·64	108·72	13,046·40
10,000	242·97	14,578·20	198·39	16,664·76	167·26	20,071·20
15,000	364·45	21,867·00	297·59	24,997·56	250·90	30,108·00

Aural exercises

Table V

CAR FERRY TARIFF

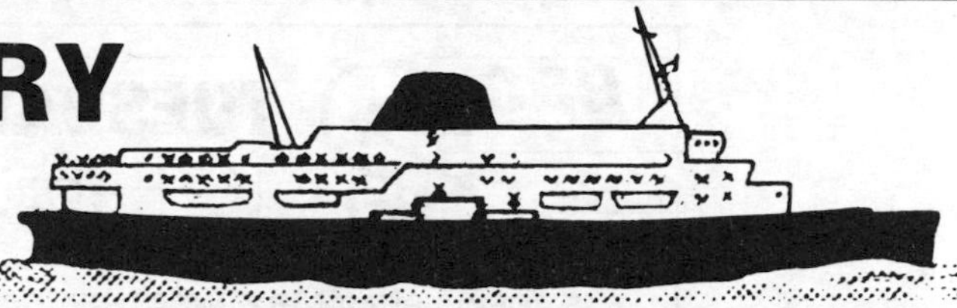

All fares are for single journeys

	Tariff A £	Tariff B £	Tariff C £	Tariff D £
Drivers and Passengers				
Adults	10·00	10·00	10·00	10·00
Senior Citizens	5·00	5·00	5·00	5·00
Children (4 and under 14 years. Under 4 free.)	5·00	5·00	5·00	5·00
Vehicles				
Cars, motor caravans, minibuses, vans:				
Overall length not exceeding 4.00m	36·00	37·00	41·00	47·00
4.50m	36·00	47·00	52·00	59·00
5.50m	36·00	59·00	64·00	74·00
over 5.50m, per additional metre or part thereof	10·00	10·00	10·00	10·00
Motorcycles, scooters and mopeds	10·00	10·50	11·00	12·00
Dogs	2·60	2·60	2·60	2·60

Diagram W

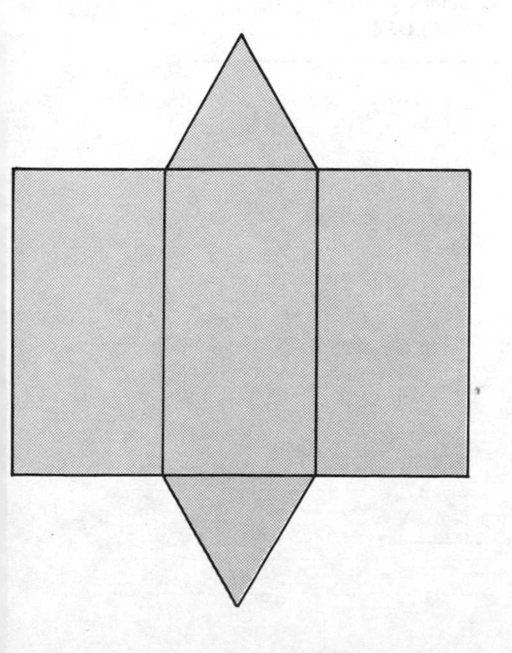

Diagram X

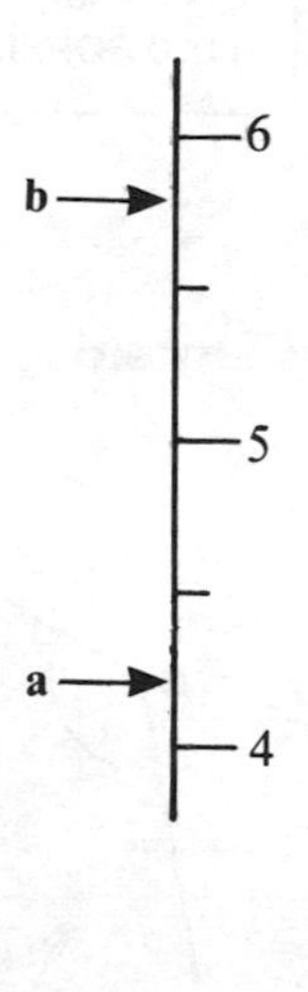

Aural exercises

Table Y

5.25	**NEWS SUMMARY: WEATHER.**
5.30	**TERRY WAITE TAKES A DIFFERENT VIEW:** The Archbishop of Canterbury's special envoy interviews top mountaineer Chris Bonington.
6.00	**WHISTLE TEST:** Visits Spandau Ballet in France, as the band prepares to re-launch.
7.00	**MAN IN THE KITCHEN:** East meets West in Tom Vernon's kitchen as he demonstrates mouth-watering Chinese dishes culled from Hong Kong.
7.30	**QUESTIONS OF DEFENCE:** The rearmament of Germany in the 1950s was America's price for its commitment to the defence of Europe. John Barry discusses this episode in NATO's history.
8.00	**WILDLIFE SHOWCASE:** The Deadliest Creature on Earth – Australia's lethal box jellyfish is feared as much as the shark on north Queensland's sun-drenched beaches. Not surprisingly this film attracted record audiences when it was first shown Down Under.
8.30	**STEAM DAYS:** Miles Kington travels by steam train from Fort William to Mallaig. Built to ferry herring through the breathtaking mountain landscape of West Scotland, the line's catch is now exclusively tourists.
9.00	**FILM:** The Club (1980). Lineker's spectacular transfer to Barcelona lends a special topicality to this week's choice.
10.35	**NEWSNIGHT. 11.20 WEATHER.**
11.25	**MUSIC AT NIGHT:** Fiona Kimm (mezzo-soprano) sings a song by Brahms.
11.30	**OPEN UNIVERSITY. 12.0 CLOSE.**

Diagram Z

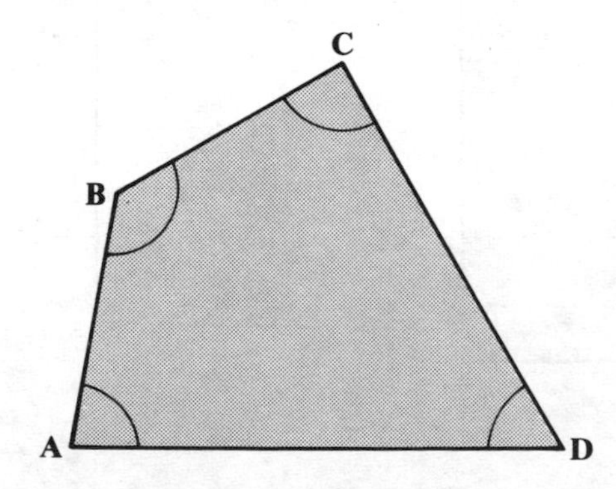